FUNDAMENTALS OF SEMICONDUCTOR MANUFACTURING AND PROCESS CONTROL

FUNDAMENTALS OF SEMICONDUCTOR MANUFACTURING AND PROCESS CONTROL

Gary S. May, Ph.D.

Georgia Institute of Technology
Atlanta, Georgia

Costas J. Spanos, Ph.D.

University of California at Berkeley
Berkeley, California

WILEY-INTERSCIENCE

A JOHN WILEY & SONS, INC., PUBLICATION

Published by John Wiley & Sons, Inc., Hoboken, New Jersey
Published simultaneously in Canada.

For general information on our other products and services or for technical support, please contact our Customer Care Department within the United States at (800) 762-2974, outside the United States at (317) 572-3993 or fax (317) 572-4002.

Wiley also publishes its books in a variety of electronic formats. Some content that appears in print may not be available in electronic formats. For more information about Wiley products, visit our web site at www.wiley.com.

Library of Congress Cataloging-in-Publication Data:

May, Gary S.
    Fundamentals of semiconductor manufacturing and process control / Gary S.
May, Costas J. Spanos.
        p. cm.
    "Wiley-Interscience."
    Includes bibliographical references and index.
    ISBN-13: 978-0-471-78406-7 (cloth : alk. paper)
    ISBN-10: 0-471-78406-0 (cloth : alk. paper)
    1. Semiconductors—Design and construction. 2. Integrated circuits—Design and construction. 3. Process control—Statistical methods. I. Spanos, Costas J. II. Title.
    TK7871.85.M379 2006
    621.3815'2—dc22
                                                                    2005028448

To my children,
Simone and Jordan,
who inspire me.

—Gary S. May

To my family,
for their love and understanding.

—Costas J. Spanos

CONTENTS

Preface   xvii

Acknowledgments   xix

1 Introduction to Semiconductor Manufacturing   1

Objectives / 1
Introduction / 1
1.1. Historical Evolution / 2
    1.1.1. Manufacturing and Quality Control / 3
    1.1.2. Semiconductor Processes / 5
    1.1.3. Integrated Circuit Manufacturing / 7
1.2. Modern Semiconductor Manufacturing / 8
    1.2.1. Unit Processes / 9
    1.2.2. Process Sequences / 11
    1.2.3. Information Flow / 12
    1.2.4. Process Organization / 14
1.3. Goals of Manufacturing / 15
    1.3.1. Cost / 15
    1.3.2. Quality / 17
    1.3.3. Variability / 17
    1.3.4. Yield / 17
    1.3.5. Reliability / 18
1.4. Manufacturing Systems / 18
    1.4.1. Continuous Flow / 19
        1.4.1.1. Batch Processes / 20
        1.4.1.2. Single Workpiece / 20
    1.4.2. Discrete Parts / 21
1.5. Outline for Remainder of the Book / 21
Summary / 22
Problems / 22
References / 23

vii

2 Technology Overview   25

Objectives / 25
Introduction / 25
  2.1. Unit Processes / 25
      2.1.1. Oxidation / 26
            2.1.1.1. Growth Kinetics / 27
            2.1.1.2. Thin Oxide Growth / 31
            2.1.1.3. Oxide Quality / 33
      2.1.2. Photolithography / 34
            2.1.2.1. Exposure Tools / 35
            2.1.2.2. Masks / 38
            2.1.2.3. Photoresist / 39
            2.1.2.4. Pattern Transfer / 41
            2.1.2.5. E-Beam Lithography / 43
            2.1.2.6. X-Ray Lithography / 45
      2.1.3. Etching / 47
            2.1.3.1. Wet Chemical Etching / 47
            2.1.3.2. Dry Etching / 48
      2.1.4. Doping / 51
            2.1.4.1. Diffusion / 52
            2.1.4.2. Ion Implantation / 56
      2.1.5. Deposition / 58
            2.1.5.1. Physical Vapor Deposition / 59
            2.1.5.2. Chemical Vapor Deposition / 60
      2.1.6. Planarization / 61
  2.2. Process Integration / 61
      2.2.1. Bipolar Technology / 63
      2.2.2. CMOS Technology / 66
            2.2.2.1. Basic NMOS Fabrication Sequence / 67
            2.2.2.2. CMOS Fabrication Sequence / 70
      2.2.3. BiCMOS Technology / 74
      2.2.4. Packaging / 75
            2.2.4.1. Die Separation / 76
            2.2.4.2. Package Types / 77
            2.2.4.3. Attachment Methods / 79
Summary / 80
Problems / 80
References / 81

3   Process Monitoring   82

Objectives / 82
Introduction / 82
3.1. Process Flow and Key Measurement Points / 83
3.2. Wafer State Measurements / 84
    3.2.1. Blanket Thin Film / 85
        3.2.1.1. Interferometry / 85
        3.2.1.2. Ellipsometry / 88
        3.2.1.3. Quartz Crystal Monitor / 91
        3.2.1.4. Four-Point Probe / 92
    3.2.2. Patterned Thin Film / 93
        3.2.2.1. Profilometry / 93
        3.2.2.2. Atomic Force Microscopy / 93
        3.2.2.3. Scanning Electron Microscopy / 95
        3.2.2.4. Scatterometry / 96
        3.2.2.5. Electrical Linewidth Measurement / 98
    3.2.3. Particle/Defect Inspection / 98
        3.2.3.1. Cleanroom Air Monitoring / 99
        3.2.3.2. Product Monitoring / 100
    3.2.4. Electrical Testing / 102
        3.2.4.1. Test Structures / 102
        3.2.4.2. Final Test / 106
3.3. Equipment State Measurements / 107
    3.3.1. Thermal Operations / 109
        3.3.1.1. Temperature / 109
        3.3.1.2. Pressure / 109
        3.3.1.3. Gas Flow / 110
    3.3.2. Plasma Operations / 111
        3.3.2.1. Temperature / 111
        3.3.2.2. Pressure / 112
        3.3.2.3. Gas Flow / 112
        3.3.2.4. Residual Gas Analysis / 112
        3.3.2.5. Optical Emission Spectroscopy / 114
        3.3.2.6. Fourier Transform Infrared
                Spectroscopy / 115
        3.3.2.7. RF Monitors / 116
    3.3.3. Lithography Operations / 116
    3.3.4. Implantation / 117

3.3.5. Planarization / 118

Summary / 118
Problems / 119
References / 120

4 Statistical Fundamentals   **122**

Objectives / 122
Introduction / 122
4.1. Probability Distributions / 123
    4.1.1. Discrete Distributions / 124
        4.1.1.1. Hypergeometric / 124
        4.1.1.2. Binomial / 125
        4.1.1.3. Poisson / 127
        4.1.1.4. Pascal / 128
    4.1.2. Continuous Distributions / 128
        4.1.2.1. Normal / 129
        4.1.2.2. Exponential / 131
    4.1.3. Useful Approximations / 132
        *4.1.3.1. Poisson Approximation to the
        Binomial / 132*
        *4.1.3.2. Normal Approximation to the
        Binomial / 132*
4.2. Sampling from a Normal Distribution / 133
    4.2.1. Chi-Square Distribution / 134
    4.2.2. *t* Distribution / 134
    4.2.3. *F* Distribution / 135
4.3. Estimation / 136
    4.3.1. Confidence Interval for the Mean with Known
        Variance / 137
    4.3.2. Confidence Interval for the Mean with Unknown
        Variance / 137
    4.3.3. Confidence Interval for Variance / 137
    4.3.4. Confidence Interval for the Difference between Two
        Means, Known Variance / 138
    4.3.5. Confidence Interval for the Difference between Two
        Means, Unknown Variances / 138
    4.3.6. Confidence Interval for the Ratio of Two
        Variances / 139
4.4. Hypothesis Testing / 140
    4.4.1. Tests on Means with Known Variance / 141
    4.4.2. Tests on Means with Unknown Variance / 142
    4.4.3. Tests on Variance / 143

Summary / 145
Problems / 145
Reference / 146

5 Yield Modeling   **147**

Objectives / 147
Introduction / 147
5.1. Definitions of Yield Components / 148
5.2. Functional Yield Models / 149

    5.2.1. Poisson Model / 151
    5.2.2. Murphy's Yield Integral / 152
    5.2.3. Negative Binomial Model / 154

5.3. Functional Yield Model Components / 156

    5.3.1. Defect Density / 156
    5.3.2. Critical Area / 157
    5.3.3. Global Yield Loss / 158

5.4. Parametric Yield / 159
5.5. Yield Simulation / 161

    5.5.1. Functional Yield Simulation / 162
    5.5.2. Parametric Yield Simulation / 167

5.6. Design Centering / 171

    5.6.1. Acceptability Regions / 172
    5.6.2. Parametric Yield Optimization / 173

5.7. Process Introduction and Time-to-Yield / 174
Summary / 176
Problems / 177
References / 180

6 Statistical Process Control   **181**

Objectives / 181
Introduction / 181
6.1. Control Chart Basics / 182
6.2. Patterns in Control Charts / 184
6.3. Control Charts for Attributes / 186

    6.3.1. Control Chart for Fraction Nonconforming / 187

        6.3.1.1. Chart Design / 188
        6.3.1.2. Variable Sample Size / 189
        6.3.1.3. Operating Characteristic and Average
            Runlength / 191

    6.3.2. Control Chart for Defects / 193
    6.3.3. Control Chart for Defect Density / 193

6.4. Control Charts for Variables / 195

    6.4.1. Control Charts for $\bar{x}$ and R / 195

        6.4.1.1. Rational Subgroups / 199

        6.4.1.2. Operating Characteristic and Average Runlength / 200

    6.4.2. Control Charts for $\bar{x}$ and s / 202

    6.4.3. Process Capability / 204

    6.4.4. Modified and Acceptance Charts / 206

    6.4.5. Cusum Chart / 208

        6.4.5.1. Tabular Cusum Chart / 210

        6.4.5.2. Average Runlength / 210

        6.4.5.3. Cusum for Variance / 211

    6.4.6. Moving-Average Charts / 212

        6.4.6.1. Basic Moving-Average Chart / 212

        6.4.6.2. Exponentially Weighted Moving-Average Chart / 213

6.5. Multivariate Control / 215

    6.5.1. Control of Means / 217

    6.5.2. Control of Variability / 220

6.6. SPC with Correlated Process Data / 221

    6.6.1. Time-Series Modeling / 221

    6.6.2. Model-Based SPC / 223

Summary / 224

Problems / 224

References / 227

7 Statistical Experimental Design   **228**

Objectives / 228

Introduction / 228

7.1. Comparing Distributions / 229

7.2. Analysis of Variance / 232

    7.2.1. Sums of Squares / 232

    7.2.2. ANOVA Table / 234

        7.2.2.1. Geometric Interpretation / 235

        7.2.2.2. ANOVA Diagnostics / 237

    7.2.3. Randomized Block Experiments / 240

        7.2.3.1. Mathematical Model / 242

        7.2.3.2. Diagnostic Checking / 243

    7.2.4. Two-Way Designs / 245

        7.2.4.1. Analysis / 245

        7.2.4.2. Data Transformation / 246

7.3. Factorial Designs / 249

    7.3.1. Two-Level Factorials / 250

        7.3.1.1. Main Effects / 251

        7.3.1.2. Interaction Effects / 251

        7.3.1.3. Standard Error / 252

        7.3.1.4. Blocking / 254

    7.3.2. Fractional Factorials / 256

        7.3.2.1. Construction of Fractional
              Factorials / 256

        7.3.2.2. Resolution / 257

    7.3.3. Analyzing Factorials / 257

        7.3.3.1. The Yates Algorithm / 258

        7.3.3.2. Normal Probability Plots / 258

    7.3.4. Advanced Designs / 260

7.4. Taguchi Method / 262

    7.4.1. Categorizing Process Variables / 263

    7.4.2. Signal-to-Noise Ratio / 264

    7.4.3. Orthogonal Arrays / 264

    7.4.4. Data Analysis / 266

Summary / 269

Problems / 269

References / 271

8 Process Modeling   **272**

Objectives / 272

Introduction / 272

8.1. Regression Modeling / 273

    8.1.1. Single-Parameter Model / 274

        8.1.1.1. Residuals / 275

        8.1.1.2. Standard Error / 276

        8.1.1.3. Analysis of Variance / 276

    8.1.2. Two-Parameter Model / 277

        8.1.2.1. Analysis of Variance / 279

        8.1.2.2. Precision of Estimates / 279

        8.1.2.3. Linear Model with Nonzero
              Intercept / 280

    8.1.3. Multivariate Models / 283

    8.1.4. Nonlinear Regression / 285

    8.1.5. Regression Chart / 287

8.2. Response Surface Methods / 289

    8.2.1. Hypothetical Yield Example / 289

8.2.1.1. Diagnostic Checking / 292

8.2.1.2. Augmented Model / 293

8.2.2. Plasma Etching Example / 294

8.2.2.1. Experimental Design / 295

8.2.2.2. Experimental Technique / 297

8.2.2.3. Analysis / 298

8.3. Evolutionary Operation / 301

8.4. Principal-Component Analysis / 306

8.5. Intelligent Modeling Techniques / 310

8.5.1. Neural Networks / 310

8.5.2. Fuzzy Logic / 314

8.6. Process Optimization / 318

8.6.1. Powell's Algorithm / 318

8.6.2. Simplex Method / 320

8.6.3. Genetic Algorithms / 323

8.6.4. Hybrid Methods / 325

8.6.5. PECVD Optimization: A Case Study / 326

Summary / 327

Problems / 328

References / 331

9 Advanced Process Control

Objectives / 333

Introduction / 333

9.1. Run-by-Run Control with Constant Term Adaptation / 335

9.1.1. Single-Variable Methods / 335

9.1.1.1. Gradual Drift / 337

9.1.1.2. Abrupt Shifts / 339

9.1.2. Multivariate Techniques / 343

9.1.2.1. Exponentially Weighted Moving-Average (EWMA) Gradual Model / 343

9.1.2.2. Predictor–Corrector Control / 343

9.1.3. Practical Considerations / 346

9.1.3.1. Input Bounds / 346

9.1.3.2. Input Resolution / 348

9.1.3.3. Input Weights / 348

9.1.3.4. Output Weights / 350

9.2. Multivariate Control with Complete Model Adaptation / 351

9.2.1. Detection of Process Disturbances via Model-Based SPC / 352

9.2.1.1. Malfunction Alarms / 352
9.2.1.2. Alarms for Feedback Control / 353
9.2.2. Full Model Adaptation / 354
9.2.3. Automated Recipe Generation / 356
9.2.4. Feedforward Control / 358
9.3. Supervisory Control / 359
9.3.1. Supervisory Control Using Complete Model
Adaptation / 359
*9.3.1.1. Acceptable Input Ranges of
Photolithographic Machines / 361*
9.3.1.2. Experimental Examples / 363
9.3.2. Intelligent Supervisory Control / 364
Summary / 373
Problems / 373
References / 378

10 Process and Equipment Diagnosis   379

Objectives / 379
Introduction / 379
10.1. Algorithmic Methods / 381
10.1.1. Hippocrates / 381
10.1.1.1. Measurement Plan / 382
10.1.1.2. Fault Diagnosis / 383
10.1.1.3. Example / 383
10.1.2. MERLIN / 384
10.1.2.1. Knowledge Representation / 385
10.1.2.2. Inference Mechanism / 387
10.1.2.3. Case Study / 390
10.2. Expert Systems / 391
10.2.1. PIES / 391
10.2.1.1. Knowledge Base / 393
10.2.1.2. Diagnostic Reasoning / 394
10.2.1.3. Examples / 395
10.2.2. PEDX / 395
10.2.2.1. Architecture / 396
10.2.2.2. Rule-Based Reasoning / 397
10.2.2.3. Implementation / 398
10.3. Neural Network Approaches / 398
10.3.1. Process Control Neural Network / 398
10.3.2. Pattern Recognition in CVD Diagnosis / 400
10.4. Hybrid Methods / 402

10.4.1. Time-Series Diagnosis / 402
10.4.2. Hybrid Expert System / 403

   10.4.2.1. Dempster–Shafer Theory / 406
   10.4.2.2. Maintenance Diagnosis / 408
   10.4.2.3. Online Diagnosis / 409
   10.4.2.4. Inline Diagnosis / 413

Summary / 414
Problems / 414
References / 415

Appendix A: Some Properties of the Error Function   417
Appendix B: Cumulative Standard Normal Distribution   420
Appendix C: Percentage Points of the χ^2 Distribution 423
Appendix D: Percentage Points of the t Distribution   425
Appendix E: Percentage Points of the F Distribution   427
Appendix F: Factors for Constructing Variables Control Charts 438

Index   441

PREFACE

In simple terms, manufacturing can be defined as the process by which raw materials are converted into finished products. The purpose of this book is to examine in detail the methodology by which electronic materials and supplies are converted into finished integrated circuits and electronic products in a high-volume manufacturing environment. This subject of this book will be issues relevant to the industrial-level manufacture of microelectronic device and circuits, including (but not limited to) fabrication sequences, process control, experimental design, process modeling, yield modeling, and CIM/CAM systems. The book will include theoretical and practical descriptions of basic manufacturing concepts, as well as some case studies, sample problems, and suggested exercises.

The book is intended for graduate students and can be used conveniently in a semester-length course on semiconductor manufacturing. Such a course may or may not be accompanied by a corequisite laboratory. The text can also serve as a reference for practicing engineers and scientists in the semiconductor industry.

Chapter 1 of the book places the manufacture of integrated circuits into its historical context, as well as provides an overview of modern semiconductor manufacturing. In the Chapter 2, we provide a broad overview of the manufacturing technology and processes flows used to produce a variety of semiconductor products. Various process monitoring methods, including those that focus on product wafers and those that focus on the equipment used to produce those wafers, are discussed in Chapter 3. As a backdrop for subsequent discussion of statistical process control (SPC), Chapter 4 provides a review of statistical fundamentals. Ultimately, the key metric to be used to evaluate any manufacturing process is cost, and cost is directly impacted by yield. Yield modeling is therefore presented in Chapter 5. Chapter 6 then focuses on the use of SPC to analyze quality issues and improve yield. Statistical experimental design, which is presented in Chapter 7, is a powerful approach for systematically varying controllable process conditions and determining their impact on output parameters which measure quality. Data derived from statistical experiments can then be used to construct process models that enable the analysis and prediction of manufacturing process behavior. Process modeling concepts are introduced in Chapter 8. Finally, several advanced process control topics, including run-by-run, supervisory control, and process and equipment diagnosis, are the subject of Chapters 9 and 10.

Each chapter begins with an introduction and a list of learning goals, and each concludes with a summary of important concepts. Solved examples are provided throughout, and suggested homework problems appear at the end of the chapter. A complete set of detailed solutions to all end-of-chapter problems has been prepared. This *Instructor's Manual* is available to all adopting faculty. The figures in the text are also available, in electronic format, from the publisher at the web site: *ftp://ftp.wiley.com/public/sci-tech-med/semiconductor-manufacturing/*

ACKNOWLEDGMENTS

G. S. May would like to acknowledge the support of the Steve W. Chaddick School Chair in Electrical and Computer Engineering at the Georgia Institute of Technology, which provided the environment that enabled the completion of this book. C. J. Spanos would like to acknowledge the contributions of the Berkeley students who, over the years, helped shape the material presented in this book.

1

INTRODUCTION TO SEMICONDUCTOR MANUFACTURING

OBJECTIVES

- Place the manufacturing of integrated circuits in a historical context.
- Provide an overview of modern semiconductor manufacturing.
- Discuss manufacturing goals and objectives.
- Describe manufacturing systems at a high level as a prelude to the remainder of the text.

INTRODUCTION

This book is concerned with the manufacturing of devices, circuits, and electronic products based on semiconductors. In simple terms, *manufacturing* can be defined as the process by which raw materials are converted into finished products. As illustrated in Figure 1.1, a manufacturing operation can be viewed graphically as a system with raw materials and supplies serving as its inputs and finished commercial products serving as outputs. In semiconductor manufacturing, input materials include semiconductor materials, dopants, metals, and insulators. The corresponding outputs include integrated circuits (ICs), IC packages, printed circuit boards, and ultimately, various commercial electronic systems and products (such as computers, cellular phones, and digital cameras). The types of processes that arise in semiconductor manufacturing include crystal

Fundamentals of Semiconductor Manufacturing and Process Control,
By Gary S. May and Costas J. Spanos
Copyright © 2006 John Wiley & Sons, Inc.

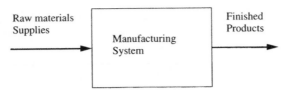

Figure 1.1. Block diagram representation of a manufacturing system.

growth, oxidation, photolithography, etching, diffusion, ion implantation, planarization, and deposition processes.

Viewed from a systems-level perspective, semiconductor manufacturing intersects with nearly all other IC process technologies, including design, fabrication, integration, assembly, and reliability. The end result is an electronic system that meets all specified performance, quality, cost, reliability, and environmental requirements. In this chapter, we provide an overview of semiconductor manufacturing, which touches on each of these intersections.

1.1. HISTORICAL EVOLUTION

Semiconductor devices constitute the foundation of the electronics industry, which is currently (as of 2005) the largest industry in the world, with global sales over one trillion dollars since 1998. Figure 1.2 shows the sales volume of the semiconductor device-based electronics industry since 1980 and projects sales to the year 2010. Also shown are the gross world product (GWP) and the sales volumes

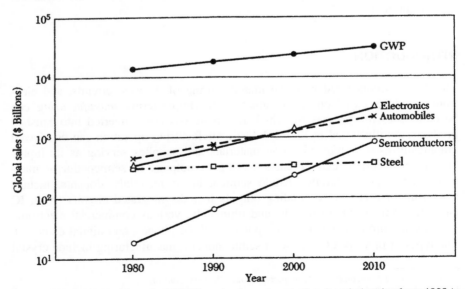

Figure 1.2. Gross world product (GWP) and sales volumes of various industries from 1980 to 2000 and projected to 2010 [1].

of the automobile, steel, and semiconductor industries [1]. If current trends continue, the sales volume of the electronic industry will reach three trillion dollars and will constitute about 10% of GWP by 2010. The semiconductor industry, a subset of the electronics industry, will grow at an even higher rate to surpass the steel industry in the early twenty-first century and to constitute 25% of the electronic industry in 2010.

The multi-trillion-dollar electronics industry is fundamentally dependent on the manufacture of semiconductor integrated circuits (ICs). The solid-state computing, telecommunications, aerospace, automotive, and consumer electronics industries all rely heavily on these devices. A brief historical review of manufacturing and quality control, semiconductor processing, and their convergence in IC manufacturing, is therefore warranted.

1.1.1. Manufacturing and Quality Control

The historical evolution of manufacturing, summarized in Table 1.1, closely parallels the industrialization of Western society, beginning in the nineteenth century. It could be argued that the key early development in manufacturing was the concept of *interchangeable parts*. Eli Whitney is credited with pioneering this concept, which he used for mass assembly of the cotton gin in the early 1800s [2]. In the late 1830s, a Connecticut manufacturer began producing cheap windup clocks by stamping out many of the parts out of sheets of brass. Similarly, in the early 1850s, American rifle manufacturers thoroughly impressed a British delegation by a display in which 10 muskets made in 10 different preceding years were disassembled, had their parts mixed up in a box, and subsequently reassembled quickly and easily. In England at that time, it would have taken a skilled craftsman the better part of a day to assemble a single unit.

The use of interchangeable parts eliminated the labor involved in matching individual parts in the assembly process, resulting in a tremendous time savings and increase in productivity. The adoption of this method required new forms of technology capable of much finer tolerances in production and measurement methods than those required by hand labor. Examples included the

Table 1.1. Major milestones in manufacturing history.

Year(s)	Event
1800–1850	Concept of interchangeable parts introduced
1850–1860	Advances in measurement and machining operations
1875	Taylor introduces scientific management principles
1900–1930	Assembly line techniques actualized by Ford
1924	Control chart introduced by Shewhart
Late 1920s	Dodge and Romig develop acceptance sampling
1950s	Computer numeric control and designed experiments introduced
1970s	Growth in the adoption of statistical experimental design
1980	Pervasive use of statistical methods in many industries

vernier caliper, which allowed workers to measure machine tolerances on small scales, and wire gauges, which were necessary in the production of clock springs. One basic machine operation perfected around this time was mechanical drilling using devices such as the turret lathe, which became available after 1850. Such devices allowed a number of tedious operations (hand finishing of metal, grinding, polishing, stamping, etc.) to be performed by a single piece of equipment using a bank of tool attachments. By 1860, a good number of the basic steps involved in shaping materials into finished products had been adapted to machine functions.

Frederick Taylor added rigor to the manufacturing research and practice by introducing the principles of *scientific management* into mass production industries around 1875 [3]. Taylor suggested dividing work into tasks so that products could be manufactured and assembled more readily, leading to substantial productivity improvements. He also developed the concept of standardized production and assembly methods, which resulted in improved quality of manufactured goods. Along with the standardization of methods came similar standardization in work operations, such as standard times to accomplish certain tasks, or a specified number of units that must be produced in a given work period.

Interchangeable parts also paved the way for the next major contribution to manufacturing: the *assembly line*. Industrial engineers had long noted how much labor is spent in transferring materials between various production steps, compared with the time spent in actually performing the steps. Henry Ford is credited for devising the assembly line in his quest to optimize the means for producing automobiles in the early twentieth century. However, the concept of the assembly line had actually been devised at least a century earlier in the flour mill industry by Oliver Evans in 1784 [2]. Nevertheless, it was not until the concept of interchangeable parts was combined with technology innovations in machining and measurement that assembly line methods were truly actualized in their ultimate form. After Ford, the assembly line gradually replaced more labor-intensive forms of production, such as custom projects or batch processing.

No matter what industry, no one working in manufacturing today can overemphasize the influence of the computer, which catalyzed the next major paradigm shift manufacturing technology. The use of the computer was the impetus for the concept of *computer numeric control* (CNC), introduced in the 1950s [4]. Numeric control was actually developed much earlier. The player piano is a good example of this technique. This instrument utilizes a roll of paper with holes punched in it to determine whether a particular note is played. The numeric control concept was enhanced considerably by the invention of the computer in 1943. The first CNC device was a spindle milling machine developed by John Parsons of MIT in 1952. CNC was further enhanced by the use of microprocessors for control operations, beginning around 1976. This made CNC devices sufficiently versatile that an existing tooling could be quickly reconfigured for different processes. This idea moved into semiconductor manufacturing more than a decade later when the machine communication standards made it possible to have factorywide production control.

The inherent accuracy and repeatability engendered by the use of the computer eventually enabled the concept of *statistical process control* to gain a foothold in manufacturing. However, the application of statistical methods actually had a long prior history. In 1924, Walter Shewhart of Bell Laboratories introduced the control chart. This is considered by many as the formal beginning of statistical quality control. In the late 1920s, Harold Dodge and Harry Romig, both also of Bell Labs, developed statistically based acceptance sampling as an alternative to 100% inspection. By the 1950s, rudimentary computers were available, and *designed experiments* for product and process improvement were first introduced in the United States. The initial applications for these techniques were in the chemical industry. The spread of these methods to other industries was relatively slow until the late 1970s, when their further adoption was spurred by economic competition between Western companies and the Japanese, who had been systematically applying designed experiments since the 1960s. Since 1980, there has been profound and widespread growth in the use of statistical methods worldwide, and particularly in the United States.

1.1.2. Semiconductor Processes

Many important semiconductor technologies were derived from processes invented centuries ago. Some of the key technologies are listed in Table 1.2 in chronological order. For the most part, these techniques were developed independently from the evolution of manufacturing science and technology. For example, the growth of metallic crystals in a furnace was pioneered by Africans living on the

Table 1.2. Major milestones in semiconductor processing history.

Year	Event
1798	Lithography process invented
1855	Fick proposes basic diffusion theory
1918	Czochralski crystal growth technique invented
1925	Bridgman crystal growth technique invented
1952	Diffusion used by Pfann to alter conductivity of silicon
1957	Photoresist introduced by Andrus; oxide masking developed by Frosch and Derrick; epitaxial growth developed by Sheftal et al.
1958	Ion implantation proposed by Shockley
1959	Kilby and Noyce invent the IC
1963	CMOS concept proposed by Wanlass and Sah
1967	DRAM invented by Dennard
1969	Self-aligned polysilicon gate process proposed by Kerwin et al.; MOCVD developed by Manasevit and Simpson
1971	Dry etching developed by Irving et al.; MBE developed by Cho; first microprocessor fabricated by Intel
1982	Trench isolation technology introduced by Rung et al.
1989	CMP developed by Davari et al.
1993	Copper interconnect introduced to replace aluminum by Paraszczak et al.

western shores of Lake Victoria more than 2000 years ago [5]. This process was used to produce carbon steel in preheated forced-draft furnaces. Another example is the lithography process, which was invented in 1798. In this first process, the pattern, or image, was transferred from a stone plate (*lithos*) [6]. The diffusion of impurity atoms in semiconductors is also important for device processing. Basic diffusion theory was described by Fick in 1855 [7].

In 1918, Czochralski developed a liquid–solid monocomponent growth technique used to grow most of the crystals from which silicon wafers are produced [8]. Another growth technique was developed by Bridgman in 1925 [9]. The Bridgman technique has been used extensively for the growth of gallium arsenide and related compound semiconductors. The idea of using diffusion techniques to alter the conductivity in silicon was disclosed in a patent by Pfann in 1952 [10]. In 1957, the ancient lithography process was applied to semiconductor device fabrication by Andrus [11], who first used photoresist for pattern transfer. Oxide masking of impurities was developed by Frosch and Derrick in 1957 [12]. In the same year, the epitaxial growth process based on chemical vapor deposition was developed by Sheftal et al. [13]. In 1958, Shockley proposed the method of using ion implantation to precisely control the doping of semiconductors [14].

In 1959, the first rudimentary integrated circuit was fabricated from germanium by Kilby [15]. Also in 1959, Noyce proposed the monolithic IC by fabricating all devices in a single semiconductor substrate and connecting the devices by aluminum metallization [16]. As the complexity of the IC increased, the semiconductor industry moved from NMOS (*n*-channel MOSFET) to CMOS (complementary MOSFET) technology, which uses both NMOS and PMOS (*p*-channel MOSFET) processes to form the circuit elements. The CMOS concept was proposed by Wanlass and Sah in 1963 [17]. In 1967, the dynamic random access memory (DRAM) was invented by Dennard [18].

To improve device reliability and reduce parasitic capacitance, the self-aligned polysilicon gate process was proposed by Kerwin et al. in 1969 [19]. Also in 1969, the metallorganic chemical vapor deposition (MOCVD) method, an important epitaxial growth technique for compound semiconductors, was developed by Manasevit and Simpson [20]. As device dimensions continued to shrink, dry etching was developed by Irving et al. in 1971 to replace wet chemical etching for high-fidelity pattern transfer [21]. Another important technique developed in the same year by Cho was molecular-beam epitaxy (MBE) [22]. MBE has the advantage of near-perfect vertical control of composition and doping down to atomic dimensions. Also in 1971, the first monolithic microprocessor was fabricated by Hoff et al. at Intel [23]. Currently, microprocessors constitute the largest segment of the industry.

Since 1980, many new technologies have been developed to meet the requirements of continuously shrinking minimum feature lengths. Trench technology was introduced by Rung et al. in 1982 to isolate CMOS devices [24]. In 1989, the chemical–mechanical polishing (CMP) method was developed by Davari et al. for global planarization of the interlayer dielectrics [25]. Although aluminum has been used since the early 1960s as the primary IC interconnect material, copper interconnect was introduced in 1993 by Paraszczak et al. to replace aluminum for minimum feature lengths approaching 100 nm [26].

1.1.3. Integrated Circuit Manufacturing

By the beginning of the 1980s, there was deep and widening concern about the economic well-being of the United States. Oil embargoes during the previous decade had initiated two energy crises and caused rampant inflation. The U.S. electronics industry was no exception to the economic downturn, as Japanese companies such as Sony and Panasonic nearly cornered the consumer electronics market. The U.S. computer industry experienced similar difficulties, with Japanese semiconductor companies beginning to dominate the memory market and establish microprocessors as the next target.

Then, as now, the fabrication of ICs was extremely expensive. A typical state-of-the-art, high-volume manufacturing facility at that time cost over a million dollars (and now costs several billion dollars) [27]. Furthermore, unlike the manufacture of discrete parts such as appliances, where relatively little rework is required and a yield greater than 95% on salable product is often realized, the manufacture of integrated circuits faced unique obstacles. Semiconductor fabrication processes consisted of hundreds of sequential steps, with potential yield loss occurring at every step. Therefore, IC manufacturing processes could have yields as low as 20–80%.

Because of rising costs, the challenge before semiconductor manufacturers was to offset large capital investment with a greater amount of automation and technological innovation in the fabrication process. The objective was to use the latest developments in computer hardware and software technology to enhance manufacturing methods. In effect, this effort in *computer-integrated manufacturing of integrated circuits* (IC-CIM) was aimed at optimizing the cost-effectiveness of IC manufacturing as *computer-aided design* (CAD) had dramatically affected the economics of circuit design.

IC-CIM is designed to achieve several important objectives, including increasing chip fabrication yield, reducing product cycle time, maintaining consistent levels of product quality and performance, and improving the reliability of processing equipment. Table 1.3 summarizes the results of a 1986 study by Toshiba that analyzed the use of IC-CIM techniques in producing 256-kbyte DRAM memory circuits [28]. This study showed that CIM techniques improved the manufacturing process on each of the four productivity metrics investigated.

Table 1.3. Results of 1986 Toshiba study.

Productivity Metric	Without CIM	With CIM
Turnaround time	1.0	0.58
Integrated unit output	1.0	1.50
Average equipment uptime	1.0	1.32
Direct labor hours	1.0	0.75

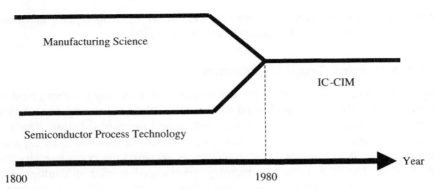

Figure 1.3. Timeline indicating convergence of manufacturing science and semiconductor processing into IC-CIM.

In addition to the demonstration of the effectiveness of IC-CIM techniques, economic concerns were so great in the early to mid-1980s that the Reagan Administration took the unprecedented step of partially funding a consortium of U.S. IC manufacturers—including IBM, Intel, Motorola, and Texas Instruments—to perform cooperative research and development on semiconductor manufacturing technologies. This consortium, SEMATECH, officially began operations in 1988 [29]. This sequence of events signaled the convergence of advances in manufacturing science and semiconductor process technology, and also heralded the origin of a more systematic and scientific approach to semiconductor manufacturing. This convergence is illustrated in Figure 1.3.

1.2. MODERN SEMICONDUCTOR MANUFACTURING

The modern semiconductor manufacturing process sequence is the most sophisticated and unforgiving volume production technology that has ever been practiced successfully. It consists of a complex series of hundreds of unit process steps that must be performed very nearly flawlessly.

This semiconductor manufacturing process can be defined at various levels of abstraction. For example, each process step has inputs, outputs, and specifications. Each step can also be modeled, either physically, empirically, or both. Much can be said about the technology of each step, and more depth in this area is provided in Chapter 2. At a higher level of abstraction, multiple process steps are linked together to form a process sequence. Between some of these links are inspection points, which merely produce information without changing the product. The flow and utilization of information occurs at another level of abstraction, which consists of various control loops. Finally, the organization of the process belongs to yet another level of abstraction, where the objective is to maximize the efficiency of product flow while reducing variability.

1.2.1. Unit Processes

It is difficult to discuss unit process steps outside the context of a process flow. Figures 1.4 and 1.5 show the major unit processes used in a simple process flow. These steps include oxidation, photolithography, etching, ion implantation, and metallization. We describe these steps briefly in this section via a simple sequence used to fabricate a $p-n$ junction [1].

The development of a high-quality silicon dioxide (SiO_2) has helped to establish the dominance of silicon in the production of commercial ICs. Generally, SiO_2 functions as an insulator in a number of device structures or as a barrier to diffusion or implantation during device fabrication. In the fabrication of a $p-n$ junction (Figure 1.4), the SiO_2 film is used to define the junction area. There are two SiO_2 growth methods, dry and wet oxidation, depending on whether

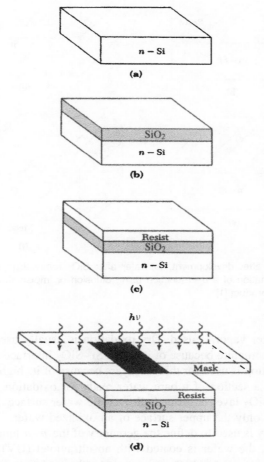

Figure 1.4. (a) A bare n-type silicon wafer; (b) an oxidized silicon wafer; (c) application of photoresist; (d) resist exposure through a mask [1].

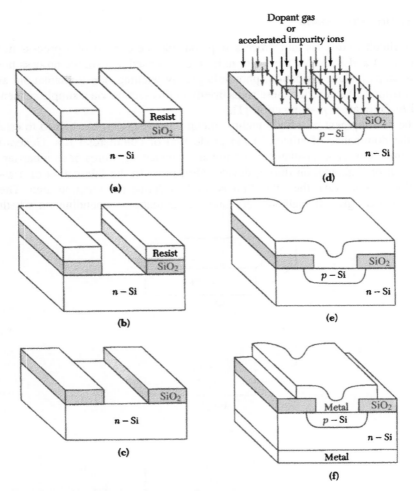

Figure 1.5. (a) Wafer after development; (b) wafer after SiO$_2$ removal; (c) result after photolithography; (d) formation of a p–n junction using diffusion or implantation; (e) wafer after metallization; (f) final product [1].

dry oxygen or water vapor is used. Dry oxidation is usually used to form thin oxides in a device structure because of its good Si–SiO$_2$ interface characteristics, whereas wet oxidation is used for thicker layers because of its higher growth rate. Figure 1.4a shows a section of a bare wafer ready for oxidation. After the oxidation process, a SiO$_2$ layer is formed all over the wafer surface. For simplicity, Figure 1.4b shows only the upper surface of an oxidized wafer.

Photolithography is used to define the geometry of the p–n junction. After the formation of SiO$_2$, the wafer is coated with an ultraviolet (UV) light-sensitive material called *photoresist*, which is spun onto the wafer surface. Afterward (Figure 1.4c), the wafer is baked to drive the solvent out of the resist and to harden the resist for improved adhesion. Figure 1.4d shows the next step, which is to expose the wafer through a patterned mask using an UV light source. The exposed region of the photoresist-coated wafer undergoes a chemical reaction. The exposed area becomes polymerized and difficult to remove in an etchant. The polymerized region remains when the wafer is placed in a developer, whereas the unexposed region dissolves away. Figure 1.5a shows the wafer after the development. The wafer is again baked to enhance the adhesion and improve the resistance to the subsequent etching process. Then, an etch using hydrofluoric acid (HF) removes the unprotected SiO_2 surface (Figure 1.5b). Last, the resist is stripped away by a chemical solution or an oxygen plasma. Figure 1.5c shows the final result of a region without oxide (a window) after the lithography process. The wafer is now ready for forming the $p-n$ junction by a diffusion or ion implantation process.

In diffusion, the wafer surface not protected by the oxide is exposed to a source with a high concentration of an opposite-type impurity. The impurity moves into the semiconductor crystal by solid-state diffusion. In ion implantation, the intended impurity is introduced into the wafer by accelerating the impurity ions to a high energy level and then implanting the ions in the semiconductor. The SiO_2 layer serves as barrier to impurity diffusion or ion implantation. After diffusion or implantation, the $p-n$ junction is formed (Figure 1.5d).

After diffusion or ion implantation, a metallization process is used to form ohmic contacts and interconnections (Figure 1.5e). Metal films can be formed by physical vapor deposition or chemical vapor deposition. The photolithography process is again used to define the front contact, which is shown in Figure 1.5f. A similar metallization step is performed on the back contact without using a photolithography process.

1.2.2. Process Sequences

Semiconductor manufacturing consists of a series of sequential process steps like the one described in the previous section in which layers of materials are deposited on substrates, doped with impurities, and patterned using photolithography to produce ICs. Figure 1.6 illustrates the interrelationship between the major process steps used for IC fabrication. Polished wafers with a specific resistivity and orientation are used as the starting material. The film formation steps include thermally grown oxide films, as well as deposited polysilicon, dielectric, and metal films. Film formation is often followed by photolithography or impurity doping. Photolithography is generally followed by etching, which in turn is often followed by another impurity doping or film formation. The final IC is made by sequentially transferring the patterns from each mask, level by level, onto the surface of the semiconductor wafer.

After processing, each wafer contains hundreds of identical rectangular chips (or dies), typically between 1 and 20 mm on each side, as shown in Figure 1.7a. The chips are separated by sawing or laser cutting; Figure 1.7b shows a separated chip. Schematic top views of a single MOSFET and a single bipolar transistor

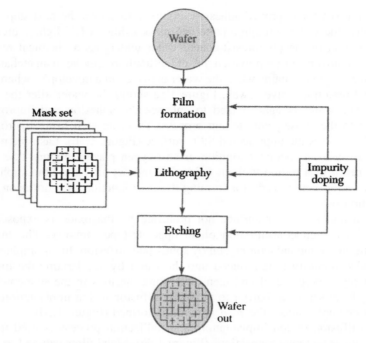

Figure 1.6. Flow diagram for generic IC process sequence [1].

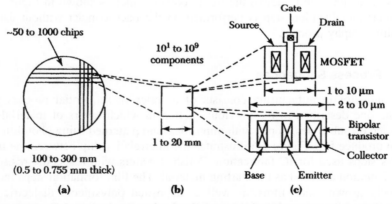

Figure 1.7. (a) Semiconductor wafer; (b) IC chip; (c) MOSFET and bipolar transistor [1].

are shown in Figure 1.7c. Inserted into this process sequence are various points at which key measurements are performed to ensure product quality.

1.2.3. Information Flow

The vast majority of quantitative evaluation of semiconductor manufacturing processes is accomplished via IC-CIM systems. The interdependent issues of ensuring high yield, high quality, and low cycle time are addressed by several critical capabilities in a state-of-the-art IC-CIM system: work-in-process (WIP) monitoring, equipment communication, data acquisition and storage, process/ equipment modeling, and process control, to name only a few. The emphasis of each of these activities is to increase throughput and prevent potential mis-processing, but each presents significant engineering challenges in their effective implementation and deployment.

A block diagram of a typical modern IC-CIM system is shown in Figure 1.8. This diagram outlines many of the key features required for efficient information flow in manufacturing operations [28]. The lower level of this two-level architecture includes embedded controllers that provide real-time control and analysis of fabrication equipment. These controllers consist of personal computers and the associated control software dedicated to each individual piece of equipment. The second level of this IC-CIM architecture consists of a distributed local-area network of computer workstations and file servers linked by a common distributed database.

Equipment communication with host computers is facilitated by an electronics manufacturing standard called the *generic equipment model* (GEM). The GEM standard is used in both semiconductor manufacturing and printed circuit board assembly. This standard is based on the *semiconductor equipment communications standard* (SECS) protocol. SECS is a standard for communication between intelligent equipment and a host. The SECS standard has two components that define the communications protocol (SECS-I) and the messages exchanged (SECS-II), respectively. SECS-I specifies point-to-point communications over a high-speed messaging service interface. GEM is a standard set of SECS capabilities that can be selected by users as needed to coordinate equipment control in an automated factory. The GEM standard defines semiconductor

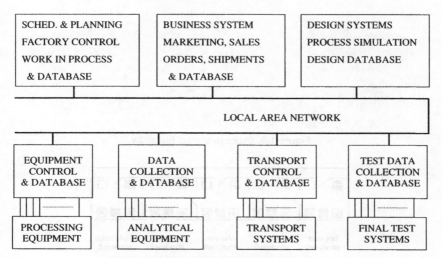

Figure 1.8. Two-level IC-CIM architecture [28].

equipment behavior as viewed through a communication link in terms of SECS-II messages communicated over that link. The GEM standard impacts equipment control and equipment–host communication and enables equipment to be integrated quickly and efficiently with a host computer [30].

The flow of information in this type of IC-CIM architecture enables equipment and process control at several levels. The highest level can be thought of as *supervisory control*, where the progression of a substrate is tracked from process to process. At this level, adjustments can be made to subsequent process steps to account for variation in previous procedures. The next lower level of control occurs on a *run-by-run* basis. For a single process, adjustments are made after each run to account for shifts and drifts that occur from wafer to wafer. Occasional shifts may occur when a new operator takes over or preventive maintenance is performed. A process may also experience drift due to equipment aging. *Real-time control* is at the lowest level of the hierarchy. In this case, adjustments are made to a process during a run to account for in situ disturbances. This hierarchy is diagramed in Figure 1.9.

1.2.4. Process Organization

As mentioned previously, the overall objective of process organization is to maximize the efficiency of product flow while minimizing variability and yield loss. Modern semiconductor factories [known as "fabs" (Fabrication Facilities)] are typically organized into *workcells*. In this approach, all the necessary equipment for completing a given process step is placed in the same room (see Figure 1.10). The workcell layout optimizes product flow, resulting in a minimal average distance traveled by semiconductor wafers as they migrate through the fabrication

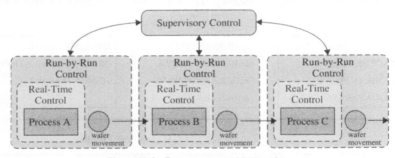

Figure 1.9. Process control hierarchy.

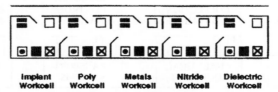

Figure 1.10. Workcell layout in a modern IC fabrication facility.

facility. This reduced distance translates into fewer chances for wafer mishandling and potential loss of product. Furthermore, the trend in equipment development since the mid-1990s has been toward single-wafer processing systems that enable enhanced reproducibility.

This modern IC factory also draws on powerful computing concepts, resulting in a highly flexible manufacturing system. In addition to the actual processing equipment, the factory consists of advanced in situ, postprocess, and end-of-the-line metrology and instrumentation necessary to quality control, equipment maintenance and diagnosis, rapid failure recovery, and inventory management. The physical factory is also augmented by simulation tools that allow various scenarios to be evaluated in a virtual manufacturing environment.

1.3. GOALS OF MANUFACTURING

From a systems-level perspective, semiconductor manufacturing intersects with design, fabrication, integration, assembly, and reliability. The fundamental goals of manufacturing are to tie all of these technologies together to achieve finished products with

- Low cost
- High quality
- High reliability

Cost is most directly impacted by yield and throughput. *Yield* is the proportion of products that meet the required performance specifications. Yield is inversely proportional to cost; that is, the higher the yield, the lower the cost. *Throughput* refers to the number of products processed per unit time. High throughput also leads to lower cost. The *quality* goal is virtually self-explanatory. It is obviously desirable to produce high-quality ICs that can be efficiently and repeatably mass-produced with a high degree of uniformity. Quality is derived from a stable and well-controlled manufacturing process. The *reliability* of electronic products is also impacted by the manufacturing process. High reliability results from the minimization of manufacturing faults. If each of the abovementioned goals is fulfilled, the end result is an IC that meets all specified performance, quality, cost, and reliability requirements.

1.3.1. Cost

Understanding the economics of IC manufacturing is important not only to the manufacturer but also to buyers and designers. A general rule of thumb is that IC fabrication, testing, and packaging each contribute about one-third of the total product cost. A variety of factors contribute to overall product costs, including the following [31]:

- Wafer processing cost

- Wafer processing yield
- Die size
- Wafer probe cost
- Probe yield
- Number of good dies
- Package cost
- Assembly yield
- Final test cost
- Final test yield

Wafer processing cost depends on wafer size, raw-wafer cost, direct labor cost, facility cost, and direct factory overhead (i.e., indirect labor costs, utilities, and maintenance). Direct costs (i.e., raw-wafer cost and direct labor) typically account for 10–15% of the wafer processing cost, with the remaining indirect costs accounted for by the equipment and facility depreciation, engineering support, facility operating costs, production control, and direct factory overhead. Increasingly, equipment costs contribute the lion's share, accounting for over 70% of the total indirect cost.

Currently, IC fabrication cost (excluding design costs) is about $4/cm^2 at mature production levels [32]. The cost per IC to produce N chips (or equivalently, N circuit functions) is proportional to e^{kN}, where k is a constant proportional to the cost of assembly and testing [33]. The interplay between these factors and their impact on cost is illustrated in Figure 1.11. Cost per IC is minimized by maximizing both the number of chips per wafer and the proportion of good chips (also known as the *yield*).

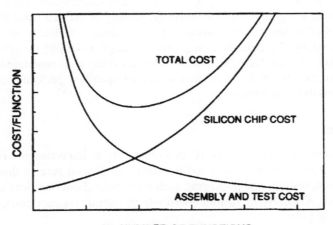

Figure 1.11. Cost per function versus number of functions [33].

1.3.2. Quality

Quality is among the most important factors in any manufacturing process. Understanding and improving quality are key ingredients to business success, growth, and enhanced competitiveness. Significant return on investment may be realized from adequate attention to continuous quality improvement as an integral part of an overall business strategy.

The term "quality" may be defined in many ways. The traditional definition is based on the notion that products must meet the requirements of those who use them. Thus, in this text, we adopt the simple definition of "fitness for use." This definition encompasses two general aspects: quality of design and quality of conformance. *Quality of design* is affected by choices in fabrication materials, component specifications, product size, and other features. *Quality of conformance* addresses how well a product conforms to the specifications required by the design. Quality of conformance is impacted by the manufacturing process, equipment performance, competence and training of the workforce, and the implementation of quality control procedures.

1.3.3. Variability

An alternative definition of quality of conformance is "the inverse of variability." This definition implies that quality may be improved by reducing variation in the various figures of merit that define product performance. This reduced variability translates directly into lower manufacturing costs due to less misprocessing, rework, and waste. Thus, processes in which the degree of quality is repeatable with a high degree of uniformity are preferred.

Unfortunately, a certain amount of variability is inherent in every product. No two products are ever completely identical. For example, the dimensions of two thin metal films used for IC interconnect will vary according to the precise conditions and equipment used to deposit and pattern the films. Small variations might have negligible impact on the final product, but large variations can lead to final products that are unacceptable. *Quality improvement* is defined as the reduction of such variability in processes and products. Since variability is usually described in statistical terms, statistical methods are necessary for quality improvement efforts.

1.3.4. Yield

As previously mentioned, IC cost is minimized by maximizing both the number of chips produced (i.e., the throughput) and the proportion of functionally operational chips per wafer. The latter parameter is known as the *yield*. As a consequence of its direct impact on manufacturing cost, yield is perhaps the most important figure of merit in semiconductor manufacturing.

Yield improvement achieved over time is referred to as "yield learning." Strategies for accelerated yield learning are critical for the economic viability of semiconductor manufacturing operations, as illustrated in Figure 1.12. Business goals

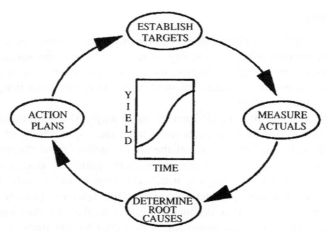

Figure 1.12. Yield learning cycle [33].

drive yield targets. The actual yield achieved is regularly monitored and tracked against those targets. The root causes of yield detractors are systematically identified and analyzed. Appropriate action plans to eliminate these causes are subsequently developed and implemented. Once target yields are achieved, they are modified (usually increased), and the learning cycle is repeated. It is of paramount importance to both reduce the duration of this cycle and optimize its efficiency.

1.3.5. Reliability

Another dimension of the quality of electronic products is their reliability. *Reliability* is a characteristic of a product that is associated with the probability that it will perform its intended function under specified conditions for a stated period of time. The enhancement of reliability is accomplished by failure-mode analysis, which is aimed at identifying the mechanisms for failure and translating this information into remedies that impact design and manufacturing processes. Reliability is usually quantified by statistical inference techniques applied to a suitable population of devices that have undergone extensive testing and failure-mode analysis. However, although the reliability of integrated circuits is often directly impacted by the manufacturing process, a detailed study of reliability and associated topics is beyond the scope of this text. Readers interested in a more thorough treatment are referred to Nash [34].

1.4. MANUFACTURING SYSTEMS

In general, manufacturing systems may be subdivided into two categories: (1) continuous-flow manufacturing and (2) discrete-parts manufacturing. *Continuous-flow* manufacturing involves chemical or physical processes that change the state of the part before the part is connected to other components to form a finished product. Most of the unit processes used in IC fabrication prior

Figure 1.13. Printed circuit board for a single-board engine controller [35].

to wafer dicing and packaging are continuous flow manufacturing operations. *Discrete-parts manufacturing*, on the other hand, refers to the assembly of distinct pieces to yield a final product. In microelectronics manufacturing, an example of a product assembled using discrete parts manufacturing is a printed circuit board (PCB) populated by individual ICs (as shown in Figure 1.13).

1.4.1. Continuous Flow

Continuous-flow manufacturing refers to processing operations that do not involve assembly of discrete parts. For continuous-flow manufacturing operations, the process inputs (see Figure 1.1) are the semiconductor substrate and raw materials such as dopants, insulators, and metals. Continuous-flow processes consist of the steps such as those described in Section 1.2. These processes may be further subdivided into *batch* and *single-workpiece* (or *single-wafer*) operations.

1.4.1.1. Batch Processes

Batch processes are those that operate on multiple products simultaneously. In IC manufacturing, the items being processed are semiconductor wafers, and the batches are called "lots." State-of-the-art semiconductor manufacturing factories employ a plethora of batch fabrication equipment, such as furnaces for high-volume wafer processing. This facilitates factory throughputs on the order of tens of thousands of wafers processed per month.

An example of a batch process in semiconductor manufacturing is chemical vapor deposition (CVD; see Chapter 2). Figure 1.14 shows a schematic of a typical CVD furnace. In this type of hot-wall, reduced-pressure reactor, the quartz furnace tube is heated in three individual zones, and reactive gas is introduced at one end and pumped out the opposite end. The wafers are placed vertically side-by-side in a container known as a "boat." Gas reacting on the surface of the wafers causes the desired thin films to be deposited.

However, the high manufacturing throughput that is characteristic of batch processing is often achieved at the expense of uniformity and process control. In the case of CVD, for example, wafers farthest from the gas inlet may exhibit lower deposition rates as a result of the reduced availability of reactant gases, which are consumed by reactions closer to the inlet. This effect can be compensated for somewhat by increasing the deposition temperature in each subsequent reaction zone from the inlet.

1.4.1.2. Single Workpiece

Single-workpiece manufacturing operations involve individual items processed one at a time. In IC manufacturing, the workpiece is the semiconductor wafer. As wafer sizes have grown over the years, single-wafer processing approaches have proliferated. This has occurred for several reasons. First, scaling up batch tools and maintaining uniformity across the wafer surface becomes more difficult for wafers 200 mm in diameter and larger. At the same time, for submicrometer features, it is nearly impossible to maintain features size control across these large wafers. In addition, when only a single wafer is processed at a time, if any flaw in a process step is detected, it can be corrected before the next wafer is

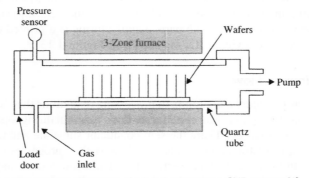

Figure 1.14. Example of a batch process: a CVD reactor [1].

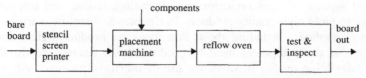

Figure 1.15. Surface mount printed circuit board assembly process.

processed. Finally, process development for large wafers has become increasingly expensive, and these costs are mitigated by the single wafer approach.

Therefore, semiconductor manufacturing has evolved from primarily a batch operation to an increasingly single-wafer operation. The advantages of single-wafer processing include (1) lower overall factory cost, (2) enhanced observability by in situ sensors for more robust process control, (3) rapid manufacturing cycle time, and (4) increased flexibility for manufacturing numerous products based on different technologies [36].

1.4.2. Discrete Parts

In discrete-parts microelectronics manufacturing, the inputs to the manufacturing system are the bare printed circuit board and the various circuit components. For example, in surface mount assembly, the manufacturing process consists of the following (Figure 1.15):

1. Screen-printing solder paste onto the bonding pads of the circuit board with a stencil printer
2. Placing the circuit components (ICs and passives) onto the pad locations using a placement machine
3. Melting the solder paste in a reflow oven to form the connection between components and the pads
4. Testing and inspecting the populated board for quality control

Following attachment of the ICs, the output of the process is the fully interconnected and populated circuit board.

The output of the process is a populated circuit board that is ready for integration into an electronic system.

1.5. OUTLINE FOR REMAINDER OF THE BOOK

In Chapter 2 we will provide a broad overview of the manufacturing technology and process flows used to produce a variety of semiconductor products. The individual unit processes used in fabricating ICs, as well as techniques for process integration and IC packaging, will be discussed. The unit processes include oxidation, photolithography, doping, etching, thin-film deposition, and planarization. The integrated process flows, which focus on silicon technology, include the complementary metal–oxide–semiconductor (CMOS), bipolar, and BiCMOS processes.

For all aspects of semiconductor manufacturing, testing and inspection are necessary to yield high-quality products. In Chapter 3, therefore, various process monitoring methods, including those that focus on product wafers and those that focus on the equipment used to produce those wafers, are discussed in detail. Maintaining quality involves the use of *statistical process control* (SPC). Since product variability is often described in statistical terms, statistical methods will necessarily play a central role in quality control and improvement efforts. Therefore, Chapter 4 will provide a review of statistical fundamentals.

Ultimately, the key metric to be used to evaluate any manufacturing process is cost, and cost is directly impacted by yield. *Yield* refers to the proportion of manufactured products that perform as required by a set of specifications. Yield is inversely proportional to the total manufacturing cost—the higher the yield, the lower the cost. *Yield modeling* is presented in Chapter 5. Chapter 6 will then focus on the use of SPC to analyze quality issues and improve yield.

A designed experiment is an extremely useful tool for discovering key variables that influence quality characteristics. *Statistical experimental design* is a powerful approach for systematically varying controllable process conditions and determining their impact on output parameters that measure quality. Data derived from such experiments can then be used to construct *process models* of various types that enable the analysis and prediction of manufacturing process behavior. Statistical experimental design is presented in Chapter 7, and process modeling concepts are introduced in Chapter 8.

Finally, several advanced process control topics are the subject of Chapters 9 and 10. These topics include run-by-run control of unit processes and supervisory control of process sequences, as well as the diagnosis of process and equipment malfunctions.

SUMMARY

In this chapter, we have provided background and motivation for the study of semiconductor manufacturing and process control. We have done so by surveying the history of integrated circuit processing, describing the attributes of manufacturing systems, and discussing the goals and objectives of modern electronics manufacturing operations. In so doing, this chapter has provided a foundation for the various issues relevant to semiconductor manufacturing that will be presented in the remainder of the book.

PROBLEMS

1.1. List the input and output parameters of a typical semiconductor manufacturing process.

1.2. What were the key milestones in the historical evolution of semiconductor manufacturing? How did the evolution of semiconductor process technology interact with and impact the development of manufacturing technology?

1.3. Describe the basic unit processes involved in IC fabrication and the sequences in which they are performed to yield products.

1.4. What is the significance of information flow within an IC factory?

1.5. Explain the differences between real-time, run-by-run, and supervisory control.

1.6. Why are IC factories organized into workcells?

1.7. List and prioritize the overall goals of IC manufacturing.

1.8. What is the difference between continuous-flow and discrete manufacturing processes? What role do each of these play in semiconductor manufacturing?

1.9. Why has semiconductor manufacturing evolved from batch operations toward single-wafer operations? What are the advantages and disadvantages of each approach?

REFERENCES

1. G. May and S. Sze, *Fundamentals of Semiconductor Fabrication*, Wiley, New York 2002.

2. G. Gunderson, *A New Economic History of America*, McGraw-Hill, New York, 1976.

3. D. Montgomery, *Introduction to Statistical Quality Control*, 3rd ed., Wiley, New York, 1997.

4. J. Stenerson and K. Curran, *Computer Numerical Control*, Prentice-Hall, Upper Saddle River, NJ, 1997.

5. D. Shore, "Steel-Making in Ancient Africa," in *Blacks in Science: Ancient and Modern*, I. Van Sertima, ed., Transaction Books, New Brunswick, NJ, 1986, p. 157.

6. M. Hepher, "The Photoresist Story," *J. Photo. Sci.* **12**, 181 (1964).

7. A. Fick, "Ueber Diffusion," *Ann. Phys. Lpz.* **170**, 59 (1855).

8. J. Czochralski, "Ein neues Verfahren zur Messung der Kristallisationsgeschwindigkeit der Metalle," *Z. Phys. Chem.* **92**, 219 (1918).

9. P. W. Bridgman, "Certain Physical Properties of Single Crystals of Tungsten, Antimony, Bismuth, Tellurium, Cadmium, Zinc, and Tin," *Proc. Am. Acad. Arts Sci.* **60**, 303 (1925).

10. W. G. Pfann, *Semiconductor Signal Translating Device*, U.S. Patent 2,597,028 (1952).

11. J. Andrus, *Fabrication of Semiconductor Devices*, U.S. Patent 3,122,817 (filed 1957; granted 1964).

12. C. J. Frosch and L. Derrick, "Surface Protection and Selective Masking during Diffusion in Silicon," *J. Electrochem. Soc.* **104**, 547 (1957).

13. N. N. Sheftal, N. P. Kokorish, and A. V. Krasilov, "Growth of Single-Crystal Layers of Silicon and Germanium from the Vapor Phase," *Bull. Acad. Sci USSR, Phys. Ser.* **21**, 140, (1957).

14. W. Shockley, *Forming Semiconductor Device by Ionic Bombardment*, U.S. Patent 2,787,564 (1958).

15. J. S. Kilby, "Invention of the Integrated Circuit," *IEEE Trans. Electron Devices* **ED-23**, 648 (1976).

16. R. N. Noyce, *Semiconductor Device-and-Lead Structure*, U.S. Patent 2,981,877 (filed 1959, granted 1961).

17. F. M. Wanlass and C. T. Sah, "Nanowatt Logics Using Field-Effect Metal-Oxide Semiconductor Triodes," *Tech. Digest IEEE Int. Solid-State Circuit Conf.*, 1963, p. 32.

18. R. M. Dennard, *Field Effect Transistor Memory*, U.S. Patent 3,387,286 (filed 1967, granted 1968).

19. R. E. Kerwin, D. L. Klein, and J. C. Sarace, *Method for Making MIS Structure*, U.S. Patent 3,475,234 (1969).

20. H. M. Manasevit and W. I. Simpson, "The Use of Metal–Organic in the Preparation of Semiconductor Materials. I. Epitaxial Gallium-V Compounds," *J. Electrochem. Soc.* **116**, 1725 (1969).

21. S. M. Irving, K. E. Lemons, and G. E. Bobos, *Gas Plasma Vapor Etching Process*, U.S. Patent 3,615,956 (1971).

22. A. Y. Cho, "Film Deposition by Molecular Beam Technique," *J. Vac. Sci. Technol.* **8**, S31 (1971).

23. R. Slater, *Portraits in Silicon*, MIT Press, Cambridge, MA, 1987, p. 175.

24. R. Rung, H. Momose, and Y. Nagakubo, "Deep Trench Isolated CMOS Devices," *Tech. Digest. IEEE Int. Electron Devices Meet.*, 1982, p. 237.

25. B. Davari et al., "A New Planarization Technique, Using a Combination of RIE and Chemical Mechanical Polish (CMP)," *Tech. Digest IEEE Int. Electron Devices Meet.*, 1989, p. 61.

26. J. Paraszczak et al., "High Performance Dielectrics and Processes for ULSI Interconnection Technologies," *Tech. Digest IEEE Int. Electron Devices Meet.*, 1993, p. 261.

27. M. Dax, "Top Fabs of 1996," *Semiconductor Intl.* **19**(5) (1996).

28. D. Hodges, L. Rowe, and C. Spanos, "Computer-Integrated Manufacturing of VLSI," *Proc. IEEE/CHMT Intl. Electron. Manuf. Tech. Symp.*, 1989.

29. IEEE Electron Devices Society, *Fifty Years of Electron Devices*, IEEE, Piscataway, NJ, 2002.

30. J. Moyne, E. del Castillo, and A. Hurwitz, *Run-to-Run Control in Semiconductor Manufacturing*, CRC Press, Boca Raton, FL, 2001.

31. M. Penn, "Economics of Semiconductor Production," *Microelectron. J.* **23**, 255–265 (1992).

32. R. Tummala (ed.), *Fundamentals of Microsystems Packaging*, McGraw-Hill, New York, 2001.

33. A. Landzberg, *Microelectronics Manufacturing Diagnostics Handbook*, Van Nostrand Reinhold, New York, 1993.

34. F. Nash, *Estimating Device Reliability: Assessment of Credibility*, Kluwer Academic Publishers, Boston, MA, 1993.

35. W. Brown (ed.), *Advanced Electronic Packaging*, IEEE Press, New York, 1999.

36. M. Moslehi, R. Chapman, M. Wong, A. Paranjpe, H. Najm, J. Kuehne, R. Yeakley, and C. Davis, "Single-Wafer Integrated Semiconductor Device Processing," *IEEE Trans. Electron Devices* **39**(1) (Jan. 1992).

2

TECHNOLOGY OVERVIEW

OBJECTIVES

- Provide an overview of the critical unit processes in semiconductor manufacturing.
- Describe the integration of such processes into sequences for fabricating specific technology families.

INTRODUCTION

Planar fabrication technology is used extensively for integrated circuit manufacturing. In Section 1.2.1 we briefly described the major steps of a planar process. We provide a more thorough description of these steps, as well as their integration for particular technology families, in this chapter. However, this treatment is in no way intended to be comprehensive. More complete and detailed discussions can be found in several other texts, such as *Fundamentals of Semiconductor Fabrication* [1], for example.

2.1. UNIT PROCESSES

Chapter 1 provided an introduction to the key unit process steps in IC fabrication, including oxidation, photolithography, etching, ion implantation, and metallization. This was accomplished using the description of the process sequence used

Fundamentals of Semiconductor Manufacturing and Process Control,
By Gary S. May and Costas J. Spanos
Copyright © 2006 John Wiley & Sons, Inc.

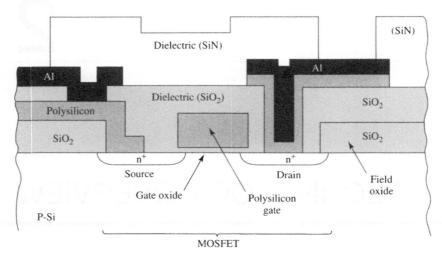

Figure 2.1. Cross section of a MOSFET [1].

to fabricate a $p-n$ junction. Here, we describe each of these steps, as well as planarization, in more detail.

2.1.1. Oxidation

Many different kinds of thin films are used to fabricate discrete devices and integrated circuits, including thermal oxides, dielectric layers, polycrystalline silicon, and metal films. For example, a silicon n-channel MOSFET (Figure 2.1) uses all four groups of films. An important oxide layer is the gate oxide, under which a conducting channel can be formed between the source and the drain. A related layer is the field oxide, which provides isolation from other devices. Both gate and field oxides generally are grown by a thermal oxidation process because only thermal oxidation can provide the highest-quality oxides having the lowest interface trap densities.

Semiconductors can be oxidized by various methods, including thermal oxidation, electrochemical anodization, and plasma-enhanced chemical vapor deposition (PECVD; see Section 2.1.5). Among these, thermal oxidation is the most important for silicon devices. It is a key process in modern silicon IC technology. The basic thermal oxidation apparatus (shown in Figure 2.2) consists of a resistance-heated furnace, a cylindrical fused-quartz tube containing the silicon wafers held vertically in a slotted quartz boat, and a source of either pure dry oxygen or pure water vapor. Oxidation temperature is generally in the range of 900–1200°C, and the typical gas flowrate is about 1 L/min. The oxidation system uses microprocessors to regulate the gas flow sequence, to control the automatic insertion and removal of silicon wafers, to ramp the temperature up (i.e., to increase the furnace temperature linearly) from a low temperature to the oxidation temperature, to maintain the oxidation temperature to within ±1°C, and to ramp the temperature down when oxidation is completed.

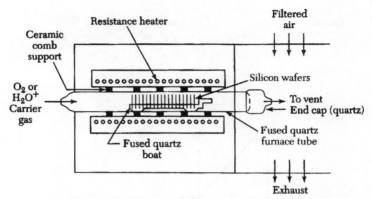

Figure 2.2. Schematic of an oxidation furnace [1].

2.1.1.1. Growth Kinetics

The following chemical reactions describe the thermal oxidation of silicon in oxygen ("dry" oxidation) and water vapor ("wet" oxidation), respectively:

$$Si(solid) + O_2(gas) \rightarrow SiO_2(solid) \tag{2.1}$$

$$Si(solid) + 2H_2O(gas) \rightarrow SiO_2(solid) + 2H_2(gas) \tag{2.2}$$

The silicon–silicon dioxide interface moves into the silicon during the oxidation process. This creates a new interface region, with surface contamination on the original silicon ending up on the oxide surface. As a result of the densities and molecular weights of silicon and silicon dioxide, growing an oxide of thickness x consumes a layer of silicon $0.44x$ thick (Figure 2.3).

The kinetics of silicon oxidation can be described on the basis of the simple model illustrated in Figure 2.4. A silicon slice contacts the oxidizing species (oxygen or water vapor), resulting in a surface concentration of C_0 molecules/cm^3 for these species. The magnitude of C_0 equals the equilibrium bulk concentration of the species at the oxidation temperature. The equilibrium concentration generally is proportional to the partial pressure of the oxidant adjacent to the oxide surface.

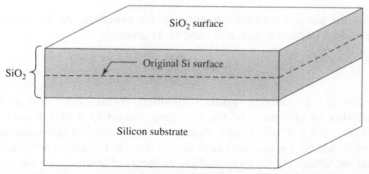

Figure 2.3. Movement of silicon–silicon dioxide interface during oxide growth [1].

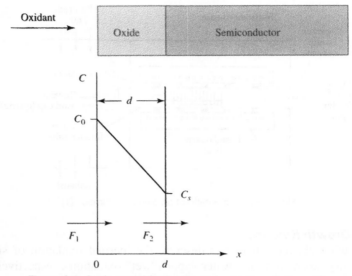

Figure 2.4. Basic model for the thermal oxidation of silicon [1].

At $1000°C$ and a pressure of 1 atm, the concentration C_0 is 5.2×10^{16} cm^{-3} for dry oxygen and 3×10^{19} cm^{-3} for water vapor.

The oxidizing species diffuses through the silicon dioxide layer, resulting in a concentration C_s at the surface of silicon. The flux F_1 can be written as

$$F_1 = D\frac{dC}{dx} \cong \frac{D(C_0 - C_s)}{x} \qquad (2.3)$$

where D is the diffusion coefficient of the oxidizing species, and x is the thickness of the oxide layer already present.

At the silicon surface, the oxidizing species reacts chemically with silicon. Assuming the rate of reaction to be proportional to the concentration of the species at the silicon surface, the flux F_2 is given by

$$F_2 = \kappa C_s \qquad (2.4)$$

where κ is the surface reaction rate constant for oxidation. At the steady state, $F_1 = F_2 = F$. Combining Eqs. (2.3) and (2.4) gives

$$F = \frac{DC_0}{x + (D/\kappa)} \qquad (2.5)$$

The reaction of the oxidizing species with silicon forms silicon dioxide. Let C_1 be the number of molecules of the oxidizing species in a unit volume of the oxide. There are 2.2×10^{22} silicon dioxide molecules/cm^3 in the oxide, and one oxygen molecule (O_2) is added to each silicon dioxide molecule, whereas we add two water molecules (H_2O) to each SiO_2 molecule. Therefore, C_1 for oxidation in dry oxygen is 2.2×10^{22} cm^{-3}, and for oxidation in water vapor it is twice this number (4.4×10^{22} cm^{-3}). Thus, the growth rate of the oxide layer thickness is given by

$$\frac{dx}{dt} = \frac{F}{C_1} = \frac{DC_0/C_1}{x + (D/\kappa)} \qquad (2.6)$$

This differential equation can be solved subject to the initial condition, $x(0) = d_0$, where d_0 is the initial oxide thickness; d_0 can also be regarded as the thickness of oxide layer grown in an earlier oxidation step. Solving Eq. (2.6) yields the general relationship for the oxidation of silicon:

$$x^2 + \frac{2D}{\kappa}x = \frac{2DC_0}{C_1}(t + \tau) \qquad (2.7)$$

where $\tau \equiv (d_0^2 + 2Dd_0/\kappa)C_1/2DC_0$, which represents a time coordinate shift to account for the initial oxide layer d_0.

The oxide thickness after an oxidizing time t is given by

$$x = \frac{D}{\kappa}\left[\sqrt{1 + \frac{2C_0\kappa^2(t + \tau)}{DC_1}} - 1\right] \qquad (2.8)$$

For small values of t, Eq. (2.8) reduces to

$$x \cong \frac{C_0\kappa}{C_1}(t + \tau) \qquad (2.9)$$

and for larger values of t, it reduces to

$$x \cong \sqrt{\frac{2DC_0}{C_1}(t + \tau)} \qquad (2.10)$$

During the early stages of oxide growth, when surface reaction is the rate limiting factor, the oxide thickness varies linearly with time. As the oxide layer becomes thicker, the oxidant must diffuse through the oxide layer to react at the silicon–silicon dioxide interface and the reaction becomes diffusion-limited. The oxide growth then becomes proportional to the square root of the oxidizing time, which results in a parabolic growth rate.

Equation (2.7) is often written in a more compact form

$$x^2 + Ax = B(t + \tau) \qquad (2.11)$$

where $A = 2D/\kappa$, $B = 2DC_0/C_1$ and $B/A = \kappa C_0/C_1$. Using this form, Eqs. (2.9) and (2.10) can be written as

$$x = \frac{B}{A}(t + \tau) \qquad (2.12)$$

for the linear region and as

$$x^2 = B(t + \tau) \qquad (2.13)$$

for the parabolic region. For this reason, the term B/A is referred to as the *linear rate constant* and B is the *parabolic rate constant*. Experimentally measured results agree with the predictions of this model over a wide range of oxidation conditions. For wet oxidation, the initial oxide thickness d_0 is very small, or $\tau \cong 0$. However, for dry oxidation, the extrapolated value of d_0 at $t = 0$ is about 25 nm. Thus, the use of Eq. (2.11) for dry oxidation on bare silicon requires a value for τ that can be generated using this initial thickness. Table 2.1 lists the values of the rate constants for wet oxidation of silicon, and Table 2.2 lists the values for dry oxidation.

The temperature dependence of the linear rate constant B/A is shown in Figure 2.5 for both dry and wet oxidation and for (111)- and (100)-oriented silicon wafers [1]. The linear rate constant varies as $\exp(-E_a/kT)$, where the activation energy E_a is about 2 eV for both dry and wet oxidation. This closely agrees with the energy required to break silicon–silicon bonds, 1.83 eV/molecule. Under a given oxidation condition, the linear rate constant depends on crystal orientation. This is because the rate constant is related to the rate of incorporation of oxygen atoms into the silicon. The rate depends on the surface bond structure of silicon atoms, making it orientation-dependent. Because the density of available bonds on the (111) plane is higher than that on the (100) plane, the linear rate constant for (111) silicon is larger.

Figure 2.6 shows the temperature dependence of the parabolic rate constant B, which can also be described by $\exp(-E_a/kT)$. The activation energy E_a is 1.24 eV for dry oxidation. The comparable activation energy for oxygen diffusion in fused silica is 1.18 eV. The corresponding value for wet oxidation, 0.71 eV, compares favorably with the value of 0.79 eV for the activation energy of diffusion of water in fused silica. The parabolic rate constant is independent of crystal orientation. This independence is expected because it is a measure of

Table 2.1. Rate constants for wet oxidation of silicon.

Temperature ($^\circ$C)	A (μm)	B (μm^2/h)	τ (h)
1200	0.05	0.72	0
1100	0.11	0.51	0
1000	0.226	0.287	0
920	0.5	0.203	0

Table 2.2. Rate constants for dry oxidation of silicon.

Temperature ($^\circ$C)	A (μm)	B (μm^2/h)	τ (h)
1200	0.04	0.045	0.027
1100	0.09	0.027	0.076
1000	0.165	0.0117	0.37
920	0.235	0.0049	1.4
800	0.37	0.0011	9.0
700	—	—	81.0

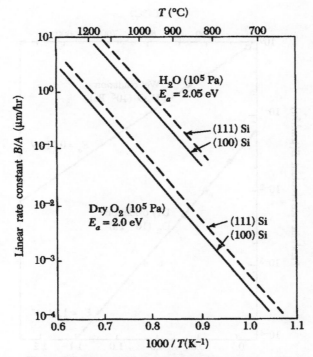

Figure 2.5. Linear rate constant versus temperature [1].

the diffusion process of the oxidizing species through a random network layer of amorphous silica.

Although oxides grown in dry oxygen have the best electrical properties, considerably more time is required to grow the same oxide thickness at a given temperature in dry oxygen than in water vapor. For relatively thin oxides such as the gate oxide in a MOSFET (typically ≤20 nm), dry oxidation is used. However, for thicker oxides such as field oxides (≥20 nm) in MOS integrated circuits, and for bipolar devices, oxidation in water vapor (or steam) is used to provide both adequate isolation and passivation.

2.1.1.2. Thin Oxide Growth

Relatively slow growth rates must be used to reproducibly grow thin oxide films of precise thickness. Approaches to achieve such slower growth rates include growth in dry O_2 at atmospheric pressure and lower temperatures (800–900°C); growth at pressures lower than atmospheric pressure; growth in a reduced partial pressures of O_2 by using a diluent inert gas, such as N_2, Ar, or He, together with the gas containing the oxidizing species; and the use of composite oxide films with the gate oxide films consisting of a layer of thermally grown SiO_2 and an overlayer of chemical vapor deposition (CVD) SiO_2. However, the mainstream approach for gate oxides 10–15 nm thick is to grow the oxide film at atmospheric pressure and lower temperatures (800–900°C). With this approach, processing

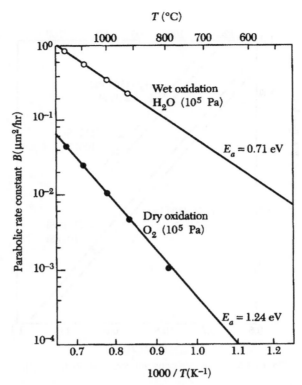

Figure 2.6. Parabolic rate constant versus temperature [1].

using modern *vertical* oxidation furnaces can grow reproducible, high-quality 10-nm oxides to within 0.1 nm across the wafer.

It was noted earlier that for dry oxidation, there is a rapid early growth that gives rise to an initial oxide thickness d_0 of about 20 nm. Therefore, the simple model given by Eq. (2.11) is not valid for dry oxidation with an oxide thickness ≤ 20 nm. For ultra-large-scale integration, the ability to grow thin (5–20 nm), uniform, high-quality reproducible gate oxides has become increasingly important.

In the early stage of growth in dry oxidation, there is a large compressive stress in the oxide layer that reduces the oxygen diffusion coefficient in the oxide. As the oxide becomes thicker, the stress will be reduced due to the viscous flow of silica and the diffusion coefficient will approach its stress-free value. Therefore, for thin oxides, the value of D/κ may be sufficiently small that we can neglect the term Ax in Eq. (2.11) and obtain

$$x^2 - d_0^2 = Bt \tag{2.14}$$

where d_0 is equal to $\sqrt{2DC_0\tau/C_1}$, which is the initial oxide thickness when time is extrapolated to zero, and B is the parabolic rate constant defined previously. We therefore expect the initial growth in dry oxidation to follow a parabolic form.

2.1.1.3. Oxide Quality

Oxides used for masking are usually grown by wet oxidation. A typical growth cycle consists of a dry–wet–dry sequence. Most of the growth in such a sequence occurs in the wet phase, since the SiO_2 growth rate is much higher when water is used as the oxidant. Dry oxidation, however, results in a higher quality oxide that is denser and has a higher breakdown voltage (5–10 MV/cm). It is for these reasons that the thin gate oxides in MOS devices (see Section 2.2) are usually formed using dry oxidation.

MOS devices are also affected by charges in the oxide and traps at the SiO_2–Si interface. The basic classification of these traps and charges, shown in Figure 2.7, are interface-trapped charge, fixed-oxide charge, oxide-trapped charge, and mobile ionic charge. Interface-trapped charges (Q_{it}) are due to the SiO_2–Si interface properties and dependent on the chemical composition of this interface. The traps are located at the SiO_2–Si interface with energy states in the silicon-forbidden bandgap. The fixed charge (Q_f) is located within approximately 3 nm of the SiO_2–Si interface. Generally, Q_f is positive and depends on oxidation and annealing conditions, as well as on the orientation of the silicon substrate. Oxide-trapped charges (Q_{ot}) are associated with defects in the silicon dioxide. These charges can be created, for example, by X-ray radiation or high-energy electron bombardment. Mobile ionic charges (Q_m), which result from contamination from sodium or other alkali ions, are mobile within the oxide under raised-temperatures (e.g., $> 100°C$) and high-electric-field operations. Trace contamination by alkali metal ions may cause stability problems in semiconductor devices operated under high-bias and high-temperature conditions. Under these

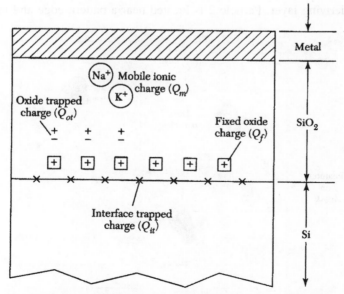

Figure 2.7. Description of charges associated with thermal oxides [1].

conditions mobile ionic charges can move back and forth through the oxide layer and cause threshold voltage shifts. Therefore, special attention must be paid to the elimination of mobile ions in device fabrication.

2.1.2. Photolithography

Photolithography is the process of transferring patterns of geometric shapes on a mask to a thin layer of photosensitive material (called *photoresist*) covering the surface of a semiconductor wafer. These patterns define the various regions in an integrated circuit, such as the implantation regions, the contact windows, and the bonding pad areas. The resist patterns defined by the lithographic process are not permanent elements of the final device, but only replicas of circuit features. To produce circuit features, these resist patterns must be transferred once more into the underlying layers of the device. Pattern transfer is accomplished by an etching process that selectively removes unmasked portions of a layer (see Section 2.1.4).

Photolithography requires a clean processing room. The need for a cleanroom arises because dust particles in the air can settle on semiconductor wafers or lithographic masks and cause defects that result in circuit failure. For example, a dust particle on a semiconductor surface can disrupt the growth of an epitaxial film, causing the formation of dislocations. A dust particle incorporated into a gate oxide can result in enhanced conductivity and cause device failure due to low breakdown voltage. The situation is even more critical in photolithography. When dust particles adhere to the surface of a photomask, they behave as opaque patterns on the mask, and these patterns will be transferred to the underlying layer along with the circuit patterns on the mask. Figure 2.8 shows three dust particles on a photomask. Particle 1 may result in the formation of a pinhole in the underlying layer. Particle 2 is located near a pattern edge and may cause

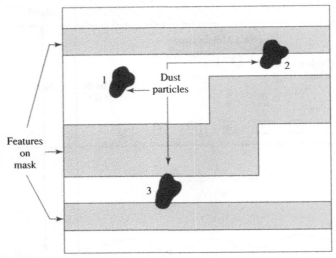

Figure 2.8. Various ways in which particles can interfere with photomask patterns [1].

a constriction of current flow in a metal runner. Particle 3 can lead to a short circuit between the two conducting regions and render the circuit useless.

In a cleanroom, the total number of dust particles per unit volume must be tightly controlled along with the temperature and humidity. There are two systems to define the classes of cleanroom. For the English system, the numerical designation of the class is taken from the maximum allowable number of particles 0.5 μm and larger per cubic foot of air. For the metric system, the class is taken from the logarithm (base 10) of the maximum allowable number of particles 0.5 μm and larger, per cubic meter. For example, a class 100 cleanroom (English system) has a dust count of 100 particles/ft^3 with particle diameters of 0.5 μm and larger, whereas a class M 3.5 cleanroom (metric system) has a dust count of $10^{3.5}$ or about 3500 particles/m^3 with particle diameters of 0.5 μm or larger.

Since the number of dust particles increases as particle size decreases, more stringent control of the cleanroom environment is required as the minimum feature lengths of ICs are reduced. For most IC fabrication areas, a class 100 cleanroom is required; that is, the dust count must be about four orders of magnitude lower than that of ordinary room air. However, for photolithography, a class 10 cleanroom or one with a lower dust count is required.

2.1.2.1. Exposure Tools

The pattern transfer process is accomplished by using a lithographic exposure tool. The performance of an exposure tool is determined by resolution, registration, and throughput. *Resolution* is the minimum feature dimension that can be transferred with high fidelity to a resist film on a semiconductor wafer. *Registration* is a measure of how accurately patterns on successive masks can be aligned (or overlaid) with respect to previously defined patterns on the wafer. *Throughput* is the number of wafers that can be exposed per unit time for a given mask level.

There are two primary optical exposure methods: shadow printing and projection printing. Shadow printing may have the mask and wafer in direct contact with one another (as in *contact printing*), or in close proximity (as in *proximity printing*). Figure 2.9a shows a basic setup for contact printing where a resist-coated wafer is brought into physical contact with a mask, and the resist is exposed by a nearly collimated beam of ultraviolet light through the back of the mask for a fixed time. The intimate contact between the resist and mask provides a resolution of ∼1 μm. However, contact printing suffers from one major drawback—a dust particle on the wafer can be embedded into the mask when the mask makes contact with the wafer. The embedded particle causes permanent damage to the mask and results in defects in the wafer with each succeeding exposure.

To minimize mask damage, the proximity exposure method is used. Figure 2.9b shows the basic setup, which is similar to contact printing except that there is a small gap (10–50 μm) between the wafer and the mask during exposure. The small gap, however, results in optical diffraction at feature edges on the photomask; that is, when light passes by the edges of an opaque mask feature, fringes are formed and some light penetrates into the shadow region. As a result, resolution is degraded to the 2–5-μm range.

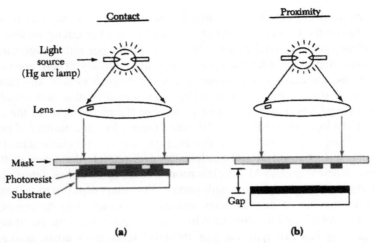

Figure 2.9. Optical shadow printing techniques: (a) contact printing; (b) proximity printing [1].

In shadow printing, the minimum linewidth, or critical dimension (CD), that can be printed is approximately

$$CD \cong \sqrt{\lambda g} \tag{2.15}$$

where λ is the wavelength of the exposure radiation and g is the gap between the mask and the wafer and includes the thickness of the resist. For $\lambda = 0.4$ μm and $g = 50$ μm, the CD is 4.5 μm. If we reduce λ to 0.25 μm (a wavelength range of 0.2–0.3 μm is in the deep-UV spectral region) and g to 15 μm, the CD becomes 2 μm. Thus, there is an advantage in reducing both λ and g. However, for a given distance g, any dust particle with a diameter larger than g potentially can cause mask damage.

To avoid the mask damage problem associated with shadow printing, projection printing tools have been developed to project an image of the mask patterns onto a resist-coated wafer many centimeters away from the mask. To increase resolution, only a small portion of the mask is exposed at a time. The small image area is scanned or stepped over the wafer to cover the entire wafer surface. Figure 2.10a shows a 1 : 1 wafer scan projection system. A narrow, arc-shaped image field ~1 mm in width serially transfers the slit image of the mask onto the wafer. The image size on the wafer is the same as that on the mask.

The small image field can also be stepped over the surface of the wafer by two-dimensional translations of the wafer only, whereas the mark remains stationary. After the exposure of one chip site, the wafer is moved to the next chip site and the process is repeated. Figures 2.10b and 2.10c show the partitioning of the wafer image by *step-and-repeat projection* with a ratio of 1 : 1 or at a demagnification ratio M : 1 (e.g., 10 : 1 for a 10 times reduction on the wafer), respectively. The 1 : 1 optical systems are easier to design and fabricate than a 10 : 1 or a 5 : 1 reduction system, but it is much more difficult to produce defect-free masks at 1 : 1 than it is at a 10 : 1 or a 5 : 1 demagnification ratio.

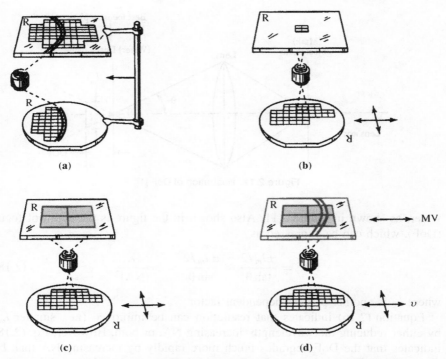

(a) (b)

(c) (d)

Figure 2.10. Image partitioning techniques for projection printing: (a) annual field wafer scan; (b) 1 : 1 step-and-repeat; (c) M : 1 step-and-repeat; and (d) M : 1 step-and-scan [1].

Reduction projection lithography can also print larger wafers without redesigning the stepper lens, as long as the field size (i.e., the exposure area onto the wafer) of the lens is large enough to contain one or more ICs. When the chip size exceeds the field size of the lens, further partitioning of the image on the reticle is necessary. In Figure 2.10d, the image field on the reticle can be a narrow, arc shape for M : 1 step-and-scan projection lithography. For the step-and-scan system, we have two-dimensional translations of the wafer with speed v, and one-dimensional translation of the mask with M times that of the wafer speed.

The resolution of a projection system is given by

$$l_m = k_1 \frac{\lambda}{\mathrm{NA}} \tag{2.16}$$

where k_1 is a process-dependent factor and NA is the numerical aperture, which is given by

$$\mathrm{NA} = \bar{n} \sin \theta \tag{2.17}$$

where $\bar{n}$ is the index of refraction in the image medium (usually air, where $\bar{n} = 1$) and θ is the half-angle of the cone of light converging to a point image at the

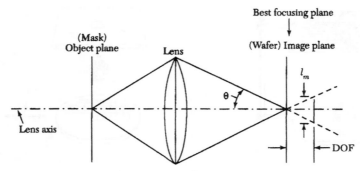

Figure 2.11. Illustration of DoF [1].

wafer, as shown in Figure 2.11. Also shown in the figure is the depth of focus (DoF), which can be expressed as

$$\text{DoF} = \frac{\pm l_m/2}{\tan \theta} \approx \frac{\pm l_m/2}{\sin \theta} = k_2 \frac{\lambda}{(\text{NA})^2} \qquad (2.18)$$

where k_2 is another process-dependent factor.

Equation (2.16) indicates that resolution can be improved (i.e., smaller l_m) by either reducing the wavelength, increasing NA, or both. However, Eq. (2.18) indicates that the DoF degrades much more rapidly by increasing NA than by decreasing λ. This explains the trend toward shorter-wavelength sources in optical lithography.

2.1.2.2. Masks

Masks used for semiconductor manufacturing are usually reduction reticles. The first step in maskmaking is to use a computer-aided design (CAD) system in which designers can completely describe the circuit patterns electrically. The digital data produced by the CAD system then drive a pattern generator, which is an electron-beam lithographic system (see Section 2.1.2.5) that transfers the patterns directly to electron-sensitized mask. The mask consists of a fused-silica substrate covered with a chrominum layer. The circuit pattern is first transferred to the electron-sensitized layer (electron resist), which is transferred once more into the underlying chrominum layer for the finished mask. The patterns on a mask represent one level of an IC design. The composite layout is broken into mask levels that correspond to the manufacturing process sequence, such as the isolation region on one level, the gate region on another, and so on. Typically, 15–20 different mask levels are required for a complete IC process cycle.

The standard-size mask substrate is a fused-silica plate 15×15 cm square, 0.6 cm thick. This size is needed to accommodate the lens field sizes for 4 : 1 or 5 : 1 optical exposure tools, whereas the thickness is required to minimize pattern placement errors due to substrate distortion. The fused-silica plate is needed for its low coefficient of thermal expansion, its high transmission at shorter

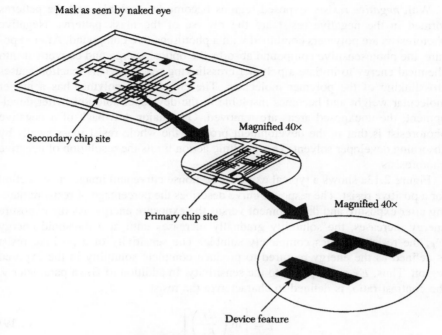

Figure 2.12. A typical IC photomask [1].

wavelengths, and its mechanical strength. Figure 2.12 shows a mask on which patterns of geometric shapes have been formed. A few secondary-chip sites, used for process evaluation, are also included in the mask.

One of the major concerns about masks is the defect density. Mask defects can be introduced during the manufacture of the mask or during subsequent lithographic processes. Even a small mask defect density has a profound effect on the final IC yield. *Yield* is defined as the ratio of good chips per wafer to the total number of chips per wafer (see Chapter 5). Inspection and cleaning of masks are important to achieve high yields on large chips. An ultraclean processing area is mandatory for photolithographic processing.

2.1.2.3. Photoresist

Photoresist is a radiation-sensitive compound that can be classified as positive or negative, depending on how they respond to radiation. For *positive resists*, the exposed regions become more soluble and thus more easily removed in the development process. The result is that the patterns formed in the positive resist are the same as those on the mask. Positive photoresists consist of three components: a photosensitive compound, a base resin, and an organic solvent. Prior to exposure, the photosensitive compound is insoluble in the developer solution. After exposure, the photosensitive compound absorbs radiation in the exposed pattern areas, changes its chemical structure, and becomes soluble in the developer solution. After development, the exposed areas are removed.

With *negative resists*, exposed regions become less soluble, and the patterns formed in the negative resist are the reverse of the mask patterns. Negative photoresists are polymers combined with a photosensitive compound. After exposure, the photosensitive compound absorbs the optical energy and converts it into chemical energy to initiate a polymer crosslinking reaction. This reaction causes crosslinking of the polymer molecules. The crosslinked polymer has a higher molecular weight and becomes insoluble in the developer solution. After development, the unexposed areas are removed. One major drawback of a negative photoresist is that in the development process, the whole resist mass swells by absorbing developer solvent. This swelling action limits the resolution of negative photoresists.

Figure 2.13a shows a typical exposure response curve and image cross section for a positive resist. The response curve describes the percentage of resist remaining after exposure and development versus the exposure energy. As the exposure energy increases, the solubility gradually increases until at a threshold energy E_T, the resist becomes completely soluble. The sensitivity of a positive resist is defined as the energy required to produce complete solubility in the exposed region. Thus, E_T corresponds to the sensitivity. In addition to E_T, a parameter γ, the contrast ratio, is defined to characterize the resist

$$\gamma \equiv \left[\ln \left(\frac{E_T}{E_1} \right) \right]^{-1} \tag{2.19}$$

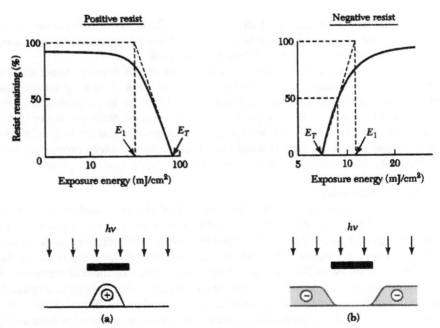

Figure 2.13. Exposure response curve and cross section of resist image after development for (a) positive photoresist and (b) negative photoresist [1].

where E_1 is the energy obtained by drawing the tangent at E_T to reach 100% resist thickness, as shown in Figure 2.13a. A larger γ implies a higher solubility of the resist with an incremental increase of exposure energy and results in sharper images.

The image cross section in Figure 2.13a illustrates the relationship between the edges of a photomask image and the corresponding edges of the resist images after development. The edges of the resist image are generally not at the vertically projected positions of the mask edges because of *diffraction*. The edge of the resist image corresponds to the position where the total absorbed optical energy equals the threshold energy E_T.

Figure 2.13b shows the exposure response curve and image cross section for a negative resist. The negative resist remains completely soluble in the developer solution for exposure energies lower than E_T. Above E_T, more of the resist film remains after development. At exposure energies twice the threshold energy, the resist film becomes essentially insoluble in the developer. The sensitivity of a negative resist is defined as the energy required to retain 50% of the original resist film thickness in the exposed region. The parameter γ is defined similarly to γ in Eq. (2.19), except that E_1 and E_T are interchanged. The image cross section for the negative resist (Figure 2.13b) is also influenced by the diffraction effect.

2.1.2.4. Pattern Transfer

Figure 2.14 illustrates the steps to transfer IC patterns from a mask to a silicon wafer that has an insulating SiO_2 layer formed on its surface. The wafer is placed in a cleanroom, which typically is illuminated with yellow light (since photoresists are not sensitive to wavelengths greater than 0.5 μm). To ensure satisfactory adhesion of the resist, adhesion promoter is then applied. The most common adhesion promoter for silicon ICs is hexamethylene–disiloxane (HMDS). After the application of this adhesion layer, the wafer is held on a vacuum spindle, and liquidous resist is applied to the center of wafer. The wafer is then rapidly accelerated up to a constant rotational speed, which is maintained for about 30 s. Spin speed is generally in the range of 1000–10,000 rpm [revolutions per minute (r/min)] to coat a uniform film about 0.5–1 μm thick, as shown in Figure 2.14a. The thickness of photoresist is correlated with its viscosity.

After spinning, the wafer is "soft-baked" (typically at 90–120°C for 60–120 s) to remove solvent from the photoresist and to increase resist adhesion to the wafer. The wafer is aligned with respect to the mask in an optical lithographic system, and the resist is exposed to ultraviolet light, as shown in Figure 2.14b. If a positive photoresist is used, the exposed resist is dissolved in the developer, as shown on the left side of Figure 2.14c. Photoresist development is usually done by flooding the wafer with the developer solution. The wafer is then rinsed and dried. After development, "postbaking" at approximately 100–180°C may be required to increase the adhesion of the resist to the substrate. The wafer is then put in an ambient that etches the exposed insulation layer but does not attack the resist, as shown in Figure 2.14d. Finally, the resist is stripped (using solvents or plasma oxidation), leaving behind an insulator image that is the same as the opaque image on the mask (left side of Figure 2.14e). For negative photoresist,

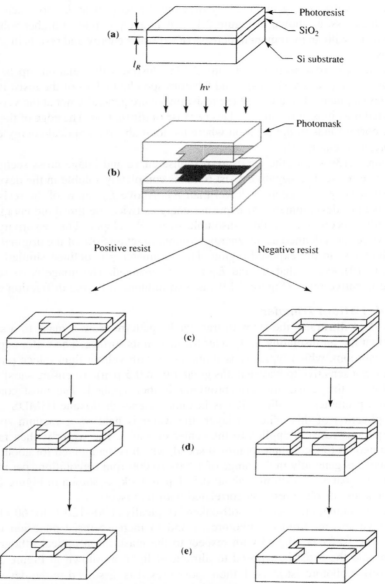

Figure 2.14. Photolithographic pattern transfer process: (a) photoresist application; (b) exposure; (c) development; (d) etching; (e) resist stripping [1].

the procedures described are also applicable, except that the unexposed areas are removed. The final insulator image (right side of Figure 2.14e) is the reverse of the opaque image on the mask.

The insulator image can be used as a mask for subsequent processing. For example, *ion implantation* (Section 2.1.3) can be done to dope the exposed semiconductor region, but not the area covered by the insulator. The dopant pattern is a duplicate of the design pattern on the photomask for a negative photoresist or is its complementary pattern for a positive photoresist. The complete circuit is fabricated by aligning the next mask in the sequence to the previous pattern and repeating the lithographic transfer process.

2.1.2.5. E-Beam Lithography

Optical lithography is so widely used because it has high throughput, good resolution, low cost, and ease of operation. However, due to deep-submicrometer IC process requirements, optical lithography has some limitations that have not yet been solved. Although we can use PSM or OPC to extend its useful lifespan, the complexity of mask production and mask inspection cannot be easily resolved. In addition, the cost of the masks is very high. Therefore, we need to find alternatives to optical lithography to process deep-submicrometer or nanometer ICs.

Electron-beam (or *e-beam*) lithography is used primarily to produce photomasks. Relatively few tools are dedicated to direct exposure of the resist by a focused electron beam without a mask. Figure 2.15 shows a schematic of an e-beam lithography system. The electron gun is a device that can generate a beam of electrons with a suitable current density. A tungsten thermionic emission cathode or single-crystal lanthanum hexaboride (LaB_6) is used for the electron gun. Condenser lenses are used to focus the electron beam to a spot size 10–25 nm in diameter. Beam blanking plates that turn the electron beam on and off, and beam deflection coils are computer-controlled and operated at MHz or higher

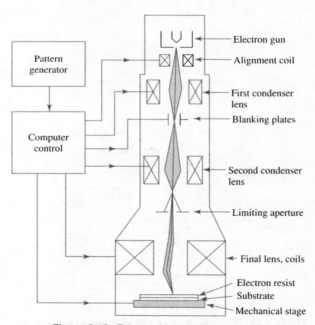

Figure 2.15. E-beam lithography system [1].

rates to direct the focused electron beam to any location in the scan field on the substrate. Because the scan field (typically 1 cm) is much smaller than the substrate diameter, a precision mechanical stage is used to position the substrate to be patterned.

The advantages of electron-beam lithography include the generation of submicrometer resist geometries, highly automated and precisely controlled operation, depth of focus greater than that available from optical lithography, and direct patterning on a semiconductor wafer without using a mask. The disadvantage is that electron-beam lithographic machines have low throughput—approximately 10 wafers per hour at less than 0.25 μm resolution. This throughput is adequate for the production of photomasks, for situations that require small numbers of custom circuits, and for design verification. However, for maskless direct writing, the machine must have the highest possible throughput, and therefore, the largest beam diameter possible consistent with the minimum device dimensions.

There are two ways to scan the focused electron beam: raster scan and vector scan. In a *raster scan* system, resist patterns are written by a beam that moves through a regular mode, vertically oriented, as shown in Figure 2.16a. The beam scans sequentially over every possible location on the mask and is blanked (turned off) where no exposure is required. All patterns on the area to be written must be

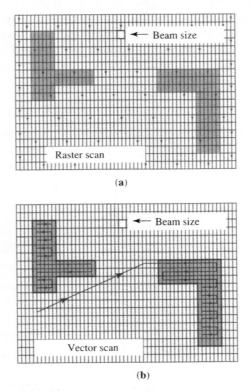

Figure 2.16. (a) Raster scan and (b) vector scan systems [1].

subdivided into individual addresses, and a given pattern must have a minimum incremental interval that is evenly divisible by the beam address size. In the *vector scan* system (Figure 2.16b), the beam is directed only to the requested pattern features and jumps from feature to feature, rather than scanning the whole chip, as in raster scan. For many chips, the average exposed region is only 20% of the chip area, which saves time.

Electron resists are polymers. The behavior of an e-beam resist is similar to that of a photoresist; that is, a chemical or physical change is induced in the resist by irradiation. This change allows the resist to be patterned. For a positive electron resist, the polymer–electron interaction causes chemical bonds to be broken (chain scission) to form shorter molecular fragments. As a result, the molecular weight is reduced in the irradiated area, which can be dissolved subsequently in a developer solution that attacks the low-molecular-weight material. Common positive electron resists include poly(methyl methacrylate) (PMMA) and poly(butene-1 sulfone) (PBS). Positive electron resists can achieve resolution of 0.1 μm or better. For a negative electron resist, the irradiation causes radiation-induced polymer linking. The crosslinking creates a complex three-dimensional structure with a molecular weight higher than that of the nonirradiated polymer. The nonirradiated resist can be dissolved in a developer solution that does not attack the high-molecular-weight material. Polyglycidylmethacrylate–coethylacrylate (COP) is a common negative electron resist. COP, like most negative photoresists, also swells during development, so the resolution is limited to about 1 μm.

While resolution is limited by diffraction of light in optical lithography, in e-beam lithography, the resolution is not impacted by diffraction (because the wavelengths associated with electrons of a few keV and higher energies are less than 0.1 nm), but by electron scattering. When electrons penetrate the resist and underlying substrate, they undergo collisions. These collisions lead to energy losses and path changes. The incident electrons spread out as they travel until either all of their energy is lost or they leave the material because of backscattering. Because of backscattering, electrons can irradiate regions several micrometers away from the center of the exposure beam. Since the dose of a resist is the sum of the irradiations from all surrounding areas, the electron-beam irradiation at one location will affect the irradiation in neighboring locations. This phenomenon is called the *proximity effect*. The proximity effect places a limit on the minimum spacings between pattern features. To correct for the proximity effect, patterns are divided into smaller segments. The incident electron dose in each segment is adjusted so that the integrated dose from all its neighboring segments is the correct exposure dose. This approach further decreases the throughput of the electron-beam system, because of the additional computer time required to expose the subdivided resist patterns.

2.1.2.6. X-Ray Lithography

X-ray lithography (XRL) is a potential candidate to succeed optical lithography for the fabrication of integrated circuits with feature sizes less than 100 nm. XRL

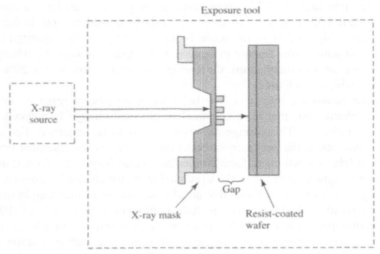

Figure 2.17. Schematic of x-ray lithography system.

uses a shadow printing method similar to optical proximity printing. Figure 2.17 shows a schematic of an XRL system. The X-ray wavelength is about 1 nm, and the printing is through a 1× mask in close proximity (10–40 μm) to the wafer. Since X-ray absorption depends on the atomic number of the material and most materials have low transparency at λ ≅ 1 nm, the mask substrate must be a thin membrane (1–2 μm thick) made of low-atomic-number material, such as silicon carbide or silicon. The pattern itself is defined in a thin (~0.5 μm), relatively high-atomic-number material, such as tantalum, tungsten, gold, or one of their alloys, which is supported by the thin membrane.

Masks are the most difficult and critical element of an XRL system, and the construction of an X-ray mask is much more complicated than that of an optical photomask. To avoid absorption of the X rays between the source and mask, the exposure generally takes place in a helium environment. The X rays are produced in a vacuum, which is separated from the helium by a thin vacuum window (usually of beryllium). The mask substrate will absorb 25–35% of the incident flux and must therefore be cooled. An X-ray resist 1 μm thick will absorb about 10% of the incident flux, and there are no reflections from the substrate to create standing waves, so antireflection coatings are unnecessary.

Electron-beam resists can be used as X-ray resists because when an X ray is absorbed by an atom, the atom goes to an excited state with the emission of an electron. The excited atom returns to its ground state by emitting an X ray having a wavelength different from that of the incident X ray. This X ray is absorbed by another atom, and the process repeats. Since all the processes result in the emission of electrons, a resist film under X-ray irradiation is equivalent to one being irradiated by a large number of secondary electrons from any of the other processes.

2.1.3. Etching

As discussed in Section 2.1.2, photolithography is the process of transferring patterns to photoresist covering the surface of a semiconductor wafer. To produce circuit features, these resist patterns must be transferred into the underlying layers of the device. Pattern transfer is accomplished by an etching process that selectively removes unmasked portions of a layer.

2.1.3.1. Wet Chemical Etching

Wet chemical etching is used extensively in semiconductor processing. Prior to thermal oxidation (Section 2.1.1) or epitaxial growth (Section 2.1.5), semiconductor wafers are chemically cleaned to remove contamination that results from handling and storing. Wet chemical etching is especially suitable for blanket etches (i.e., over the whole wafer surface) of polysilicon, oxide, nitride, metals, and III–V compounds.

The mechanisms for wet chemical etching involve three essential steps, as illustrated in Figure 2.18; the reactants are transported by diffusion to the reacting surface, chemical reactions occur at the surface, and the products from the surface are removed by diffusion. Both agitation and the temperature of the etchant solution will influence the etch rate, which is the amount of film removed by etching per unit time. In IC processing, most wet chemical etches proceed by immersing the wafers in a chemical solution or by spraying the wafers with the etchant solution. For immersion etching, the wafer is immersed in the etch solution, and mechanical agitation is usually required to ensure etch uniformity and a consistent etch rate. Spray etching has gradually replaced immersion etching

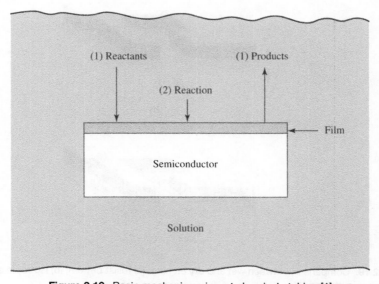

Figure 2.18. Basic mechanisms in wet chemical etching [1].

because it greatly increases the etch rate and uniformity by constantly supplying fresh etchant to the wafer surface.

In semiconductor production lines, highly uniform etch rates are important. Etch rates must be uniform across a wafer, from wafer to wafer, from run to run, and for any variations in feature sizes and pattern densities. Etch rate uniformity is given by

$$\text{Etch rate uniformity } (\%) = \frac{(\text{max. etch rate} - \text{min. etch rate})}{\text{max. etch rate} + \text{min. etch rate}} \times 100\% \quad (2.20)$$

2.1.3.2. Dry Etching

In pattern transfer operations, a resist pattern is defined by a photolithographic process to serve as a mask for etching of its underlying layer (Figure 2.19a). Most of the layer materials (e.g., SiO_2, Si_3N_4, and deposited metals) are amorphous or polycrystalline thin films. If they are etched in a wet etchant, the etch rate

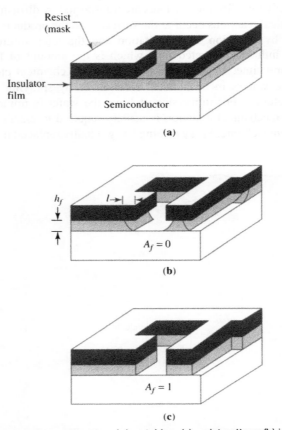

Figure 2.19. Comparison between wet and dry etching: (a) resist pattern; (b) isotropic etching; (c) anisotropic etching [1].

is generally isotropic (i.e., the lateral and vertical etch rates are the same), as illustrated in Figure 2.19b. If h_f is the thickness of the layer material and l the lateral distance etched underneath the resist mask, we can define the degree of anisotropy (A_f) by

$$A_f \equiv 1 - \frac{l}{h_f} = 1 - \frac{R_l t}{R_v t} = 1 - \frac{R_l}{R_v} \tag{2.21}$$

where t is time and R_l and R_v are the lateral and vertical etch rates, respectively. For isotropic etching, $R_l = R_v$ and $A_f = 0$.

The major disadvantage of wet etching in pattern transfer is the undercutting of the layer underneath the mask, resulting in a loss of resolution in the etched pattern. In practice, for isotropic etching, the film thickness should be about one-third or less of the resolution required. If patterns are required with resolutions much smaller than the film thickness, anisotropic etching (i.e., $1 \geq A_f > 0$) must be used. In practice, the value of A_f is chosen to be close to unity. Figure 2.19c shows the limiting case where $A_f = 1$, corresponding to $l = 0$ (or $R_l = 0$).

To achieve a high-fidelity transfer of the resist patterns required for ultra-large-scale integration (ULSI) processing, dry etching methods have been developed. Dry etching is synonymous with plasma-assisted etching, which denotes several techniques that use plasma in the form of low-pressure discharges. Dry-etch methods include plasma etching, reactive-ion etching (RIE), sputter etching, magnetically enhanced RIE (MERIE), reactive-ion-beam etching, and high-density plasma (HDP) etching.

A *plasma* is a fully or partially ionized gas composed of equal numbers of positive and negative charges and a different number of un-ionized molecules. Plasma is produced when an electric field of sufficient magnitude is applied to a gas, causing the gas to break down and become ionized. The plasma is initiated by free electrons that are released by some means, such as field emission from a negatively biased electrode. The free electrons gain kinetic energy from the electric field. In the course of their travel through the gas, the electrons collide with gas molecules and lose their energy. The energy transferred in the collision causes the gas molecules to be ionized (i.e., to free electrons). The free electrons gain kinetic energy from the field, and the process continues. Therefore, when the applied voltage is larger than the breakdown potential, a sustained plasma is formed throughout the reaction chamber.

Plasma etching is a process in which a solid film is removed by a chemical reaction with ground-state or excited-state neutral species. The process is often enhanced or induced by energetic ions generated in a gaseous discharge. Plasma etching proceeds in five steps, as illustrated in Figure 2.20: (1) the etchant species is generated in the plasma, (2) the reactant is then transported by diffusion through a stagnant gas layer to the surface, (3) the reactant is adsorbed on the surface, (4) a chemical reaction (along with physical effects such as ion bombardment) follows to form volatile compounds, and (5) the compounds are desorbed from the surface, diffused into the bulk gas, and pumped out by the vacuum system.

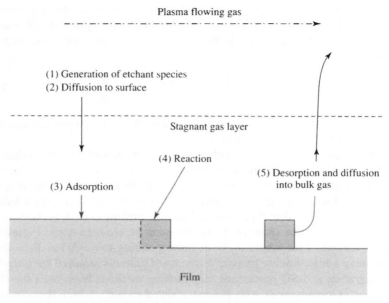

Plasma flowing gas

(1) Generation of etchant species
(2) Diffusion to surface

Stagnant gas layer

(4) Reaction

(5) Desorption and diffusion
into bulk gas

(3) Adsorption

Film

Figure 2.20. Basic steps in dry etching [1].

Plasma etching is based on the generation of plasma in a gas at low pressure. Two basic methods are used: physical methods and chemical methods. The former includes sputter etching, and the latter includes pure chemical etching. In physical etching, positive ions bombard the surface at high speed; small amounts of negative ions formed in the plasma cannot reach the wafer surface and therefore play no direct role in plasma etching. In chemical etching, neutral reactive species generated by the plasma interact with the material surface to form volatile products. Chemical and physical etch mechanisms have different characteristics. Chemical etching exhibits a high etch rate, and good selectivity (i.e., the ratio of etch rates for different materials) produces low ion bombardment–induced damage but yields isotropic profiles. Physical etching can yield anisotropic profiles, but it is associated with low etch selectivity and severe bombardment-induced damage. Combinations of chemical and physical etching give anisotropic etch profiles, reasonably good selectivity, and moderate bombardment-induced damage. An example is reactive-ion etching (RIE), which uses a physical method to assist chemical etching or creates reactive ions to participate in chemical etching.

Plasma reactor technology in the IC industry has changed dramatically since the first application of plasma processing to photoresist stripping. A reactor for plasma etching contains a vacuum chamber, pump system, power supply generators, pressure sensors, gas flow control units, and *endpoint detector* (see Chapter 3). Each etch tool is designed empirically and uses a particular combination of pressure, electrode configuration and type, and source frequency to control the two primary etch mechanisms—chemical and physical. Higher etch rates and tool automation are required for most etchers used in manufacturing.

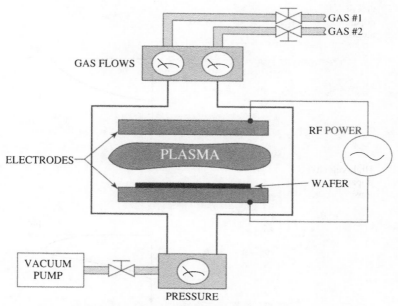

Figure 2.21. Typical reactive-ion etching system.

RIE has been extensively used in the microelectronic industry. In a parallel-plate diode system (Figure 2.21), a radio frequency (RF), capacitively coupled bottom electrode holds the wafer. This allows the grounded electrode to have a significantly larger area because it is, in fact, the chamber itself. The larger grounded area combined with the lower operating pressure (<500 mTorr) causes the wafers to be subjected to a heavy bombardment of energetic ions from the plasma as a result of the large, negative self-bias at the wafer surface. The etch selectivity of this system is relatively low compared with traditional barrel etch systems because of strong physical sputtering. However, selectivity can be improved by choosing the proper etch chemistry.

2.1.4. Doping

Impurity doping is the introduction of controlled amounts of impurities into semiconductors to change their electrical properties. Diffusion and ion implantation are the two key methods of impurity doping. Both diffusion and ion implantation are used for fabricating discrete devices and integrated circuits because these processes generally complement each other. For example, diffusion is used to form a deep junction (e.g., a twin well in CMOS), whereas ion implantation is used to form a shallow junction (e.g., a source–drain junction of a MOSFET).

Until the early 1970s, impurity doping was performed by diffusion at elevated temperatures, as shown in Figure 2.22a. In this method the dopant atoms are placed on or near the surface of the wafer by deposition from the gas phase of the dopant or by using doped-oxide sources. The doping concentration decreases

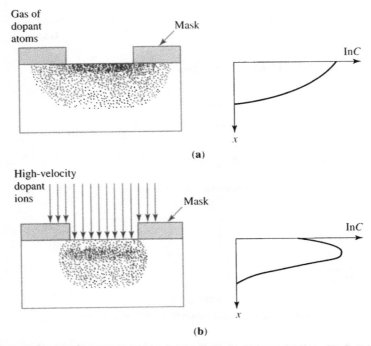

Figure 2.22. (a) Diffusion and (b) ion implantation techniques for impurity doping [1].

monotonically from the surface, and the profile of the dopant distribution is determined mainly by the temperature and the diffusion time. Since the 1970s, doping operations have been performed chiefly by ion implantation, as shown in Figure 2.22b. In this process the dopant ions are implanted into the semiconductor by means of an ion beam. The doping concentration has a peak distribution inside the semiconductor, and the profile of the dopant distribution is determined mainly by the ion mass and the implanted ion energy.

2.1.4.1. Diffusion

Diffusion of impurities is accomplished by placing semiconductor wafers in a carefully controlled, high-temperature quartz-tube furnace and passing a gas mixture that contains the desired dopant through it. The number of dopant atoms that diffuse into the semiconductor is related to the partial pressure of the dopant impurity in the gas mixture. For diffusion in silicon, boron is the most popular dopant for introducing a p-type impurity, whereas arsenic and phosphorus are used extensively as n-type dopants. These dopants can be introduced in several ways, including solid sources (e.g., BN for boron, As_2O_3 for arsenic, and P_2O_5 for phosphorus), liquid sources (BBr_3, $AsCl_3$, and $POCl_3$), and gaseous sources (B_2H_6, AsH_3, and PH_3). However, liquid sources are most commonly used. A schematic diagram of the furnace and gas flow arrangement for a liquid source is shown in Figure 2.23. This arrangement is similar to that used for thermal

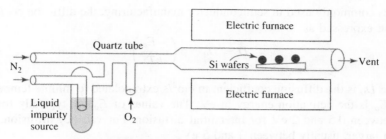

Figure 2.23. Schematic of a diffusion system [1].

oxidation. An example of the chemical reaction for phosphorus diffusion using a liquid source is

$$4POCl_3 + 3O_2 \rightarrow 2P_2O_5 + 6Cl_2 \uparrow \tag{2.22}$$

The P_2O_5 is then reduced to phosphorus by silicon according to

$$2P_2O_5 + 5Si \rightarrow 4P + 5SiO_2 \tag{2.23}$$

and the phosphorus is released and diffuses into the silicon, and Cl_2 is vented.

Diffusion in a semiconductor is the atomic movement of the diffusant (dopant atoms) in the crystal lattice by vacancies or interstitials. The diffusion process is similar to that of charge carriers (electrons and holes). Let a flux F be defined as the number of dopant atoms passing through a unit area in a unit time and C as the dopant concentration per unit volume. Then

$$F = -D\frac{\partial C}{\partial x} \tag{2.24}$$

where the proportionality constant D is the *diffusion coefficient* or *diffusivity*. Note that the basic driving force of the diffusion process is the concentration gradient dC/dx. The flux is proportional to the concentration gradient, and the dopant atoms will move (diffuse) away from a high-concentration region toward a lower-concentration region.

If Eq. (2.24) is substituted into the one-dimensional continuity equation under the condition that no materials are formed or consumed in the host semiconductor, the result is

$$\frac{\partial C}{\partial t} = -\frac{\partial F}{\partial x} = \frac{\partial}{\partial x}\left(D\frac{\partial C}{\partial x}\right) \tag{2.25}$$

When the concentration of the dopant atoms is low, the diffusion coefficient can be considered to be independent of doping concentration, and Eq. (2.25) becomes

$$\frac{\partial C}{\partial t} = D\frac{\partial^2 C}{\partial x^2} \tag{2.26}$$

Equation (2.26) is often referred to as *Fick's diffusion equation* or *Fick's law*. Note that the diffusion coefficient varies with temperature. Over the temperature ranges commonly used in semiconductor manufacturing, the diffusion coefficient can be expressed as

$$D = D_0 \exp\left(\frac{-E_a}{kT}\right) \tag{2.27}$$

where D_0 is the diffusion coefficient in cm^2/s extrapolated to infinite temperature and E_a is the activation energy in eV. The values of E_a are typically found to be between 0.5 and 2 eV for interstitial diffusion. For vacancy diffusion, E_a is much larger, usually between 3 and 5 eV.

The diffusion profile of the dopant atoms is dependent on the initial and boundary conditions. There are two important cases to consider: (1) constant-surface-concentration diffusion and (2) constant-total-dopant diffusion. In the first case, impurity atoms are transported from a vapor source onto the semiconductor surface and diffused into the semiconductor wafers. The vapor source maintains a constant level of surface concentration during the entire diffusion period. In the second case, a fixed amount of dopant is deposited onto the semiconductor surface and is subsequently diffused into the wafers.

For case 1, the initial condition at $t = 0$ is

$$C(x, 0) \tag{2.28}$$

which indicates that the dopant concentration in the host semiconductor is initially zero. The boundary conditions are

$$C(0, t) = C_s \tag{2.29a}$$

$$C(\infty, t) = 0 \tag{2.29b}$$

where C_s is the surface concentration (at $x = 0$), which is independent of time. The second boundary condition states that at long distances from the surface there are no impurity atoms.

The solution of Fick's equation that satisfies the initial and boundary conditions is

$$C(x, t) = C_s \, \text{erfc}\left(\frac{x}{2\sqrt{Dt}}\right) \tag{2.30}$$

where *erfc* is the complementary error function and $\sqrt{Dt}$ is the diffusion length. The diffusion profile for the constant surface concentration condition is shown in Figure 2.24a, where, on both linear (upper) and logarithmic (lower) scales, the normalized concentration as a function of depth for three values of the diffusion length corresponding to three consecutive diffusion times and a fixed D for a given diffusion temperature are plotted. Note that as the time progresses, the dopant penetrates deeper into the semiconductor.

The total number of dopant atoms per unit area of the semiconductor is given by

$$Q(t) = \int_0^\infty C(x, t)dx \tag{2.31}$$

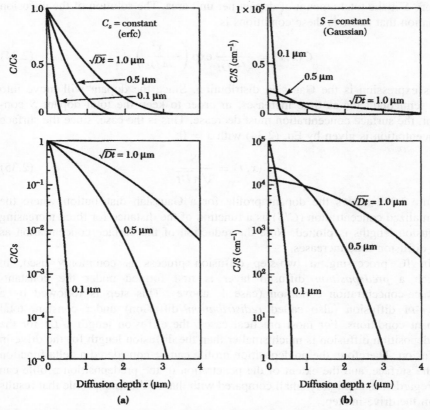

Figure 2.24. Diffusion profiles: (a) normalized erfc versus distance for successive diffusion times; (b) normalized Gaussian function versus distance [1].

Substituting Eq. (2.30) into Eq. (2.31) yields

$$Q(t) = \frac{2}{\sqrt{\pi}} C_s \sqrt{Dt} \cong 1.13 C_s \sqrt{Dt} \qquad (2.32)$$

The quantity $Q(t)$ represents the area under one of the diffusion profiles of the linear plot in Figure 2.24a. These profiles can be approximated by triangles with height C_s and base $2\sqrt{Dt}$. This leads to $Q(t) \cong C_s \sqrt{Dt}$, which is close to the exact result obtained from Eq. (2.32).

Now consider constant total dopant diffusion. For this case, a fixed (or constant) amount of dopant is deposited onto the semiconductor surface in a thin layer, and the dopant subsequently diffuses into the semiconductor. The initial condition is the same as in Eq. (2.28). The boundary conditions are

$$\int_0^\infty C(x, t) = S \qquad (2.33a)$$

$$C(\infty, t) = 0 \qquad (2.33b)$$

where S is the total amount of dopant per unit area. The solution of the diffusion equation that satisfies these conditions is

$$C(x, t) = \frac{S}{\sqrt{\pi Dt}} \exp\left(-\frac{x^2}{4Dt}\right) \qquad (2.34)$$

This expression is the Gaussian distribution. Since the dopant will move into the semiconductor as time increases, in order to keep the total dopant S constant, the surface concentration must decrease. This is the case, since the surface concentration is given by Eq. (2.34) with $x = 0$:

$$C(x, t) = \frac{S}{\sqrt{\pi Dt}} \qquad (2.35)$$

Figure 2.24b shows the dopant profile for a Gaussian distribution where the normalized concentration (C/S) as a function of the distance for three increasing diffusion lengths is plotted. Note the reduction of the surface concentration as the diffusion time increases.

In IC processing, a two-step diffusion process is commonly used, in which a *predeposition* diffused layer is first formed under the constant-surface-concentration condition (case 1, above). This step is followed by a *drive-in* diffusion (also called *redistribution* diffusion) under constant total dopant conditions. For most practical cases, the diffusion length $\sqrt{Dt}$ for the predeposition diffusion is much smaller than the diffusion length for the drive-in diffusion. Therefore, the predeposition profile can be considered a delta function at the surface, and the extent of the penetration of the predeposition profile can be regarded as negligibly small compared with that of the final profile that results from the drive-in step.

2.1.4.2. Ion Implantation

As discussed above, diffusion and ion implantation are the two key methods of impurity doping. Since the early 1970s, many doping operations have been performed by ion implantation, which is shown in Figure 2.22b. In this process the energetic dopant ions are implanted into the semiconductor by means of an ion beam. The doping concentration has a peak distribution inside the semiconductor and the profile of the dopant distribution is determined mainly by the ion mass and energy.

Implantation energies are typically between 1 keV and 1 MeV, resulting in ion distributions with average depths ranging from 10 nm to 10 μm. Ion doses vary from 10^{12} ions/cm^2 for threshold voltage adjustment in MOSFETs to 10^{18} ions/cm^2 for the formation of buried insulating layer. Note that the dose is expressed as the number of ions implanted into 1 cm^2 of the semiconductor surface area. The main advantages of ion implantation are its more precise control and reproducibility of impurity dopings and its lower processing temperature compared with those of the diffusion process.

Figure 2.25 shows schematically a medium-energy ion implantor. The ion source has a heated filament to break up source gas, such as BF_3 or AsH_3, into

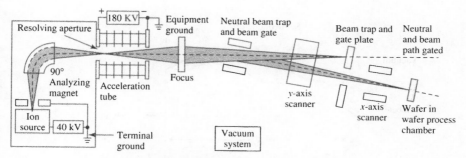

Figure 2.25. Schematic of ion implantor [1].

charged ions (B^+ or As^+). An extraction voltage (around 40 kV) causes the charged ions to move out of the ion source chamber into a mass analyzer. The magnetic field of the analyzer is chosen such that only ions with the desired mass : charge ratio can travel through it without being filtered. The selected ions then enter the acceleration tube, where they are accelerated to the implantation energy as they move from high voltage to ground. Apertures ensure that the ion beam is well collimated. The pressure in the implantor is kept below 10^{-4} Pa to minimize ion scattering by gas molecules. The ion beam is then scanned over the wafer surface using electrostatic deflection plates and is implanted into the semiconductor substrate.

The energetic ions lose their energies through collision with electrons and nuclei in the substrate and finally come to rest at some depth within the lattice. The average depth can be controlled by adjusting the acceleration energy. The dopant dose can be controlled by monitoring the ion current during implantation. The principal side effect is the disruption or damage of the semiconductor lattice due to ion collisions. Therefore, a subsequent annealing treatment is needed to remove these damages.

The total distance that an ion travels in coming to rest is called its *range* (R) and is illustrated in Figure 2.26a. The projection of this distance along the axis of incidence is called the *projected range* (R_p). Because the number of collisions per unit distance and the energy lost per collision are random variables, there will be a spatial distribution of ions having the same mass and the same initial energy. The statistical fluctuations in the projected range are called the *projected straggle* (σ_p). There is also a statistical fluctuation along an axis perpendicular to the axis of incidence, which is called the *lateral straggle* ($\sigma_\perp$).

Figure 2.26b shows the ion distribution. Along the axis of incidence, the implanted impurity profile can be approximated by a Gaussian distribution function

$$n(x) = \frac{S}{\sqrt{2\pi}\sigma_p} \exp\left[-\frac{(x - R_p)^2}{2\sigma_p^2}\right] \tag{2.36}$$

where S is the ion dose per unit area. This equation is similar to Eq. (2.34) for constant total dopant diffusion, except that the quantity $4Dt$ is replaced by $2\sigma_p^2$

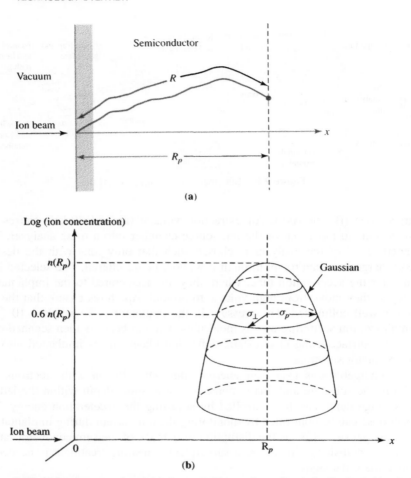

Figure 2.26. (a) Ion range and projected range; (b) two-dimensional distribution of implanted ions [1].

and the distribution is shifted along the x axis by R_p. Thus, for diffusion, the maximum concentration is at $x = 0$, whereas for ion implantation the maximum concentration is at the projected range. The ion concentration is reduced by 40% from its peak value at $(x - R_p) = \pm\sigma_p$, by one decade at $\pm 2\sigma_p$, by two decades at $\pm 3\sigma_p$, and by five decades at $\pm 4.8\sigma_p$.

2.1.5. Deposition

Many different types of thin films are used to manufacture integrated circuits, including thermal oxides, dielectric layers, epitaxial layers, polycrystalline silicon, and metal films. This section addresses two of the various techniques for depositing such films: physical vapor deposition and chemical vapor deposition.

2.1.5.1. Physical Vapor Deposition

The most common methods of physical vapor deposition (PVD) of metals are evaporation, electron-beam evaporation, plasma spray deposition, and sputtering. Metals and metal compounds can be deposited by PVD. Evaporation occurs when a source material is heated above its melting point in an evacuated chamber. The evaporated atoms then travel at high velocity in straight-line trajectories. The source can be melted by resistance heating, by radio frequency (RF) heating, or with a focused electron beam (or e-beam). Evaporation and e-beam evaporation were used extensively in earlier generations of integrated circuits, but they have been replaced by sputtering for modern ICs.

In ion-beam sputtering, a source of ions is accelerated toward the target and impinges on its surface. Figure 2.27a shows a standard sputtering system. The sputtered material deposits on a wafer that is placed facing the target. The ion current and energy can be independently adjusted. Since the target and wafer are placed in a chamber that has lower pressure, more target material and less contamination are transferred to the wafer.

One method to increase the deposition rate in sputtering is to use a third electrode that provides more electrons for ionization. Another method is to use a magnetic field, such as in *electron cyclotron resonance* (ECR) systems, to capture and spiral electrons, increasing their ionizing efficiency in the vicinity of the sputtering target. This technique, referred to as *magnetron sputtering*, has found widespread applications for the deposition of aluminum and its alloys at a rate that can approach 1 μm/min.

Long-throw sputtering is another technique used to control the angular distribution. Figure 2.27b shows a long-throw sputtering system. In standard sputtering configurations, there are two primary reasons for a wide angular distribution of incident flux at the surface: (1) the use of a small target to substrate separation d_{ts} and (2) scattering of the flux by the working gas as the flux travels from the target to the substrate. These two factors are linked because a small d_{ts} is needed to achieve good throughput, uniformity, and film properties when there is substantial gas scattering. A solution to this problem is to sputter at very low

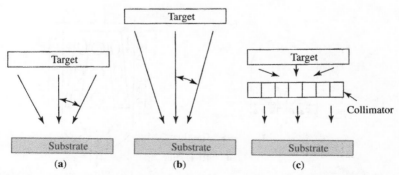

Figure 2.27. (a) Standard sputtering; (b) long-throw sputtering; (c) sputtering through a collimator [1].

pressures, a capability that has been developed using a variety of systems, which can sustain the magnetron plasma under more rarefied conditions. These systems allow for sputtering at working pressures of less than 0.1 Pa. At these pressures, gas scattering is less important, and the target–substrate distance can be greatly increased. From a simple geometric argument, this allows the angular distribution to be greatly narrowed, which permits more deposition at the bottom of high-aspect features such as contact holes.

Contact holes with large aspect ratio are difficult to fill with material, mainly because scattering events cause the top opening of the hole to seal before appreciable material has deposited on its floor. This problem can be overcome by collimating the sputtered atoms by placing an array of collimating tubes just above the wafer to restrict the depositing flux to normal ±5°. Sputtering with a collimator is shown in Figure 2.27c. Atoms whose trajectory is more than 5° from normal are deposited on the inner surface of the collimators.

2.1.5.2. Chemical Vapor Deposition

Chemical vapor deposition (CVD), also known as *vapor-phase epitaxy* (VPE), is a process whereby an epitaxial layer is formed by a chemical reaction between gaseous compounds. CVD can be performed at atmospheric pressure (APCVD) or at low pressure (LPCVD). Figure 2.28 shows three common susceptors for epitaxial growth. Note that the geometric shape of the susceptor provides the name for the reactor: horizontal, pancake, and barrel susceptors—all made from graphite blocks. Susceptors in epitaxial reactors are analogous to crucibles in the

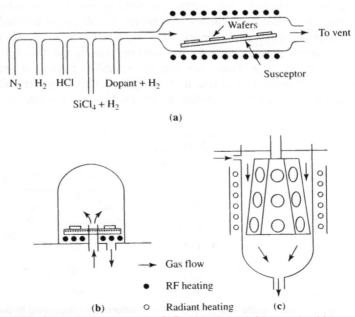

Figure 2.28. Common susceptors for CVD: (a) horizontal; (b) pancake; (c) barrel [1].

crystal growing furnaces. Not only do they mechanically support the wafer, but in induction-heated reactors, they also serve as the source of thermal energy for the reaction. The mechanism of CVD involves a number of steps: (1) the reactants (gases and dopants) are transported to the substrate region; (2) they are transferred to the substrate surface, where they are adsorbed; (3) a chemical reaction occurs, catalyzed at the surface, followed by growth of the epitaxial layer; (4) the gaseous products are desorbed into the main gas stream; and (5) the reaction products are transported out of the reaction chamber.

CVD is attractive for metallization because it offers coatings that are conformal, has good step coverage, and can coat a large number of wafers at a time. The basic CVD setup is the same as that used for deposition of dielectrics and polysilicon (see Figure 1.14). Low-pressure CVD (LPCVD) is capable of producing conformal step coverage over a wide range of topographical profiles, often with lower electrical resistivity than that from PVD. One of the major new applications of CVD metal deposition for integrated circuit production is in the area of refractory metal deposition. For example, tungsten's low electrical resistivity (5.3 $\mu\Omega \cdot$ cm) and refractory nature make it a desirable metal for use in IC fabrication.

2.1.6. Planarization

The development of chemical–mechanical polishing (CMP) has become important for multilevel interconnection technology because it is the only method that allows global planarization (i.e., a flat surface across the whole wafer). It also offers other advantages, including reduced defect density and the avoidance of plasma damage (which would occur in an RIE-based planarization system).

The CMP process consists of moving the sample surface against a pad that carries slurry between the sample surface and the pad. Abrasive particles in the slurry cause mechanical damage on the sample surface, loosening the material for enhanced chemical attack or fracturing off the pieces of surface into a slurry where they dissolve or are swept away. The process is tailored to provide an enhanced material removal rate from high points on surfaces. Mechanical grinding alone may theoretically achieve the desired planarization, but is undesirable because of extensive associated damage to the material surface. There are three main parts of the process: (1) the surface to be polished; (2) the pad, which is the key medium enabling the transfer of mechanical action to the surface being polished; and (3) the slurry, which provides both chemical and mechanical effects. Figure 2.29 shows a typical CMP setup.

2.2. PROCESS INTEGRATION

An integrated circuit is an ensemble of active (e.g., transistors) and passive devices (e.g., resistors, capacitors, and inductors) formed on and within a single-crystal semiconductor substrate and interconnected by a metallization pattern.

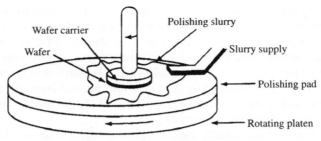

Figure 2.29. CMP schematic [1].

ICs have enormous advantages over discrete devices, including (1) reduction of the interconnection parasitics, (2) full utilization of a semiconductor wafer's area, and (3) drastic reduction in processing cost. In this section, we discuss the manner in which the basic processes described in previous portions of this chapter are combined to fabricate ICs. We consider three major IC technologies associated with two transistor families (viz., bipolar junction transistors and metal–oxide–semiconductor field-effect transistors, or MOSFETs): bipolar, CMOS, and BiCMOS. In addition, we will discuss the packaging of ICs by various techniques.

Figure 2.30 illustrates the interrelationship between the major process steps used for IC fabrication. Polished wafers with a specific resistivity and orientation are used as the starting material. The film formation steps include thermally grown

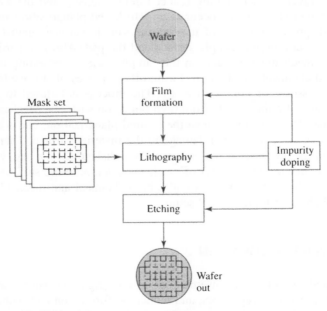

Figure 2.30. Schematic diagram of IC fabrication [1].

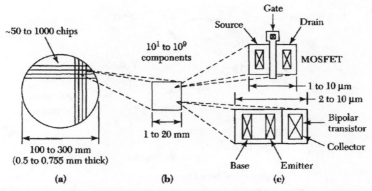

Figure 2.31. (a) Semiconductor wafer; (b) IC chip; (c) MOSFET and bipolar transistor [1].

oxide films (Section 2.1.1), deposited polysilicon, dielectric, and metal films (Section 2.1.5). Film formation is often followed by lithography (Section 2.1.2) or impurity doping (Section 2.1.4). Lithography is generally followed by etching (Section 2.1.3), which in turn is often followed by another impurity doping or film formation. The final IC is made by sequentially transferring the patterns from each mask, level by level, onto the surface of the semiconductor wafer.

After processing, each wafer contains hundreds of identical rectangular chips (or dies), typically between 1 and 20 mm on each side, as shown in Figure 2.31a. The chips are separated by sawing or laser cutting. Figure 2.31b shows a separated chip. Schematic top views of a single MOSFET and a single bipolar transistor are shown in Figure 2.31c to give some perspective of the relative size of a component in an IC chip. Prior to chip separation, each chip is electrically tested. Good chips are selected and packaged to provide an appropriate thermal, electrical, and interconnection environment for electronic applications.

2.2.1. Bipolar Technology

The majority of bipolar transistors used in ICs are of the $n-p-n$ type because the higher mobility of minority carriers (electrons) in the base region results in higher-speed performance than can be obtained with $p-n-p$ types. Figure 2.32 shows a perspective view of an $n-p-n$ bipolar transistor in which lateral isolation is provided by oxide walls and vertical isolation is provided by the $n^{+}-p$ junction. The lateral oxide isolation approach reduces not only the device size but also the parasitic capacitance because of the smaller dielectric constant of silicon dioxide (3.9, compared with 11.9 for silicon).

For an $n-p-n$ bipolar transistor, the starting material is a p-type, lightly doped ($\sim 10^{15}$ cm^{-3}), $\langle 111 \rangle$- or $\langle 100 \rangle$-oriented, polished silicon wafer. Because the junctions are formed inside the semiconductor, the choice of crystal orientation is not as critical as for MOS devices (see Section 2.2.2). The first step is to form a buried layer. The main purpose of this layer is to minimize the series resistance of the collector. A thick oxide (0.5–1 μm) is thermally grown

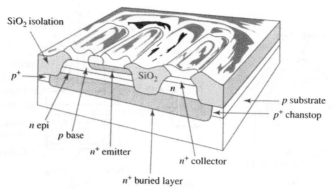

SiO$_2$ isolation

p^+

SiO$_2$

n

p substrate

p^+ chanstop

n epi

p base

n^+ emitter

n^+ collector

n^+ buried layer

Figure 2.32. Oxide-isolated bipolar transistor [1].

on the wafer, and a window is then opened in the oxide. A precisely controlled amount of low-energy arsenic ions ($\sim$30 keV, $\sim10^{15}$ cm^{-2}) is implanted into the window region to serve as a predeposit (Figure 2.33a). Next, a high-temperature ($\sim$1100°C) drive-in step forms the n^+-buried layer, which has a typical sheet resistance of 20 $\Omega/\Box$.

The second step is to deposit an n-type epitaxial layer. The oxide is removed and the wafer is placed in an epitaxial reactor for epitaxial growth. The thickness and the doping concentration of the epitaxial layer are determined by the ultimate use of the device. Analog circuits (with their higher voltages for amplification) require thicker layers ($\sim$10 μm) and lower dopings ($\sim5 \times 10^{15}$ cm^{-3}), whereas digital circuits (with their lower voltages for switching) require thinner layers ($\sim$3 μm) and higher dopings ($\sim2 \times 10^{16}$ cm^{-3}). Figure 2.33b shows a cross-sectional view of the device after the epitaxial process.

The third step is to form the lateral oxide isolation region. A thin oxide pad ($\sim$50 nm) is thermally grown on the epitaxial layer, followed by a silicon nitride deposition ($\sim$100 nm). If nitride is deposited directly onto the silicon without the thin oxide pad, the nitride may cause damage to the silicon surface during subsequent high-temperature steps. Next, the nitride–oxide layers and about half of the epitaxial layer are etched using a photoresist as mask (Figures 2.33c and 2.33d). Boron ions are then implanted into the exposed silicon areas (Figure 2.33d).

The photoresist is removed, and the wafer is placed in an oxidation furnace. Since the nitride layer has a very low oxidation rate, thick oxides will be grown only in the areas not protected by the nitride layer. The isolation oxide is usually grown to a thickness such that the top of the oxide becomes coplanar with the original silicon surface to minimize the surface topography. This oxide isolation process is called *local oxidation of silicon* (LOCOS). Figure 2.34a shows the cross section of the isolation oxide after removal of the nitride layer. Because of segregation effects, most of the implanted boron ions are pushed underneath the isolation oxide to form a p^+ layer. This is called the p^+-*channel stop* (or *chanstop*), because the high concentration of p-type semiconductor will prevent surface inversion and eliminate possible high-conductivity paths (or channels) among neighboring buried layers.

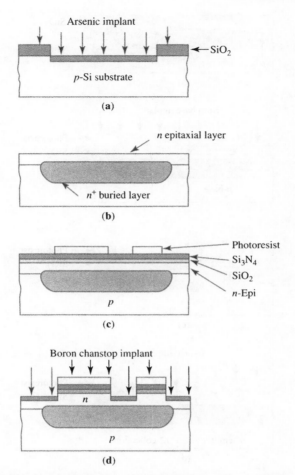

Figure 2.33. Cross-sectional views of bipolar transistor fabrication: (a) buried-layer implantation; (b) epitaxial layer; (c) photoresist mask; (d) channel-stop layer [1].

The fourth step is to form the base region. A photoresist is used as a mask to protect the right half of the device. Then, boron ions ($\sim 10^{12}$ cm^{-2}) are implanted to form the base regions, as shown in Figure 2.34b. Another lithographic process removes all the thin pad oxide except for a small area near the center of the base region (Figure 2.34c). The fifth step is to form the emitter region. As shown in Figure 2.34d, the base contact area is protected by a photoresist mask. Then, a low-energy, high-arsenic-dose ($\sim 10^{16}$ cm^{-2}) implantation forms the n^+-emitter and n^+-collector contact regions. The photoresist is removed, and a final metallization step forms the contacts to the base, emitter, and collector, as shown in Figure 2.32.

In this basic bipolar process, there are six film formation operations, six lithographic operations, four ion implantations, and four etching operations. Each operation must be precisely controlled and monitored. Failure of any one of the

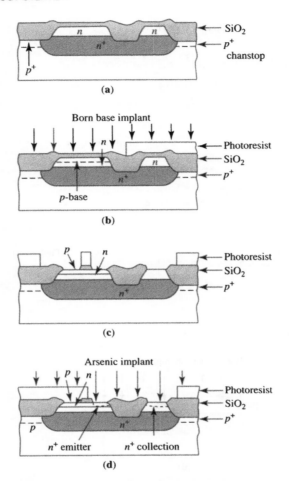

Figure 2.34. Cross-sectional views of bipolar transistor fabrication: (a) oxide isolation; (b) base implantation; (c) removal of thin oxide; (d) emitter–collector implant [1].

operations generally will render the wafer useless. The doping profiles of the completed transistor along a coordinate perpendicular to the surface and passing through the emitter, base, and collector are shown in Figure 2.35. The emitter profile is abrupt because of the concentration-dependent diffusivity of arsenic. The base doping profile beneath the emitter can be approximated by a Gaussian distribution for limited-source diffusion. The collector doping is given by the epitaxial doping level ($\sim 2 \times 10^{16}$ cm^{-3}) for a representative switching transistor.

2.2.2. CMOS Technology

The MOSFET is the dominant device used in modern integrated circuits because it can be scaled to smaller dimensions than other types of devices. The dominant technology for MOSFET is *complementary MOSFET* (CMOS) technology, in

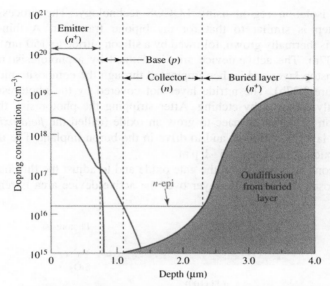

Figure 2.35. n–p–n bipolar transistor doping profile [1].

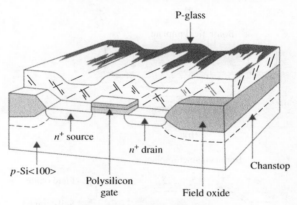

Figure 2.36. n-channel MOSFET [1].

which both n-channel and p-channel devices (NMOS and PMOS, respectively) are provided on the same chip. CMOS technology is particular attractive because it has the lowest power consumption of all IC technology. Figure 2.36 shows a perspective view of an n-channel MOSFET prior to final metallization. The top layer is a phosphorus-doped silicon dioxide (P-glass) that is used as an insulator between the polysilicon gate and the gate metallization and also as a gettering layer for mobile ions.

2.2.2.1. Basic NMOS Fabrication Sequence

In an NMOS process, the starting material is a p-type, lightly doped ($\sim 10^{15}$ cm^{-3}), $\langle 100 \rangle$-oriented, polished silicon wafer. The first step is to form the oxide isolation region using LOCOS technology. The process sequence for this step is similar to that for the bipolar transistor. A thin-pad oxide (~35 nm) is thermally grown, followed by a silicon nitride (~150 nm) deposition (Figure 2.37a). The active-device area is defined by a photoresist mask and a boron chanstop layer and is then implanted through the composite nitride–oxide layer (Figure 2.37b). The nitride layer not covered by the photoresist mask is subsequently removed by etching. After stripping the photoresist, the wafer is placed in an oxidation furnace to grow an oxide (called the *field oxide*), where the nitride layer is removed, and to drive in the boron implant. The thickness of the field oxide is typically 0.5–1 μm.

The second step is to grow the gate oxide and to adjust the threshold voltage. The composite nitride–oxide layer over the active-device area is removed, and

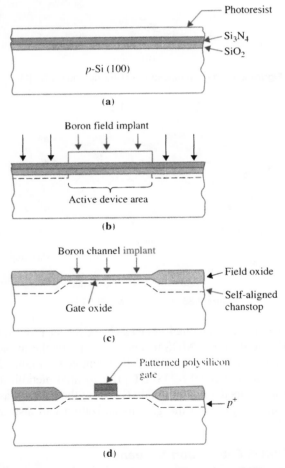

Figure 2.37. NMOS fabrication sequence: (a) formation of SiO_2, Si_3N_4, and photoresist layers; (b) boron implant; (c) field oxide; (d) gate [1].

a thin-gate oxide layer (less than 10 nm) is grown. For an enhancement-mode
n-channel device, boron ions are implanted in the channel region, as shown in
Figure 2.37c, to increase the threshold voltage to a predetermined value (e.g.,
+0.5 V). For a depletion-mode n-channel device, arsenic ions are implanted in
the channel region to decrease the threshold voltage (e.g., −0.5 V).

The third step is to form the gate. A polysilicon is deposited and is heavily
doped by diffusion or implantation of phosphorus to a typical sheet resistance of
20–30 $\Omega/\square$. This resistance is adequate for MOSFETs with gate lengths larger
than 3 μm. For smaller devices, polycide, a composite layer of metal silicide and
polysilicon such as W-polycide, can be used as the gate materials to reduce the
sheet resistance to about 1 $\Omega/\square$.

The fourth step is to form the source and drain. After the gate is patterned
(Figure 2.37d), it serves as a mask for the arsenic implantation ($\sim$30 keV, $\sim$5 $\times$
10^{15} cm^{-2}) to form the source and drain (Figure 2.38a), which are self-aligned

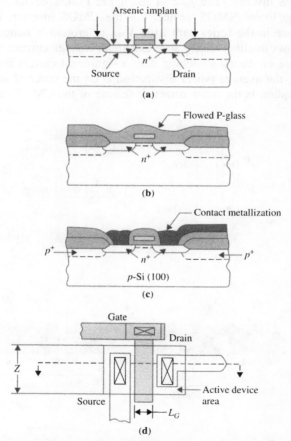

Figure 2.38. NMOS fabrication sequence: (a) source and drain; (b) P-glass deposition;
(c) MOSFET cross section; (d) MOSFET top view [1].

with respect to the gate. At this stage, the only overlapping of the gate is due to lateral straggling of the implanted ions (for 30 keV As, $\sigma_\perp$ is only 5 nm). If low-temperature processes are used for subsequent steps to minimize lateral diffusion, the parasitic gate–drain and gate–source coupling capacitances can be much smaller than the gate–channel capacitance.

The last step is metallization. P-glass is deposited over the entire wafer and is flowed by heating the wafer to give a smooth surface topography (Figure 2.38b). Contact windows are defined and etched in the P-glass. A metal layer, such as aluminum, is then deposited and patterned. A cross-sectional view of the completed MOSFET is shown in Figure 2.38c, and the corresponding top view is shown in Figure 2.38d. The gate contact is usually made outside the active-device area to avoid possible damage to the thin-gate oxide.

2.2.2.2. CMOS Fabrication Sequence

The MOS process forms the foundation for CMOS technology. Figure 2.39a shows a CMOS inverter. The gate of the upper PMOS device is connected to the gate of the lower NMOS device. For the CMOS inverter, in either logic state, one device in the series path from V_{DD} to ground is nonconductive. The current that flows in either steady state is a small leakage current, and only when both devices are on during switching does a significant current flow through the inverter. Thus, the average power dissipation is on the order of nanowatts. Low power consumption is the most attractive feature of the CMOS circuit.

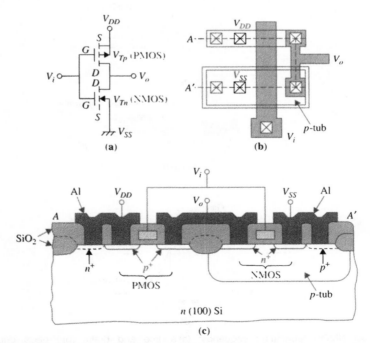

Figure 2.39. CMOS inverter: (a) circuit diagram; (b) layout; and (c) cross section [1].

Figure 2.39b shows a layout of the CMOS inverter, and Figure 2.39c shows the device cross section along the $A-A'$ line. In the processing, a p tub (also called a p *well*) is first implanted and subsequently driven into the n substrate. The p-type dopant concentration must be high enough to overcompensate the background doping of the n substrate. The subsequent processes for the n-channel MOSFET in the p tub are identical to those described previously. For the p-channel MOSFET, $^{11}B^+$ or $^{49}(BF_2)^+$ ions are implanted into the n substrate to form the source and drain regions. A channel implant of $^{75}As^+$ ions may be used to adjust the threshold voltage and a n^+ chanstop is formed underneath the field oxide around the p-channel device. Because of the p tub and the additional steps needed to make the p-channel MOSFET, the number of steps to make a CMOS circuit is essentially double that to make an NMOS circuit. Thus, there is a trade-off between the complexity of processing and a reduction in power consumption.

Instead of the p tub described above, an alternate approach is to use an n tub formed in p-type substrate, as shown in Figure 2.40a. In this case, the n-type dopant concentration must be high enough to overcompensate for the background doping of the p substrate (i.e., $N_D > N_A$). In both the p-tub and the n-tub approaches, the channel mobility will be degraded because mobility is determined by the total dopant concentration ($N_A + N_D$). A more recent approach using two separated tubs implanted into a lightly doped substrate is shown in Figure 2.40b. This structure is called a "twin tub." Because no overcompensation is needed in either of the twin tubs, higher channel mobility can be obtained.

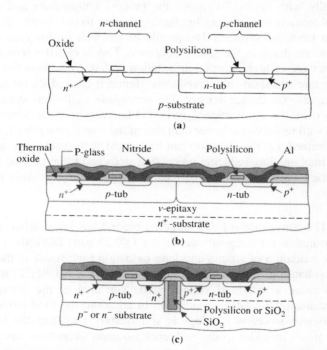

Figure 2.40. Various CMOS structures: (a) n-tub; (b) twin-tub; (c) refilled trench [1].

All CMOS circuits have the potential for a problem called *latchup* that is associated with parasitic bipolar transistors. These parasitic devices consist of the *npn* transistor formed by the NMOS source–drain regions, *p* tub, and *n*-type substrate, as well as the *pnp* transistor formed by the PMOS source-drain regions, *n*-type substrate, and *p* tub. Under appropriate conditions, the collector of the *pnp* device supplies base current to the *npn* and vice versa in a positive feedback arrangement. This latchup current can have serious negative repercussions in a CMOS circuit.

An effective processing technique to eliminate latchup is to use deep-trench isolation, as shown in Figure 2.40c. In this technique, a trench with a depth deeper than the well is formed in the silicon by anisotropic reactive sputter etching. An oxide layer is thermally grown on the bottom and walls of the trench, which is then refilled by deposited polysilicon or silicon dioxide. This technique can eliminate latchup because the *n*-channel and *p*-channel devices are physically isolated by the refilled trench. The detailed steps for trench isolation and some related CMOS processes are now considered.

Well Formation. The well of a CMOS circuit can be a single well, a twin well, or a retrograde well. The twin-well process exhibits some disadvantages. For example, it needs high-temperature processing (above 1050°C) and a long diffusion time (longer than 8 h) to achieve the required well depth of 2–3 μm. In this process, the doping concentration is highest at the surface and decreases monotonically with depth. To reduce the process temperature and time, high-energy implantation is used (i.e., implanting the ion to the desired depth instead of diffusion from the surface). The profile of the well in this case can have a peak at a certain depth in the silicon substrate. This is called a *retrograde well.*

The advantage of high-energy implantation is that it can form the well under low-temperature and short-time conditions. Hence, it can reduce the lateral diffusion and increase the device density. The retrograde well offers some additional advantages over the conventional well: (1) because of high doping near the bottom, the well resistivity is lower than that of the conventional well, and latchup can be minimized; (2) the chanstop can be formed at the same time as the retrograde well implantation, reducing processing steps and time; and (3) higher well doping in the bottom can reduce the chance of punchthrough from the drain to the source.

Isolation. The conventional MOS isolation process has some disadvantages that make it unsuitable for deep-submicrometer ($\leq$0.25-μm) fabrication. The high-temperature oxidation of silicon and long oxidation time result in the encroachment of the chanstop implantation (usually boron for *n*-MOSFET) to the active region and cause a threshold voltage shift. The area of the active region is reduced because of the lateral oxidation. In addition, the field oxide thickness in submicrometer-isolation spacings is significantly less than the thickness of field oxide grown in wider spacings. Trench isolation technology can avoid these problems.

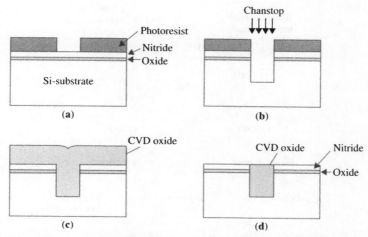

Figure 2.41. Shallow-trench isolation: (a) patterning on nitride–oxide films; (b) dry etching and chanstop implantation; (c) CVD oxide to refill; (d) surface after CMP [1].

An example is *shallow trench* (depth less than 1 μm) isolation, shown in Figure 2.41. After patterning (Figure 2.41a), the trench area is etched (Figure 2.41b) and then refilled with oxide (Figure 2.41c). Before refilling, a channel stop implantation can be performed. Since the oxide has overfilled the trench, the oxide on the nitride should be removed. Chemical–mechanical polishing is used to remove the oxide on the nitride and to get a flat surface (Figure 2.41d). Because of its high resistance to polishing, the nitride acts as a stop layer for the CMP process. After the polishing, the nitride layer and the oxide layer can be removed by H_3PO_4 and HF, respectively. This initial planarization step at the beginning is helpful for the subsequent polysilicon patterning and planarizations of the multilevel interconnection processes.

Gate Engineering. If n^+-polysilicon is used for both PMOS and NMOS gates, the threshold voltage for PMOS has to be adjusted by boron implantation. This makes the channel of the PMOS a buried type, as shown in Figure 2.42a. The buried-type PMOS suffers serious short-channel effects as the device size shrinks below 0.25 μm. The most noticeable phenomena for short-channel effects are threshold voltage rolloff, drain-induced barrier lowering, and the large leakage current at the OFF state. To alleviate these problems, the n^+-polysilicon can be changed to p^+-polysilicon for the PMOS devices. Due to the workfunction difference (1.0 eV from n^+- to p^+-polysilicon), a surface p-type channel device can be achieved without the boron V_T adjustment implantation. Hence, as the technology shrinks to 0.25 μm and less, dual-gate structures are required: p^+-polysilicon gate for PMOS and n^+-polysilicon for NMOS (Figure 2.42b).

To form the p^+-polysilicon gate, ion implantation of BF_2 is commonly used. However, boron penetrates easily from the polysilicon through the oxide into the silicon substrate at high temperatures, resulting in a V_T shift. This penetration

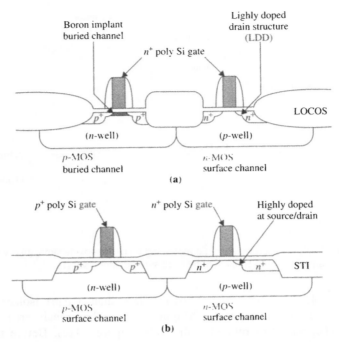

Figure 2.42. (a) Conventional CMOS structure with a single polysilicon gate; (b) advanced CMOS structure with dual polysilicon gates [1].

is enhanced in the presence of a F atom. There are methods to reduce this effect: use of rapid thermal annealing to reduce the time at high temperatures and, consequently, the diffusion of boron; use of nitrided oxide to suppress the boron penetration, since boron can easily combine with nitrogen and becomes less mobile; and the creation of a multilayer of polysilicon to trap the boron atoms at the interface of the two layers.

2.2.3. BiCMOS Technology

BiCMOS is a technology that combines both CMOS and bipolar device structures in a single IC. The reason to combine these two different technologies is to create an IC chip that has the advantages of both CMOS and bipolar devices. We know that CMOS exhibits advantages in power dissipation, noise margin, and packing density, whereas bipolar technology shows advantages in switching speed, current drive capability, and analog capability. As a result, for a given design rule, BiCMOS can have a higher speed than CMOS, better performance in analog circuits than CMOS, a lower power dissipation than bipolar, and a higher component density than bipolar.

BiCMOS has been widely used in many applications. Early on, it was used in static random access memory (SRAM) circuits. Currently, BiCMOS technology has been successfully developed for transceiver, amplifier, and oscillator

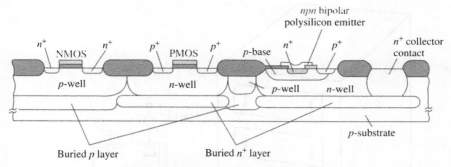

Figure 2.43. BiCMOS device structure [1].

applications in wireless communication equipment. Most BiCMOS processes are based on standard CMOS process with some modifications, such as adding masks for bipolar transistor fabrication. The example shown in Figure 2.43 is for a high-performance BiCMOS process based on the twin-well CMOS approach.

The initial material is a p-type silicon substrate. An n^+-buried layer is formed to reduce collector resistance. The buried p layer is formed by ion implantation to increase the doping level and prevent punchthrough. A lightly doped n-epi layer is grown on the wafer, and a twin-well process for the CMOS is performed. To achieve high performance for the bipolar transistor, four additional masks are needed: the buried n^+ mask, the collector deep n^+ mask, the base p mask, and the polyemitter mask. The p^+ region for base contact can be formed with the p^+ implant in the source–drain implantation of the PMOS, and the n^+ emitter can be formed with the source/drain implantation of the NMOS. The additional masks and longer processing time compared with a standard CMOS process are the main drawbacks of BiCMOS.

2.2.4. Packaging

Before finished ICs can be put to their intended use in various commercial electronic systems and products (such as computers, cellular phones, and digital cameras), several other key processes must take place. These include both *electrical testing* and *packaging*. Testing, which is discussed in detail in Chapter 3, is clearly necessary to ensure high-quality products. The term *packaging* refers to the set of technologies and processes that connect ICs with electronic systems. A useful analogy is to consider an electronic product as the human body. Like the body, these products have "brains," which are analogous to ICs. Electronic packaging provides the "nervous system," as well as the "skeletal system." The package is responsible for interconnecting, powering, cooling, and protecting the IC.

Overall, electronic systems consist of several levels of packaging, each with distinctive types of interconnection devices. Figure 2.44 depicts this packaging hierarchy. Level 0 consists of on-chip interconnections. Chip-to-printed circuit board or chip-to-module connections constitute level 2, and board-to-board

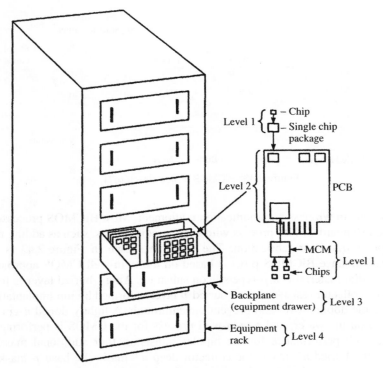

Figure 2.44. Electronic packaging hierarchy [2].

interconnections make up level 3. Levels 4 and 5 consist of connections between subassemblies and between systems (such as computer to printer), respectively.

2.2.4.1. Die Separation

After functional testing, individual ICs (or dies) must be separated from the substrate. This is the first step in the packaging process. In a common method that has been used for many years, the substrate wafer is mounted on a holder and scribed in both the x and y directions using a diamond scribe. This is done along scribe borders of 75–250 μm in width that are formed around the periphery of the dies during fabrication. These borders are aligned with the crystal planes of this substrate if possible. After scribing, the wafer is removed form the holder and placed upside-down on a soft support. A roller is then used to apply pressure, fracturing the wafer along the scribe lines. This must be accomplished with minimal damage to the individual die.

More modern die separation processes use a diamond saw, rather than a diamond scribe. In this procedure, the wafer is attached to an adhesive sheet of mylar film. The saw is then used to either scribe the wafer or to cut completely through it. After separation, the dies are removed from the mylar. The separated dice are then ready to be placed into packages.

Plastic dual-in-line-package

Figure 2.45. Dual-inline package [2].

2.2.4.2. Package Types

There are a number of approaches to the packaging of single ICs. The *dual-inline package* (DIP) (Figure 2.45), is the package most people envision when they think of integrated circuits. The DIP was developed in the 1960s, quickly became the primary package for ICs, and has long dominated the electronics packaging market. The DIP can be made of plastic or ceramic; the latter is called the *CerDIP*. The CerDIP consists of a DIP constructed of two pieces of sandwiched ceramic with leads protruding from between the ceramic plates.

In the 1970s and 1980s, *surface-mount packages* were developed in response to a need for higher-density interconnect than the DIP approach could provide. In contrast to DIPs, the leads of a surface-mounted package do not penetrate the printed circuit board (PCB) on which it is mounted. This means that the package can be mounted on both sides of the board, thereby allowing higher density. One example of such a package is the quad flatpack (QFP) (Figure 2.46), which has leads on all four sides to further increase the number of input/output (I/O) connections.

More recently, the need for rapidly increasing numbers of I/O connections has led to the development of pin-grid array (PGA) and ball-grid array (BGA) packages (Figures 2.47 and 2.48, respectively). PGAs have an I/O density of about 600, and BGAs can have densities greater than 1000, as compared to

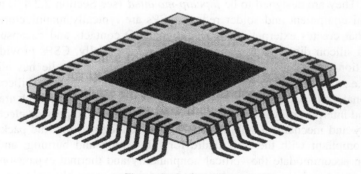

Figure 2.46. Quad flatpack [2].

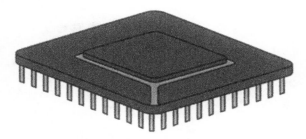

Figure 2.47. Pin-grid array [2].

Figure 2.48. Ball-grid array [2].

~200 for QFPs. BGAs can be identified by the solder bumps on the bottom of the package. With QFPs, as the spacing between leads becomes tighter, the manufacturing yield decreases rapidly. The BGA allows higher density and takes up less space than the QFP, but its manufacturing process is inherently more expensive.

The most recent development in packaging is the chip-scale package (CSP), which is shown in Figure 2.49. CSPs, defined as packages no larger than 20% greater than the size of the IC die itself, often take the form of miniaturized ball-grid arrays. They are designed to be *flipchip-mounted* (see Section 2.2.4.3) using conventional equipment and solder reflow. CSPs are typically manufactured in a process that creates external power and signal I/O contacts and encapsulates the finished silicon die prior to dicing the wafer. Essentially, CSPs provide an interconnection framework for ICs so that before dicing, each die has all the functions (i.e., external electrical contacts, encapsulation of the finished silicon) of a conventional, fully packaged IC. Two essential features of this approach are that the leads and interposer layer (an added layer on the IC used to provide electrical functionality and mechanical stability) are flexible enough so that the packaged device is compliant with the test fixture for full testing and burning, and the package can accommodate the vertical nonplanarity and thermal expansion and contraction of the underlying printed circuit board during assembly and operation.

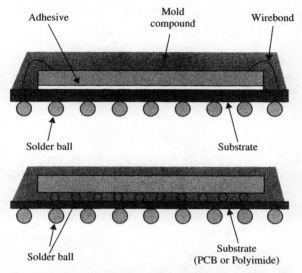

Figure 2.49. Two typical chip-scale packages [3].

2.2.4.3. Attachment Methods

An IC must be mounted and bonded to a package, and that package must be attached to a printed circuit board before the IC can be used in an electronic system. Methods of attaching ICs to PCBs are referred to as *level 1 packaging*. The technique used to bond a bare die to a package has a significant effect on the ultimate electrical, mechanical and thermal properties of electronic system being manufactured. Chip-to-package interconnection is generally accomplished by either *wire bonding, tape-automated bonding* (TAB), or *flipchip bonding* (see Figure 2.50).

Wire bonding is the oldest attachment method and is still the dominant technique for chips with fewer that 200 I/O connections. Wire bonding requires connecting gold or aluminum wires between chip bonding pads and contact points on the package. ICs are first attached to the substrate using a thermally conductive adhesive with their bonding pads facing upward. The Au or Al wires are then attached between the pads and substrate using *ultrasonic, thermosonic*, or *thermocompression* bonding [1]. Although automated, this process is still time-consuming since each wire must be attached individually.

Tape-automated bonding (TAB) was developed in the early 1970s and is often used to bond packages to PCBs. In TAB, chips are first mounted on a flexible polymer tape (usually polyimide) containing repeated copper interconnection patterns. The copper leads are defined by lithography and etching, and the lead pattern can contain hundreds of connections. After the IC pads have been aligned to metal interconnection stripes on the tape, attachment takes place by thermocompression. Gold bumps are formed on either side of the die or tape and are used to bond the die to the leads on the tape.

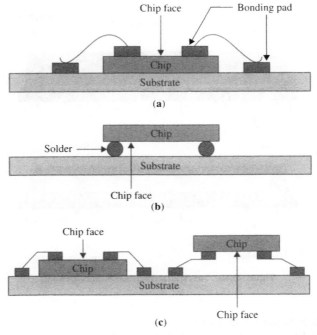

Figure 2.50. (a) Wire; (b) flipchip; (c) tape-automated bonding [2].

Flipchip bonding is a direct interconnection approach in which the IC is mounted upside-down onto a module or printed circuit board. Electrical connections are made via solder bumps (or solderless materials such as epoxies or conductive adhesives) located over the surface of the chip. Since bumps can be located anywhere on the chip, flipchip bonding ensures that the interconnect distance between the chip and package is minimized. The I/O density is limited only by the minimum distance between adjacent bond pads.

SUMMARY

In this chapter, we have provided an overview of the critical unit processes in IC fabrication and described the integration of these unit processes into sequences for fabricating and packaging ICs. In the next chapter, we will discuss how these processes are monitored to facilitate quality control.

PROBLEMS

2.1. Assuming that a silicon oxide layer of thickness x is grown by thermal oxidation, show that the thickness of silicon being consumed is $0.44x$. The molecular weight of Si is 28.9 g/mol, and the density of Si is 2.33 g/cm^3. The corresponding values for SiO$_2$ are 60.08 g/mol and 2.21 g/cm^3.

2.2. A silicon sample is oxidized in dry O$_2$ at 1200°C for one hour.

    (a) What is the thickness of the oxide grown?

    (b) How much additional time is required to grow 0.1 μm more oxide in wet O$_2$ at 1200°C?

2.3. Find the parameter γ for the photoresists shown in Figure 2.13.

2.4. Calculate the Al average etch rate and etch rate uniformity on a 200-mm-diameter silicon wafer, assuming that the etch rates at the center, left, right, top, and bottom of the wafer are 750, 812, 765, 743, and 798 nm/min, respectively.

2.5. The electron densities in RIE and HDP systems range within $10^9 - 10^{10}$ and $10^{11} - 10^{12}$ cm^{-3}, respectively. Assuming that the RIE chamber pressure is 200 mTorr and HDP chamber pressure is 5 mTorr, calculate the ionization efficiency in RIE reactors and HDP reactors at room temperature. The ionization efficiency is the ratio of the electron density to the density of molecules.

2.6. For a boron diffusion in silicon at 1000°C, the surface concentration is maintained at 10^{19} cm^{-3} and the diffusion time is 1 h. If the diffusion coefficient of boron at 1000°C is 2×10^{14} cm^2/s, find $Q(t)$ and the gradient at $x = 0$ and at a location where the dopant concentration reaches 10^{15} cm^{-3}.

2.7. Arsenic was predeposited by arsine gas, and the resulting total amount of dopant per unit area was 1×10^{14} atoms/cm^2. How long would it take to drive the arsenic in to a junction depth of 1 μm? Assume a background doping of $C_B = 1 \times 10^{15}$ atoms/cm^3, and a drive-in temperature of 1200°C. For As diffusion, $D_0 = 24$ cm^2/s, and $E_a = 4.08$ eV.

2.8. Assume 100-keV boron implants on a 200-mm silicon wafer at a dose of 5×10^{14} ions/cm^2. The projected range and project straggle are 0.31 and 0.07 μm, respectively. Calculate the peak concentration and the required ion-beam current for 1 min of implantation.

2.9. In a CMP process, the oxide removal rate and the removal rate of a layer underneath the oxide (called a *stop layer*) are $1r$ and $0.1r$, respectively. To remove 1 μm of oxide and a 0.01-μm stop layer, the total removal time is 5.5 min. Find the oxide removal rate.

REFERENCES

1. G. May and S. Sze, *Fundamentals of Semiconductor Fabrication*, Wiley, New York, 2003.

2. W. Brown, ed., *Advanced Electronic Packaging*, IEEE Press, New York, 1999.

3. R. Tummala, ed., *Fundamentals of Microsystems Packaging*, McGraw-Hill, New York, 2001.

3

PROCESS MONITORING

OBJECTIVES

- Survey various sensor metrology and methods of monitoring IC fabrication processes.
- Place this metrology in the context of process needs.
- Identify key measurement points in the process flow.
- Differentiate between wafer state and equipment state measurements.

INTRODUCTION

In Chapter 2, the basic unit processes used in fabricating an integrated circuit, as well as the process flows for several major IC technologies, were discussed in detail. In order for these processes to repeatably produce reliable, high-quality devices and circuits, each unit process must be strictly controlled. Many diagnostic tools are used to maintain systematic control. Such control requires that the key output variables for each process step (i.e., those that are correlated with product functionality and performance) be carefully monitored.

Process monitoring enables operators and engineers to detect problems early on to minimize their impact. The economic benefit of effective monitoring systems increases with the complexity of the manufacturing process. Manufacturing line monitors consist of extremely sophisticated metrology equipment that can be divided into tools characterizing the state of features on the semiconductor

Fundamentals of Semiconductor Manufacturing and Process Control,
By Gary S. May and Costas J. Spanos
Copyright © 2006 John Wiley & Sons, Inc.

wafers themselves and those that describe the status of the fabrication equipment operating on those wafers. The issues involved in understanding and implementing both wafer state and equipment state measurements will be discussed in detail in this chapter.

3.1. PROCESS FLOW AND KEY MEASUREMENT POINTS

When we monitor a physical system, we observe that system's behavior. On the basis of these observations, we take appropriate actions to influence that behavior in order to guide the system to some desirable state. Semiconductor manufacturing systems consist of a series of sequential process steps in which layers of materials are deposited on substrates, doped with impurities, and patterned using photolithography to produce sophisticated integrated circuits and devices.

As an example of such a system, Figure 3.1 depicts a typical CMOS process flow (refer to Section 2.2.1 for more details). Inserted into this flow diagram in various places are symbols denoting key measurement points. Clearly, CMOS technology involves many unit processes with high complexity and tight tolerances. This necessitates frequent and thorough inline process monitoring to assure high-quality final products.

The measurements required may characterize physical parameters, such as film thickness, uniformity, and feature dimensions; or electrical parameters, such as resistance and capacitance. These measurements may be performed directly on product wafers, either directly or using test structures, or alternatively, on nonfunctional monitor wafers (or "dummy" wafers). In addition to these, some measurements are actually performed "in situ," or *during* a fabrication step. When a process sequence is complete, the product wafer is diced, packaged, and subjected to final electrical and reliability testing.

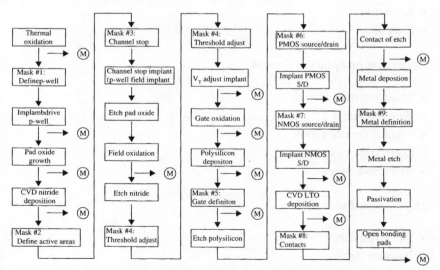

Figure 3.1. CMOS process flow showing key measurement points (denoted by "M").

3.2. WAFER STATE MEASUREMENTS

There is no substitute for regular inspection of products during manufacturing to ensure high quality. Inspections can reveal contamination, structural flaws, or other problems. Such investigations must not be limited to visual inspections, however, since not all processes have a visible effect on electronic products. Thin-film deposition and ion implantation are two important examples of this. In addition, with ever-increasing levels of integration, features on wafers become smaller and more difficult to inspect. As a result, visual inspection must be supplemented by sophisticated physical and electrical measurements of various characteristics that describe the state of a wafer.

Wafer state characterization includes the measurement of the physical parameters related to each manufacturing process step. Examples include

Lithography

- Linewidth
- Overlay
- Print bias
- Resist profiles

Etch

- Etch rate
- Selectivity
- Uniformity
- Anisotropy
- Etch bias

Deposition or Epitaxial Growth

- Sheet resistance
- Film thickness
- Surface concentration
- Dielectric constant
- Refractive index

Diffusion or Implantation

- Sheet resistance
- Junction depth
- Surface concentration

The total collection of such measurements relate to the physical characteristics of product wafers, and these physical characteristics can be correlated with the electrical performance of devices and circuits. The following sections describe wafer state measurements and the corresponding measurement apparatus in greater detail.

3.2.1. Blanket Thin Film

We begin the discussion of wafer state measurements with those measurements that are performed on blanket thin films. The term "blanket" is used to differentiate wafers that have been uniformly coated by a thin film from those in which the film has been patterned using photolithography and etching.

3.2.1.1. Interferometry

Optical metrology provides fast and precise measurements of film thickness and optical constants. In semiconductor manufacturing, *interferometry* (sometimes called *reflectometry*) is a widely used optical method for measuring such parameters. Single- or multiple wavelength interferometers are commonly used for both in situ and postprocess measurements of film thickness. In this method, a light source, usually a laser, is focused on a semiconductor wafer while a detector measures the reflected light intensity. The wafer consists of a parallel stack of partially transparent thin films. The reflected light intensity varies as a function of time depending on the thickness of the top layer due to constructive and destructive interference caused by multiple reflections.

To illustrate the basic concept, consider Figure 3.2, which shows a film of uniform thickness d and index of refraction n, with the eye of the observer focused on point a. The film is illuminated by broad source of monochromatic light S. There is a point P on the source such that two rays (represented by the single and double arrows) can leave P and enter the eye after traveling through point a. These two rays follow different paths, one reflected from the upper surface of the film and the other from the lower surface. Whether point a appears bright or dark depends on the nature of the interference (i.e., constructive or destructive) between the two waves that diverge from a.

The two factors that impact the nature of the interference are differences in optical path length and phase changes on reflection. For the two rays to combine to give maximum intensity, we must have

$$2dn \cos \theta = (m + 0.5)\lambda \tag{3.1}$$

where $m = 0, 1, 2, \ldots$ and θ is the angle of the refracted beam relative to the surface normal. The term 0.5λ accounts for the phase change that occurs on reflection since a phase change of $180°$ is equivalent to half a wavelength. The condition for minimum intensity is

$$2dn \cos \theta = m\lambda \tag{3.2}$$

Equations (3.1) and (3.2) hold when the index of refraction of the film is either greater or less than the indices of the media on *each* side of the film. Therefore,

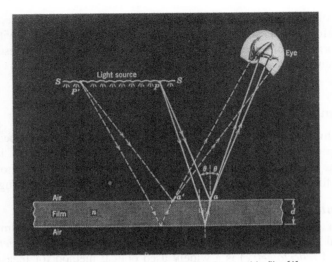

Figure 3.2. Interference by reflection from a thin film [1].

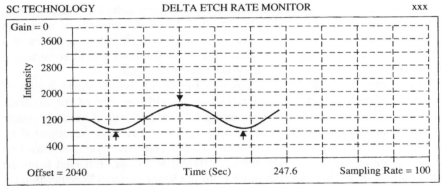

Figure 3.3. Sample interferogram used for plasma etch monitoring [2].

if the index of refraction is known, the thickness of the film may be computed by simply counting peaks or valleys in the reflected waveform. An example of such a waveform (or *interferogram*) appears in Figure 3.3.

Interferometry becomes more complex when applied to stacks of several thin films. The overall goal, however, is still to obtain film thickness information from the time-varying reflected intensity signal. The reflected light intensity is given by [7]

$$I_r(d, \lambda) = I_0(\lambda) r(d, \lambda, \phi_1, \phi_2, \dots, \phi_N) \tag{3.3}$$

where I_0 is the incident light intensity, r is the reflection coefficient, d is the thickness of the top layer, and ϕ_i are physical constants (i.e., thicknesses and refractive indices) associated with the lower films in the film stack.

The reflected intensity is monitored using a detector consisting of a light-sensitive transducer, such as a photodiode, in conjunction with an optical filter or diffraction grating to select the wavelength(s) of interest. The output of the detector corresponding to a particular wavelength is of the form

$$y_\lambda(kT) = \alpha(\lambda, kT)A(\lambda, kT)I_0(\lambda, kT)r(d(kT), \lambda) + e_\lambda(kT) \qquad (3.4)$$

where T denotes the sampling period, k is an integer, α represents losses in the optical system, A is the gain of the detector, and e_λ is measurement noise. The physical parameters ϕ_i are considered to be fixed and known in this formulation and are not shown. For multiple-wavelength (or *spectroscopic*) measurements, this expression is repeated for each wavelength used. For p wavelengths, in matrix form, this is written as

$$\mathbf{y}(kT) = \text{diag}(\mathbf{h}(kT)\mathbf{r}(d(kT)) + e(kT) \qquad (3.5)$$

where $\text{diag}(\mathbf{x})$ represents a matrix with the elements of the vector $\mathbf{x}$ along the diagonal and

$$\mathbf{y}(kT) = [y_{\lambda_1}(kT) \cdots y_{\lambda_p}(kT)] \qquad (3.6)$$

$$\mathbf{h}(kT) = [\alpha(\lambda_1, kT)A(\lambda_1, kT)I_0(\lambda_1, kT) \cdots \alpha(\lambda_p, kT)$$

$$A(\lambda_p, kT)I_0(\lambda_p, kT)] \qquad (3.7)$$

$$\mathbf{r}(d(kT)) = [r(d(kT), \lambda_1) \cdots r(d(kT), \lambda_p)]^T \qquad (3.8)$$

$$\mathbf{e}(kT) = [e_{\lambda_1}(kT) \cdots e_{\lambda_p}(kT)]^T \qquad (3.9)$$

where the superscript T represents the transpose operation.

To obtain film thickness or the rate of change of thickness (i.e., etch or deposition rate), the detector output is processed in one of two ways: (1) extrema counting or (2) least-squares fitting. Extrema counting takes advantage of the fact that the reflected light intensity varies approximately periodically with both the wavelength of the incident light and the thickness of the top film. The distance between peaks and valleys is a known function of the top film thickness. Thus, if many wavelengths are available, thickness can be determined by counting the peaks in a plot of reflectance versus wavelength. If only a single wavelength is available, the movement of peaks and valleys over time during in situ measurements indicates that a specific amount of material has been etched or deposited. This provides the average etch or deposition rate between successive minima and maxima.

To use the least-squares approach, at each timepoint, the following nonlinear optimization problem is posed:

$$\min_d[(\mathbf{y}(kT) - \text{diag}(\mathbf{h})\mathbf{r}(d))^T (\mathbf{y}(kT) - \text{diag}(\mathbf{h})\mathbf{r}(d))] \qquad (3.10)$$

The film thickness is then that for which the minimum is achieved. Etch rate or deposition rate is then calculated from the resulting thickness versus time curve.

The final variation of interferometry we will discuss briefly is one that is particularly applicable to thickness monitoring during plasma etching. During etching, the emission from the plasma itself may be used as the light source. As this light is reflected from the etched film and underlying film surfaces while the thickness of the etched film decreases, the optical path difference between light rays varies and the changing constructive and destructive interference results in periodic signals in the same manner as previously described. If a charge-coupled device (CCD) camera is placed in such a way that it can view these signals (see Figure 3.4), each pixel of the CCD camera then acts as an individual interferometer monitoring a different part of the wafer. This arrangement is called *full-wafer interferometry* [5].

3.2.1.2. Ellipsometry

Ellipsometry is a widely used measurement technique based on the polarization changes that occur when light is reflected from or transmitted through a medium. Changes in polarization are a function of the optical properties of the material (i.e., its complex refractive indices), its thickness, and the wavelength and angle of incidence of the light beam relative to the surface normal. When multiple light beams of varying wavelength are used, the technique is referred to as *spectroscopic ellipsometry* (SE). SE, which can be used to make in situ or postprocess measurements, is a fundamentally more accurate technique than interferometry for obtaining film thickness and optical dielectric function information. In general, SE measurements are performed at an off-normal angle with respect to the sample. In this configuration, the measurement is sensitive to the polarization state of both the incident and reflected waves.

Figure 3.5 shows an unpolarized beam of light falling on a dielectric surface. In this case, the dielectric is glass. The electric field vector for each wavetrain

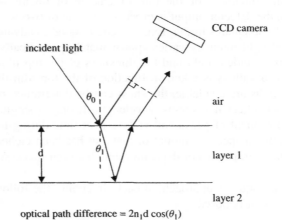

optical path difference = $2n_1 d \cos(\theta_1)$

Figure 3.4. Schematic of full-wafer interferometry [5].

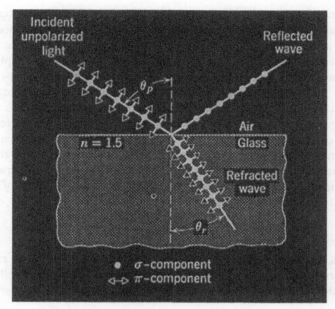

Figure 3.5. Illustration of components of polarization [1].

in the beam can be resolved into two components—one perpendicular to the plane of incidence (i.e., the plane of the figure) and another parallel to this plane. The perpendicular component, represented by the dots, is the σ component (or "s component"). The parallel component, represented by the arrows, is the π component (or "p component"). On average, for completely unpolarized incident light, these two components are of equal amplitude. However, if the incident beam is polarized (as is the case in ellipsometry), this is no longer true.

In the most common configuration, linearly polarized light is incident on the surface, and the elliptical polarization status of the reflected light is analyzed. Measured ellipsometry data are usually written in the form of the ratio (ρ) of the *total reflection coefficients* for s and p polarization (R^s and R^p, respectively). In other words

$$\rho = R^p/R^s = \tan(\psi)e^{i\Delta} \tag{3.11}$$

where $\tan(\psi)$ is the ratio of the magnitude of the p-polarized light to the s-polarized reflected light and Δ is the difference in phase shifts on reflection for the p and s polarizations, respectively.

Another set of expressions called the *Fresnel equations* relate [Eq. (3.11)] to the bulk complex dielectric function (ε). The dielectric function represents the degree to which the material may be polarized by an applied external electric field, and as a complex number, it is expressed as

$$\varepsilon = \varepsilon_1 + j\varepsilon_2 \tag{3.12}$$

where ε_1 and ε_2 are the real and imaginary parts, respectively. For heterogeneous samples consisting of multiple layers, the dielectric function determined by ellipsometry is an average over the region penetrated by the incident light called the *effective dielectric function*, $\langle \varepsilon \rangle$. If the sample structure is not too complicated, $\langle \varepsilon \rangle$ can be simulated by appropriate models (such as the "ambient–film–substrate" model). In this case, film and substrate properties can be separated, and film properties (i.e., thickness or dielectric function) can be determined as follows.

Because there are a maximum of two independent optical parameters (Ψ and Δ) measured at each wavelength, the maximum number of unknowns that can be determined from a single spectral measurement is $2w$, where w is the number of wavelengths scanned. Thus far, we have discussed the index of refraction as if it were a single parameter. However, in general, the *complex index of refraction* (N) consists of a real part (n) and an imaginary part (k), or

$$N = n - jk \tag{3.13}$$

where k is the *extinction coefficient*, which is a measure of how rapidly the intensity decreases as light passes through a material. The dielectric function is related to the complex index of refraction by the relationship

$$\varepsilon = N^2 \tag{3.14}$$

Therefore, we can obtain values for n and k in terms of ε_1 and ε_2 using

$$n = \sqrt{\tfrac{1}{2}\left[(\varepsilon_1^2 + \varepsilon_2^2)^{1/2} + \varepsilon_1\right]} \tag{3.15}$$

$$k = \sqrt{\tfrac{1}{2}\left[(\varepsilon_1^2 + \varepsilon_2^2)^{1/2} - \varepsilon_1\right]} \tag{3.16}$$

As mentioned above, the complex index of refraction is related to the total reflection coefficients by the Fresnel equations, which are given by [6]

$$R^p = \frac{r_{12}^p + r_{23}^p \exp(-j2\beta)}{1 + r_{12}^p r_{23}^p \exp(-j2\beta)} \tag{3.17}$$

$$R^s = \frac{r_{12}^s + r_{23}^s \exp(-j2\beta)}{1 + r_{12}^s r_{23}^s \exp(-j2\beta)} \tag{3.18}$$

where the Fresnel reflection coefficients at the individual interfaces are of the form

$$r_{12}^p = \frac{N_2 \cos \phi_1 - N_1 \cos \phi_2}{N_2 \cos \phi_1 + N_1 \cos \phi_2} \tag{3.19}$$

$$r_{12}^s = \frac{N_1 \cos \phi_1 - N_2 \cos \phi_2}{N_1 \cos \phi_1 + N_2 \cos \phi_2} \tag{3.20}$$

and

$$\beta = 2\pi \left(\frac{d}{\lambda}\right) N_2 \cos \phi_2 \tag{3.21}$$

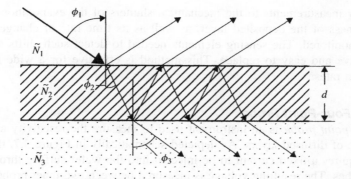

Figure 3.6. Reflections and transmissions in ambient (1), film (2), and substrate (3) [6].

All subscripts and angles mentioned in Eqs. (3.17)–(3.21) are described in Figure 3.6.

Thus, materials with finite light absorption have two unknowns (ε_1 and ε_2, or equivalently, n and k), at each wavelength and one additional unknown in the film thickness. Thus, the total number of unknowns is $2w + 1$. Because this number of unknowns is one too many to be determined from spectroscopic ellipsometry data, it is necessary to employ a dispersion model. Such a model describes the functional dependence of n and k on λ based on P fitting parameters. Therefore, the total number of unknowns becomes $P + 1$. As long as $2w > P + 1$, film thickness and the optical constants may be determined simultaneously by numerically iterating the $P + 1$ fitting parameters to fit spectra [7].

For example, for a thin film on a substrate, the usual objective is to determine thickness d for a known substrate and film dielectric function. To do so, the value of d is found that minimizes the function

$$\sum_{\lambda} |\langle \varepsilon \rangle - \langle \varepsilon \rangle_{calc}|^2 \tag{3.22}$$

(or similar functions using ρ, or ψ and Δ) [8]. Here, the first term represents measured values, and the second term represents theoretically calculated values. This expression can be minimized using well-known procedures such as Newton's method or the Levenberg–Marquardt algorithm [9].

3.2.1.3. Quartz Crystal Monitor

As described in Chapter 2, the deposition of metals such as aluminum is often accomplished using the evaporation technique. The deposition rate during evaporation operations is commonly measured using a device known as a *quartz crystal monitor*. This device is a vibrating crystal sensor that is allowed to oscillate at its resonant frequency as the frequency is monitored. This resonant frequency then shifts as a result of mass loading as additional mass from the evaporated metal is deposited on top of the crystal. When enough material has been added, the resonant frequency shifts by several percent. By feeding the frequency measurements to the mechanical shutters of the evaporation system, the thickness of the deposited layer, as well as its time rate of change, can be readily monitored. The sensing elements needed to detect such shifts are quite inexpensive and easy to replace. This method is effective for a wide range of deposition rates.

3.2.1.4. Four-Point Probe

The *four-point probe* is an instrument used to measure the resistivity and sheet resistance of diffused layers. As depicted schematically in Figure 3.7, this technique requires a fixed current to be injected into the wafer surface through two outer probes. The resulting voltage is measured between two inner probes. If the probes have a uniform spacing (s, in cm), and the sample is infinite, then the resistivity in $\Omega\cdot$cm is given by [11]

$$\rho = 2\pi s V/I \tag{3.23}$$

for $t \gg s$ and

$$\rho = (\pi t/\ln 2)V/I \tag{3.24}$$

for $s \gg t$. For shallow layers such as this, Eq. (3.24) means that the sheet resistance (R_s) is then given by

$$R_s = \rho/t = (\pi/\ln 2)V/I = 4.53 V/I \tag{3.25}$$

Although the approximations used in Eqs. (3.24) and (3.25) are valid for shallow diffused layers in silicon, different correction factors must be used for sheet resistance measurements on bulk wafers.

It should be noted that monitor wafers used for sheet resistance measurements can also be used to determine junction depth (x_j). After the wafers are diffused or implanted with dopants, the thickness of the diffused region is defined as the junction depth. This parameter may be determined from sheet resistance measurements by replacing t with x_j in Eq. (3.25).

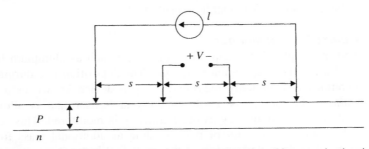

Figure 3.7. Schematic of four-point probe measurement [11]. In this example, the sheet resistance of a p-type epitaxial layer of thickness t on an n-type substrate is measured.

3.2.2. Patterned Thin Film

We continue our discussion of wafer state measurements with those measurements that are performed primarily on wafers that have previously been patterned using photolithography and etching to form specific structures or devices.

3.2.2.1. Profilometry

Profilometry is a very common method of postprocess film thickness measurement. In this technique, a step feature in the grown or deposited film is first created, either by masking during deposition or by etching afterward. The profilometer then drags a fine stylus across the film surface (see Figure 3.8). When the stylus encounters a step, a signal variation (based on a differential capacitance or inductance technique) indicates the step height. This information is then displayed on a chart recorder or cathode-ray tube (CRT) screen. Films of thicknesses greater than 100 nm can be measured with this instrument. The measurement of thinner films is difficult because of vibration, surface roughness, and the precision required in leveling the instrument. Some more recently developed surface profilometers use atomic force microscopy (see Section 3.2.2.2 below).

3.2.2.2. Atomic Force Microscopy

Atomic force microscopy (AFM) is a method for measuring surface properties and/or profiles with atomic-scale topographical definition. In this technique, a sharp tip built at the end of a soft cantilever arm is vibrated perpendicular to

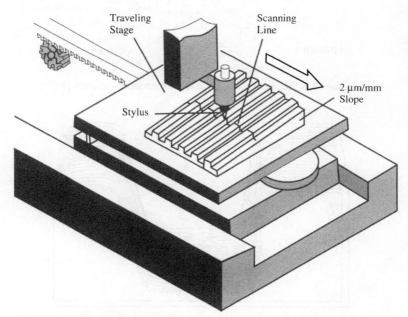

Figure 3.8. Schematic of surface profilometer [12].

the surface at close to the resonant frequency of the cantilever–tip mass as the probe tip traverses laterally across the feature to be characterized. The tip is in atomically close proximity to the surface, so a van der Waals electrostatic force is created between them. This force, which has a strong dependence on the gap between the tip and surface topography, modifies the resonant frequency of the system. The changes in resonance are monitored by an interferometric detection technique that provides a corresponding displacement signal, resulting in a direct measure of the atomic-scale surface topography. A schematic of an AFM system is shown in Figure 3.9. Figure 3.10 shows a typical AFM scan of a

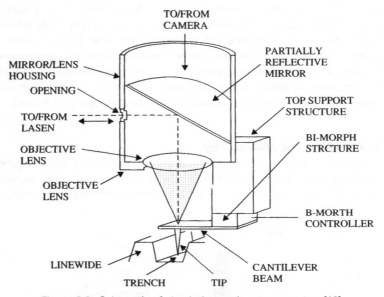

Figure 3.9. Schematic of atomic force microscopy system [13].

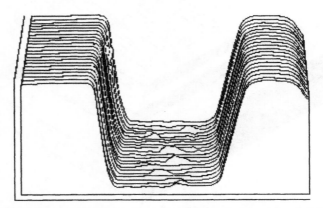

Figure 3.10. Typical AFM image of a surface feature (in this case, a trench) [13].

surface structure. One disadvantage of this technique compared to conventional methods is its low throughput.

3.2.2.3. Scanning Electron Microscopy

Scanning electron microscopy (SEM) is a key technique for assessing minimum feature size in semiconductor manufacturing. The minimum feature size is often expressed in terms of the critical dimension (CD) or minimum *linewidth* that can be resolved by the photolithography system. The decrease in linewidths toward the scale of fractions of a micrometer has rendered conventional optical microscopes nearly obsolete. However, linewidth measurements based on SEM can overcome the limitations of optical techniques for submicrometer geometry features.

The fine imaging capability of the SEM is due to the fact that the wavelength of electrons is four orders of magnitude less than that of optical systems. At such small wavelengths, diffraction effects are usually negligible and spatial resolution is excellent. Features as small as 100 nm can be readily resolved [13]. The electron beam may be based on thermionic or field emission sources. A schematic of a typical field emission SEM is shown in Figure 3.11.

As shown in this figure, the electron gun consists of a tip, first anode, and second anode. A voltage is established between the tip and first anode to facilitate field emission from the tip. An accelerating voltage is then applied between the tip and the second anode to accelerate the electrons. The electron beam emitted from the tip passes through the aperture provided at the center of the first anode, is accelerated, and passes through the center aperture of the second

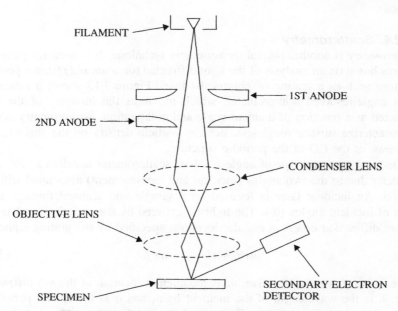

Figure 3.11. Schematic of field emission SEM optics [13].

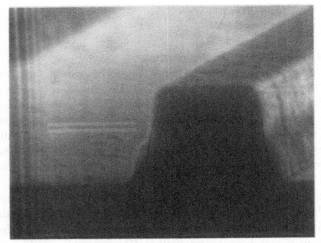

Figure 3.12. Sample SEM output (the parallel lines are calibration marks).

anode to the condenser lens. Electron beams are collected by the condenser lens and aggregated into a small spot on the objective lens. Figure 3.12 shows a typical digital photo output of an SEM. The CD of the feature is usually determined by an arbitrary edge criterion. While lateral resolution offers a tremendous benefit, it must be pointed out that SEM still suffers from several disadvantages, including high cost, low throughput (only ~30 wafers per hour), and the destructive nature of the measurement (i.e., wafers must be cleaved to expose the feature to be imaged).

3.2.2.4. Scatterometry

Scatterometry is another optical measurement technique. It is used for patterned features based on an analysis of the light diffracted (or *scattered*) from a periodic structure such as a grating of photoresist lines. Figure 3.13 shows a schematic of an angle-resolved scatterometer, which measures the intensity of the light diffracted as a function of incident angle and polarization. Scatterometry is used to characterize surface roughness, defects, particle density on the surface, film thickness, or the CD of the periodic structure.

The most common type of angle-resolved scatterometer is called a "2θ" scatterometer due to the two angles (incident and measurement) associated with the method. An incident laser is focused on a sample and scanned through some range of incident angles (θ_i). The light is scattered by the periodic patterns into distinct diffraction orders at angular locations specified by the grating equation

$$\sin \theta_i + \sin \theta_n = n\lambda/d \qquad (3.26)$$

where θ_i is taken to be negative, θ_n is the angular location of the nth diffraction order, λ is the wavelength of the incident light, and d is the spatial period (or pitch) of the periodic structure. Because of the complex interaction between the

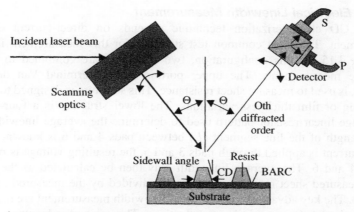

Figure 3.13. Schematic of a 2θ angle-resolved scatterometer [14].

incident light and the periodic features, the fraction of power diffracted into each order is a function of the dimensions of the structure and thus may be used to characterize them.

Capturing diffracted light "signatures" (such as those depicted in Figure 3.14) is just the first phase of scatterometry. In the subsequent analytical phase, a diffraction model is used to interpret the experimental signatures in terms of key parameters such as CD or film thickness. Doing so requires a library of theoretical signatures for comparison to the measured data. The generation of such a library is accomplished by first specifying nominal film stack dimensions and the expected variation of each parameter to be measured. A computerized diffraction model is then used to produce the library of scatter signatures that encompasses all combinations of these parameters for subsequent analysis.

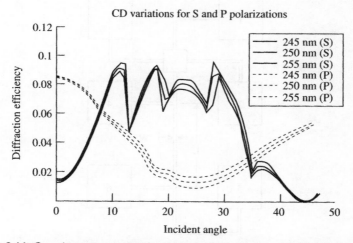

Figure 3.14. Sample scatterometry signatures for 5-nm photoresist CD variations [14].

3.2.2.5. Electrical Linewidth Measurement

Another CD characterization technique depends on direct-current electrical measurement. The most common test structure for this measurement is shown in Figure 3.15. In this configuration, two structures are combined to perform resistance measurements. The upper portion, a four-terminal Van der Pauw structure, is used to measure sheet resistance. This structure is designed to account for doping or film thickness variations. The lower structure is a four-terminal crossbridge linear resistor pattern used to determine the average linewidth (W).

The length of the line segment (L) between pads 4 and 6 is known. When a known current is applied through pads 3 and 5, the resulting voltage is measured at pads 4 and 6. The average linewidth may then be calculated as the product of the measured sheet resistance and length, divided by the measured resistance (V_{46}/I_{35}). The key advantages of electrical linewidth measurement are resolutions on the order of 1 nm and short cycle time. The main disadvantages are the requirement that the film be conductive and the need for physical contact with the wafer.

3.2.3. Particle/Defect Inspection

Contamination is a major concern in semiconductor manufacturing, and billions of dollars are spent annually by manufacturers in order to reduce it. Contamination often takes the form of particles that can appear on the surface of wafers and cause defects in devices or circuits. The fraction of the product that is sensitive to particles depends in part on the particle size. A general rule of thumb is that particles as small as one-tenth the size of a structure can cause the structure to fail. With the industry currently immersed in manufacturing devices with submicrometer features, even nanometer-scale particles are of great concern. Inspection and characterization of particles are therefore critical.

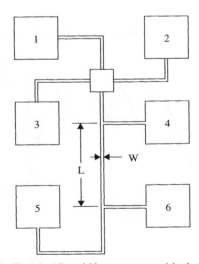

Figure 3.15. Electrical linewidth measurement test structure [13].

3.2.3.1. Cleanroom Air Monitoring

One method of controlling particulate contamination is performing manufacturing operations in a *cleanroom* environment, such as the one schematically depicted in Figure 3.16. Air enters the cleanroom through high-efficiency particulate air (HEPA) or ultra-high-efficiency particulate air (ULPA) filters. The air is forced to flow laminarly (as opposed to turbulently) so that lateral dispersion of contaminants generated in the room is minimized. Cleanrooms are categorized by their "class," which quantifies the number of particles of a given size per cubic foot of air. Various aspects of cleanroom performance affect product quality, and as feature sizes continue to decrease, cleanroom specifications are likewise becoming progressively tighter.

Despite the use of cleanrooms, semiconductor fabrication processes, as well as manufacturing personnel themselves, still generate materials that can contaminate products. Such contamination may originate from process gases and vapors, process liquids, processes that break up bulk material (such as sputtering), deposition processes, metallic impurities, wafer handling, or tool wear, to name just a few. The usual methods for quantitatively determining cleanroom air quality involve sampling via optical particle counters and sampling onto "witness plates" that are later read by surface particle counters.

In the latter approach, a preinspected clean silicon wafer is placed in a location to be monitored. After a fixed time period, the plate is removed and reinspected. The particles per unit area added to the plate are counted. Surface particle counters can inspect an entire plate within minutes with nearly complete detection of particles of sizes a low as fractions of micrometers.

For gases, liquids, and many types of surfaces, optical particle counters are used. Using these devices, particles are illuminated as they pass through a focused

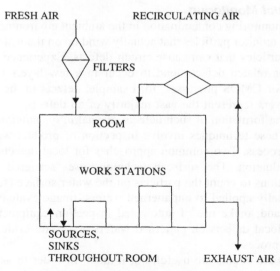

Figure 3.16. Cleanroom schematic [13].

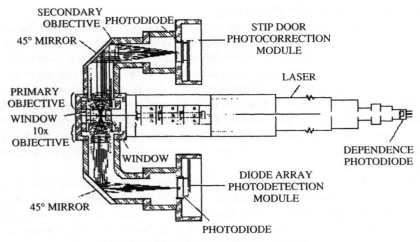

Figure 3.17. Optical particle counter [13].

laser beam (see Figure 3.17). The light scattered from the particles is then measured and correlated with the number of particles present. The amount of light scattered into the sensing element will depend on the light (intensity, wavelength, polarization), the characteristics of the particles (size, shape, orientation, refractive index), and the measurement geometry (position and solid angle subtended by the optics with respect to the beam and the particle). In addition to cleanroom monitoring, this technique is also used for *in situ particle monitoring* (ISPM) inside of processing equipment that produces particles, such as ion implantation or sputtering equipment.

3.2.3.2. Product Monitoring
In addition to monitoring contamination in the ambient environment, it is perhaps more crucial to monitor particles that actually wind up on the wafer surface, since these are the particles that can cause circuit defects. Experience has shown that most processing-related defects tend to occur in a few layers of the complete process [15]. For CMOS processes, for example, defects in the gate oxide and interconnect layers represent the vast majority of all defects.

To control the formation of such defects, special *inline monitoring* techniques are required. These techniques involve inspection of product wafers at various stages in the process. Two common approaches for local defects are "surfscan" and image evaluation. The surfscan technique uses scattered laser light and analyzes reflections to count the particles on the wafer surface (see Figure 3.18). Surfscan is usually applied to unpatterned wafers. Image evaluation techniques, on the other hand, make use of automated inspection equipment to check the occurrence of local defects on patterned wafers at several critical points in the manufacturing process.

Generic particle counts are useful, but limited. In order to assess the impact of the presence of defects caused by particles, specially designed test structures

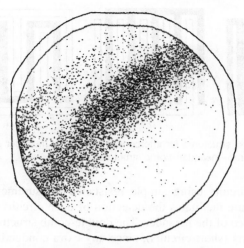

Figure 3.18. Sample surfscan [15].

are used. These structures, also known as *process control monitors* (PCMs), include single transistors, single lines of conducting material, MOS capacitors, via chains, and interconnect monitors. Product wafers typically contain several PCMs distributed across the surface, either in die sites or in the scribe lines between die (see Figure 3.19).

Process quality can be checked at various stages of manufacturing through inline measurements on PCM structures. Three typical interconnect test structures are shown in Figure 3.20. Using such test structures, measurements are performed to assess the presence of defects, which can be inferred by the presence of short

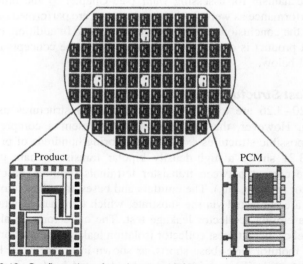

Product   PCM

Figure 3.19. Configuration of products and PCMs on a typical wafer [15].

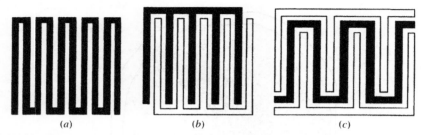

Figure 3.20. Basic test structures for interconnect layers: (a) meander structure; (b) double-comb structure; (c) comb–meander–comb structure [15].

circuits or open circuits using simple resistance measurements. For example, the meander structure facilitates the detection of open circuits through increased end-to-end resistance of the meander. The double-comb structure can likewise be used to detect shorts (short circuits), since any extra conducting material bridging the two combs will reduce the resistance between combs significantly. The comb–meander–comb structure combines the capabilities of the other two structures and permits the detection of both shorts and opens. Various combinations of widths of lines and spaces in these test structures allow the collection of statistics on defects of various sizes.

3.2.4. Electrical Testing

In the preceding section, the concept of test structures for process monitoring was introduced. Although this introduction was presented in the context of particle and defect monitoring, it should not be construed that this is the only use of test structures. In fact, electrical measurements performed on test structures are a major mechanism for assessing *yield* (see Chapter 5) and other indicators of product performance as well. Such measurements are performed on an inline basis and also at the conclusion of the fabrication process. In addition, electrical testing of the final product is crucial to ensure quality. These concepts are discussed in more detail below.

3.2.4.1. Test Structures

Figures 3.20–3.26 are examples of electrical test structures used for process monitoring. However, these by no means represent a comprehensive set, as dozens of possible structures exist for monitoring hundreds of process variables.

Figure 3.21 shows a high-density bipolar transistor chain used to monitor the leakage current between transistor terminals (emitter–base leakage, emitter–collector leakage, etc.). The emitters and bases are wired into parallel chains. Collectors are contacted via the substrate, which eliminates metal short interference in the emitter–collector leakage test. The collectors are also wired to the second level of metal to test collector isolation leakage. Transistor chains can also be used to monitor base–base shorts, as shown in Figure 3.22. In this example, shorts due to polysilicon bridging can be detected by forming the polysilicon

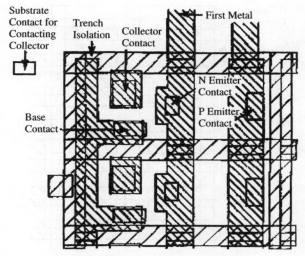

Figure 3.21. Bipolar transistor chain [13].

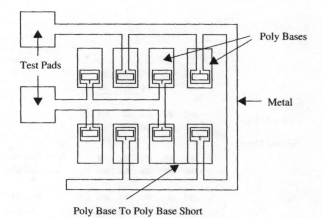

Figure 3.22. Polysilicon base to polysilicon base short chain [13].

bases on field oxide to eliminate the possibility of shorts through the substrate. It is also important to monitor transistor contacts for open circuits. Series-type chains similar to the meander structure in Figure 3.20 can be used for this purpose by connecting the contacts for the various transistor terminals. Figure 3.23 shows an example of a collector contact chain. Note that although the structures depicted in Figures 3.21–3.23 were designed for bipolar circuits, analogous structures can be fabricated to evaluate MOS circuits by wiring up chains connecting their source, drain, and gate terminals in similar chains.

Figure 3.24 is a typical example of a via chain structure used to test connectivity between metal layers. This chain also includes a first-level metal stripe

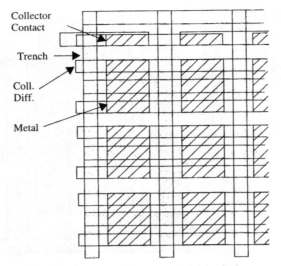

Figure 3.23. Collector contact chain [13].

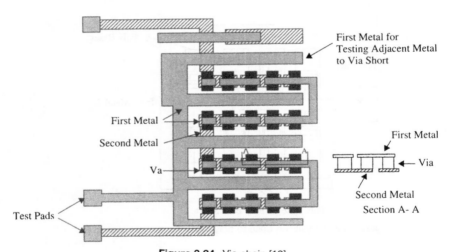

Figure 3.24. Via chain [13].

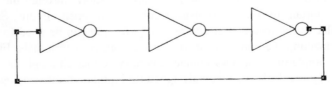

Figure 3.25. Ring oscillator.

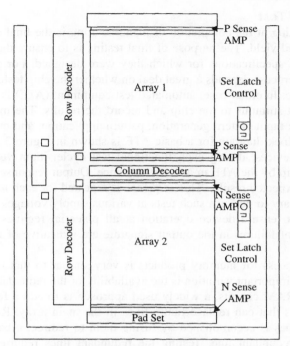

Figure 3.26. Array diagnostic monitor [13].

running parallel to the chain as a mask misalignment monitor. An adjacent metal stripe runs on every level, but never along the full length of the chain. They instead appear at certain sections, alternating with each other.

In addition to defect monitoring, test structures are also used to assess functional characteristics of the semiconductor devices and their dependence on processing conditions. These can be individual devices or simple subcircuits. A common example of such a structure is a *ring oscillator*, which is used to measure speed and capacitive loading effects. A ring oscillator is essentially a chain of inverters (see Figure 3.25). It is formed by connecting an odd number of inverters in a loop. In general, a ring with N inverters will oscillate with a period of $2N\tau_p$ and a frequency of $1/2N\tau_p$, where τ_p is the propagation delay through a single inverter. Inverter chains can also be used to monitor transistor current gain or voltage drops across transistors [13].

An example of a more elaborate functional test structure is the array diagnostic monitor (ADM) shown in block diagram form in Figure 3.26. The ADM, which is used to assess CMOS DRAM circuits, has DC and AC diagnostic capabilities. It is essentially a simplified, yet fully functional duplicate version of a memory array. ADM testing allows for rapid process feedback and ultimately translates into accelerated process improvement.

3.2.4.2. Final Test

Functional testing at the completion of manufacturing is the final arbiter of process quality and yield. The purpose of final testing is to ensure that all products perform to the specifications for which they were designed. For integrated circuits, the test process depends a great deal on whether the chip tested is a logic or memory device. In either case, automated test equipment (ATE) is used to apply a measurement stimulus to the chip and record the results. The major functions of the ATE are input pattern generation, pattern application, and output response detection. A block diagram for a basic ATE is shown in Figure 3.27.

For logic devices, during each functional test cycle, input vectors are sent through the chip by the ATE in a timed sequence. Output responses are read and compared to expected results. This sequence is repeated for each input pattern. It is often necessary to perform such tests at various supply voltages and operating temperatures to ensure device operation at all potential regimes. The number and sequence of failures in the output signature are indicative of manufacturing process faults.

The test process for memory products is very similar to that used for logic. However, one important variation is the availability of the redundancy technique. For dynamic RAM circuits, a widely used approach is to add a few extra word and/or bit lines that can replace faulty lines in the main array. Replacement of these faulty lines is accomplished by fusing them to redirect a bad word or bit address to a redundant line. Testing the redundant lines requires two passes. During the first pass, the addresses of errors are recorded and stored. As long as the number of faults is less than the number of extra lines, the chip is repairable. Although redundancy adds considerable cost and complexity to testing, the yield benefit achieved more than compensates for this.

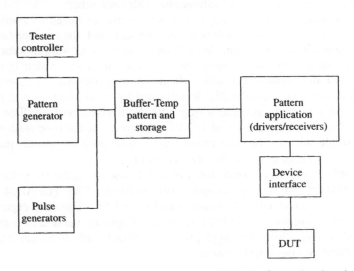

Receivers - Output data detection

Figure 3.27. Block diagram of basic test system (DUT = "device under test") [13].

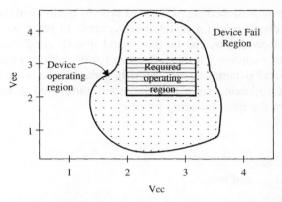

Figure 3.28. Example of two-dimensional voltage *shmoo* plot for hypothetical bipolar chip [13].

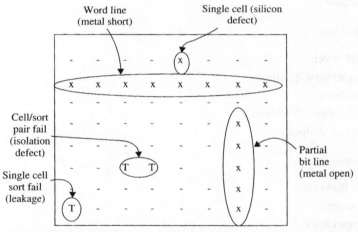

Figure 3.29. Cell map showing examples of failure patterns and defect types [13].

Test results may be expressed in a variety of ways. A couple of examples are shown in Figures 3.28 and 3.29. Figure 3.28 shows a plot of a two-dimensional plot called a "shmoo" plot for a hypothetical bipolar product. In a shmoo plot, the outlined shaded region is where the device is intended to operate, while the blank area outside represents the failure region. Another typical test output is the cell map shown in Figure 3.29. Cell maps are very useful in identifying and isolating device failures, particularly in memory arrays. In addition, the patterns generated in the cell map may be compiled, catalogued, and later compared to a library of existing defect types, thereby aiding in the diagnosis of faults.

3.3. EQUIPMENT STATE MEASUREMENTS

Rather than characterizing the state of the product wafers themselves, equipment monitors measure the status of tools while they are processing these wafers. Such monitors are the most immediate measure of process quality and therefore provide the shortest feedback loop for maintaining control. In other words, the impact of out-of-control conditions can be minimized if such conditions are promptly identified by tool monitors and immediate corrective action is taken.

Certain physical parameters are routinely measured as a part of equipment monitoring. The following are a few commonly monitored process variables at various stages of the manufacturing process:

Lithography

- Exposure energy
- Exposure dose and intensity
- Time
- Magnification
- Aperture

Wet Stations

- Fluid level
- Temperature gradients
- Flowrates
- Development/etch rates
- Time to endpoint

Deposition

- Gas flowrate
- Pressure
- Temperature

Implantation

- Accelerating voltage
- Beam current

Diffusion

- Source composition
- Pressure
- Flowrate
- Temperature

The combined effects of these tool variables eventually lead to measurable impact on the characteristics of product wafers. The process engineer must therefore have available reliable methods for monitoring these variables in order to facilitate process control. The following sections describe several equipment state measurements used for monitoring such characteristics.

3.3.1. Thermal Operations

Thermal operations refer to any process step that occurs at an elevated temperature. Examples include epitaxial growth, chemical vapor deposition, evaporation, and annealing. This subsection describes the measurement of key process variables during these operations.

3.3.1.1. Temperature

In situ measurements of conditions such as temperature can be used to infer the quality of the wafers being produced in thermal processes. In many types of thermal processing equipment, temperature is measured using a thermocouple embedded in the wafer holder (or *susceptor*). A *thermocouple* is a circuit consisting of a pair of wires made of different metals joined at one end (the "sensing junction") and terminated at the other end (the "reference junction") in such a way that the terminals are both at a known reference temperature. Leads from the reference junction to a load resistance (i.e., an indicating meter) complete the thermocouple circuit. Due to the *thermoelectric effect* (or Seebeck effect), a current is induced in the circuit whenever the sensing and reference junctions are different temperatures. This current varies linearly with the temperature difference between the junctions.

In some cases (such as in rapid thermal processes), the use of a thermocouple is not possible because there is no susceptor. Alternative temperature sensors used in such situations include thermopiles and optical pyrometers. A *thermopile*, which also operates via the Seebeck effect, consists of several sensing junctions made of the same material pairs located in close proximity and connected in series in order to multiply their output.

The second alternative method to the thermocouple is pyrometry. Pyrometers operate by measuring the radiant energy received in a certain band of energies, assuming that the source is a graybody of known emissivity. The input energy can then be converted to a source temperature using the Stefan–Boltzmann relationship [16]. Most commercial systems monitor the mid-infrared band (3–6 μm). One major issue in using pyrometry is that the effective emissivity of the source must be accurately known. The effective emissivity includes both intrinsic and extrinsic contributions. *Intrinsic emissivity* is a function of the material, surface finish, temperature, and wavelength. *Extrinsic emissivity* is affected by the amount of radiant energy from other sources reflected back to the spot being measured (which can increase the apparent temperature). In addition, the presence of multiple layers of different thin-film materials can also alter the apparent emissivity due to interference effects.

3.3.1.2. Pressure

Pressure in vacuum systems used in thermal operations can be measured using a variety of transducers, including capacitance manometers, thermal conductivity gauges, and ionization gauges. Capacitance manometers are mechanical gauges that sense the deflection caused by the pressure difference between the chamber to be measured and a reference volume. These devices detect the movement of a thin metal diaphragm to do so. Although they can be used to detect pressures as low as 1 mTorr, they are also often used to measure pressures as high as 1 Torr.

Thermal conductivity gauges derive the thermal conductivity of the ambient gas by passing a current through a wire and measuring its temperature. Pressure may then be inferred from the conductivity measurement.

However, neither mechanical deflection nor thermal conductivity gauges are able to measure pressures below 1 mTorr. This type of application requires an *ionization gauge*, which operates using an electron stream to ionize the gas in the gauge and an electric field to collect the ions. The ion current is a function of the pressure in the chamber. The pressure that can be measured in this way is limited only by the ability to sense small ion currents, so ionization gauges can detect pressures as low as 10^{-12} Torr.

3.3.1.3. Gas Flow

Thermal systems of various types, as well as plasma etchers, require controlled rates of introduction of process gases into the reaction chambers. This is most commonly achieved using an instrument called a *mass flow controller*, which consists of a flowmeter, a controller, and a valve, and it is located between the gas source and the chamber itself. Gas flow is measured in units of volume/time. The most common unit is the *standard cubic centimeter per minute* (sccm), defined as the flux of one cm^3 of gas per minute at 273 K.

There are two primary types of mass flowmeters: (1) the differential pressure type and (2) the thermal type. The differential pressure flowmeter relates a pressure drop at a physical flow restriction to rate of mass flow. The thermal type, which is more widely used in semiconductor manufacturing, relies on the ability of a flowing gas to transfer heat. As shown in Figure 3.30, the thermal flowmeter consists of a larger gas flow tube in parallel with a small sensor. A heating coil is wrapped around the sensor midway along its length, and temperature sensors are placed both upstream and downstream of the heated point. Flowing gas causes the temperature distribution in the sensor tube to change as a result of thermal transfer between the heated wall and the gas stream. The temperature downstream from the heated region becomes higher than the upstream temperature as the flowing gas conducts heat away. It can be shown that the rate of mass flow (m_f) is then given by

$$m_f = (\kappa W_h \, \Delta T)^{1.25} \qquad (3.27)$$

where W_h is the heater power, ΔT is the temperature difference between the two sensors, and κ is a constant that depends on the heat transfer coefficients and the specific heat, density, and thermal conductivity of the gas. Assuming that the remaining parameters remain constant over the flow range of interest, the mass flowrate can be obtained by measuring the temperature difference. The two temperature sensors, which are usually *resistance thermometers* (see Section 3.3.2.1),

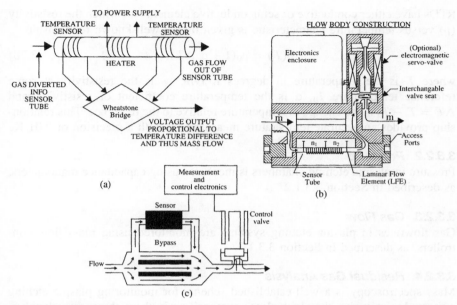

Figure 3.30. Mass flow controller: (a) operational principles; (b) cross-sectional drawing; (c) schematic diagram [12].

are connected to one port of an unbalanced Wheatstone bridge, and the temperature difference is converted into a voltage signal. As the flowrate is determined, its value is compared to a setpoint value and adjusted as necessary to maintain that value by the controller.

3.3.2. Plasma Operations

As discussed in Chapter 2, plasma etching has emerged as a critical process in the production of integrated circuits. This emergence has stemmed from a continuous need to fabricate devices with extremely small dimensions. However, without sufficient online monitoring and control, etch equipment can produce unacceptably large volumes of defective products, leading to millions of dollars lost as a result of misprocessing. As a result, in addition to the measurements described above for thermal operations, plasma etching systems often employ some unique supplemental equipment monitoring devices.

3.3.2.1. Temperature

In many plasma etching systems, the process temperature is controlled by means of a system that removes heat from the lower electrode by circulating deionized water. This closed-loop recirculation system is sometimes referred to as a "chiller." The chiller maintains a preset temperature, often room temperature. This temperature is monitored using a standard resistance thermometer device (RTD).

RTDs have either conductive or semiconductive elements for which the resistivity (ρ) versus temperature characteristic is given by the well-known relationship

$$\rho(T) = \rho_0(1 + \alpha \, \Delta T) \tag{3.28}$$

where T is the temperature in degrees Kelvin, ρ_0 is the resistivity at some reference temperature T_0, α is the temperature coefficient of resistivity, and $\Delta T = T - T_0$ is the change in temperature relative to the reference. This relationship provides an accurate temperature measurement with a precision of 0.01 K.

3.3.2.2. Pressure

Pressure in plasma etching chambers is measured using capacitance manometers, as described in Section 3.3.1.2.

3.3.2.3. Gas Flow

Gas flowrates in plasma etching systems are monitored using mass flow controllers, as described in Section 3.3.1.3.

3.3.2.4. Residual Gas Analysis

Mass spectroscopy is a well-established scheme for monitoring plasma etching systems by analyzing the residual gas composition in the etch process chamber. The fundamental principle by which a mass spectrometer operates is based on the separation of gas molecules by atomic mass. An etch system continuously depletes its chamber gases during processing. At the beginning of an etch, the gas in the chamber consists of a mixture of process gas and that resulting from the etch. Toward the end, the gas in the chamber will resemble its mixture prior to etching. This information may be used to detect the etch endpoint using *residual gas analysis* (RGA).

There are two main methods for mass spectroscopic monitoring of plasmas: flux analysis and partial-pressure analysis. *Flux analysis* involves sampling the plasma directly by coupling the emission of plasma particles through a small aperture into the ion optics of a mass spectrometer. This method is primarily a research tool and is best suited for plasma species and energy analysis. On the other hand, *partial-pressure analysis* is accomplished by simple vacuum connections between the spectrometer and the plasma chamber. Because of its simplicity, partial-pressure analysis is the method of choice in most production systems used in semiconductor manufacturing.

Figure 3.31 is a schematic diagram of a *quadrupole mass spectrometer* (QMS), the main apparatus used for partial-pressure analysis. Depending on the operating pressure of the plasma system, there is either a high or low conductance connection between the etch and QMS chamber, which is usually differentially pumped. This results in the dynamic response of a pressure change in the QMS chamber ($\Delta P_Q(t)$) differing from the pressure change in the etch chamber ($\Delta P_D(t)$). For dynamic measurements, it can be shown that [17]

$$P_D(t) = \left(1 + \frac{S_Q}{C_T}\right) P_Q(t) + \frac{V_Q}{C_T} \frac{d[P_Q(t)]}{dt} - \left(1 + \frac{S_Q}{C_T}\right) P_B \tag{3.29}$$

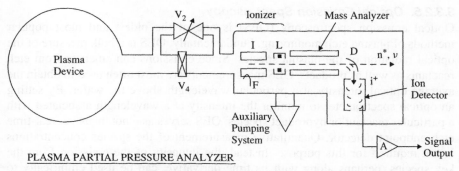

PLASMA PARTIAL PRESSURE ANALYZER

Figure 3.31. Schematic diagram of QMS system used for partial-pressure analysis [17].

where a QMS chamber with volume V_Q is pumped with a pump speed of S_Q and is connected to the etch chamber through a tube with conductance C_T. The last term represents the background pressure (P_B) correction in the QMS. The quantities S_Q and C_T are a function of the gas temperature, pressure, mass, and viscosity of the chamber gas mixture. Equation (3.29) usually must be solved numerically.

Figure 3.32 is an example of the results of RGA using a QMS system for the etching of a GaAs/AlGaAs metal–semiconductor–metal structure in a BCl_3/Cl_2 plasma [18]. The time evolution of the RGA signals from the various reaction product species are clearly evident, indicating the usefulness of this technique for etch process monitoring.

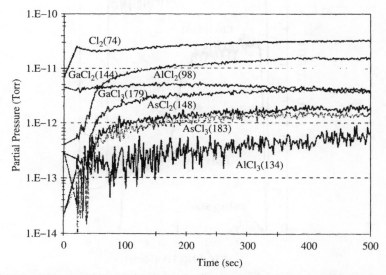

Figure 3.32. RGA signals from a BCl_3/Cl_2 etch of a GaAs/AlGaAs structure (numbers in parentheses represent the atomic mass of the species).

3.3.2.5. Optical Emission Spectroscopy

Optical emission spectroscopy (OES) is one of the oldest and most popular methods of plasma etch monitoring. Fundamentally, OES is a bulk measure of the optical radiation of the plasma species. Since emissions can emanate from etch reactants as well as products, OES measurements are most often used to obtain the average optical intensity at a particular wavelength above the wafer. By setting an optical spectrometer to monitor the intensity at a wavelength associated with a particular reactant or byproduct species, OES serves as a noninvasive, real-time etch endpoint detector. Quantitative measurement of the species concentrations is not required for this purpose. Instead, the intensity of the emission from the key species, perhaps along with its time derivative, can be used empirically to determine the proper point to discontinue the etch process.

A series of such measurements for a particular etch process is referred to as an "endpoint trace," a curve representing the intensity of the optical emission of the key species over time. An example of such a trace is illustrated in Figure 3.33, which depicts fluorine and CN emission intensities during silicon nitride etching. At the beginning of the etch, the gas in the chamber consists of a mixture of process gas and that resulting from the etch. At the end of the etch, the gas mixture again resembles its mixture prior to the start of the process. Therefore, the etch endpoint is characterized by a sharp change in the intensity of the endpoint trace.

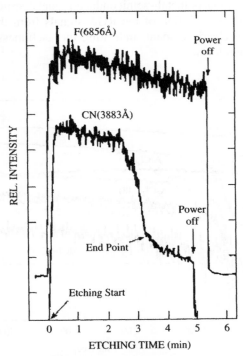

Figure 3.33. OES endpoint trace showing the intensity of the emission of key species in a silicon nitride etch process [17].

OES measurements not only reflect the chemistry of the plasma but also inherently have embedded in them information concerning the operational status of the plasma equipment, pattern density on the substrate, and nonideal fluctuations in the processing conditions (gas flow, pressure, etc.). It is therefore also possible to use OES signals to monitor and diagnose etch equipment problems.

3.3.2.6. Fourier Transform Infrared Spectroscopy

Infrared (IR) spectroscopy is a widely used method for identifying organic compounds, such as those that may result from the etching of polymer films. This method is based on the absorption of infrared radiation by molecules at characteristic wavelengths. Radiation causes various components of such molecules to vibrate and rotate. Since the frequency of vibration/oscillation is dependent on the nature of the chemical bonds present, the presence or absence of absorption in certain well-defined regions of the IR spectrum can be used to determine the presence or absence of chemical groups. The intensity of the absorption peaks is proportional to the amount of material present. Computer databases and search routines are usually used to identify compounds.

In Fourier transform infrared (FTIR) spectroscopy, an infrared source is sent through a beamsplitter to the surface of the wafer being etched and to a movable mirror. The reflected radiation from both surfaces is added and sent to a detector. The distance of the mirror path is swept, and the intensity of the reflected beam as a function of the position of the mirror is monitored. The intensity of the IR peaks can then be used to determine the composition of the film on the wafer surface. An example of typical FTIR output is provided in Figure 3.34.

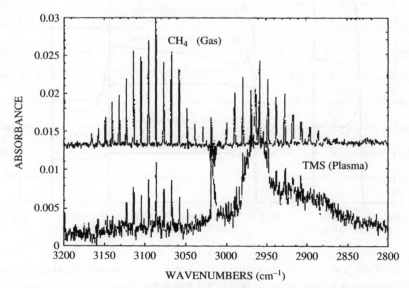

Figure 3.34. IR spectra of CH_4 and tetramethylsilane (TMS) in an electron cyclotron resonance plasma system [19].

3.3.2.7. RF Monitors

Historically, the only aspect of radiofrequency (RF) power monitored in plasma systems is the power delivered from the RF supply to the matching network. This is typically expressed in terms of forward and reflected RF power. The addition of an RF plasma impedance sensor between the matching network and plasma electrode, however, allows new electrical variables to be monitored and controlled. This allows problems such as poor RF connections, electrode condition, and changes in process gas mixture to be detected more easily [20].

Monitoring these parameters facilitates inferences regarding the state of the etch system, such as the degree of ionization of the presence of chamber wall coatings. Etch endpoint can also be detected using changes in RF impedance during the etch cycle. Figure 3.35 shows an example of RF data that can be gathered by a plasma impedance sensor.

3.3.3. Lithography Operations

The success of pattern transfer in photolithographic operations is determined by interactions between four constituents. Those constituents (and examples of the relevant process variables in each) are (1) the wafer (reflectivity, pattern density,

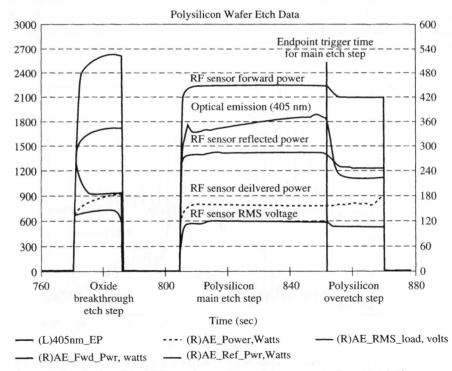

Figure 3.35. Set of RF waveforms gathered during a polysilicon etch [20].

topography), (2) the photoresist (thickness, uniformity, age), (3) the exposure tool (mask variance, wavelength, exposure dose, lens characteristics, barometric pressure), and (4) the developer (concentration, temperature).

The key output measurement in photolithography is linewidth or critical dimension, which is a wafer state variable (see Section 3.2.2). Nevertheless, the CD is significantly impacted by several equipment state variables that must also be monitored to ensure quality. For example, resist thickness and uniformity are controlled in part by the spin speed and ramp of spin coaters (as well as the coating solvent and viscosity of the resist). The primary equipment state measurements, however, are related to the exposure tool. In most modern exposure tools, the monitoring of equipment state variables occurs internally. For example, barometric pressure and lens characteristic changes are now monitored as part of the tool package.

3.3.4. Implantation

In modern ion implantation systems, it is important to monitor and carefully control the dose of the implant. This is accomplished in the end station by placing the wafer undergoing implantation in a Faraday cup, which is simply a cage that captures all the charge that enters it. The ion current into the wafer is measured by connecting an ammeter between the Faraday cup and ground. The dose is the time integral of this current divided by the wafer area.

Accurate measurement of the dose requires that precautions be taken against errors due to secondary-electron ejection. This process involves the creation of large numbers of electrons, many of which have sufficient energy to escape the wafer when a high-energy ion strikes the wafer surface. To prevent secondary-electron dose measurement errors, the wafer is biased with a small positive voltage. This bias (usually tens of volts) is sufficient to attract all the secondary electrons back to the surface of wafer, where they are absorbed.

Another problem often seen in implanting through a photoresist mask is outgassing. Ions striking the surface break apart organic molecules in the resist, leading to the formation of gaseous hydrogen that evolves from the surface and leaves behind carbon. Not only can this hardened carbonize layer be difficult to remove, but outgassing can raise the pressure in the end station enough to cause neutralization of the ion beam through impact with the H_2 molecules, which can result in significant dose rate measurement errors. Modern cryopumps are very effective in pumping away H_2 and other photoresist outgassing products, but these cryopumps must be regenerated at regular intervals to maintain adequate vacuum levels, and this impacts throughput. The beam neutralization effect of outgassing is controlled in some implant systems by the use of a feedback loop that corrects the observed signal at the Faraday cup in response to changes in the beamline pressure. Other systems avoid these problems by monitoring the ion-beam current during those portions of the implant operation when the ion beam is not impinging on the wafer surface and the photoresist outgassing rate is low.

3.3.5. Planarization

As discussed in Chapter 2, planarization operations employ chemical–mechanical polishing (CMP) systems. In CMP systems, some of the key equipment state measurements that must be performed include characterization of polishing pads, determining the condition of the slurry, and endpoint detection.

The condition of CMP polishing pads is a key indicator of removal rate since the porosity of the pad determines the slurry arrival rate at the surface of the wafer. Glazing of the pad tends to occur after several runs, which slows the removal rate. The solution of this problem is frequent conditioning of the pad to obtain consistent roughness. Care must be taken in pad conditioning, however, since processed wafers show greatly increased particle counts immediately after pad conditioning [16].

Slurries for CMP applications consist of particles suspended in various liquids (depending on the specific material being polished). By measuring and controlling the pH of the slurry, particle agglomeration is minimized. In addition, for oxide CMP, the polishing rate increases with increasing pH, particle concentration, and particle size. Therefore monitoring each of these qualities of the slurry is important.

Since CMP is a process for reducing thickness at selected locations on the wafer, it is necessary to identify when a suitable degree of overall planarization has been achieved and the process has reached its endpoint. One method of endpoint detection involves monitoring the current supplied to the motor of the wafer carrier. This motor current monitoring technique is a production-proven method that works well when polishing down to a stop layer (such as polishing a $CVD-SiO_2$ film on a silicon nitride stop layer in shallow-trench isolation processes) [12]. Circuitry such as a current shunt or Hall effect probe is used to monitor the current supplied to the motor that rotates the wafer carrier. Since the carrier is driven at a constant rotational speed to maintain a constant polishing rate, the drive current is varied to compensate for any load changes on the motor. This makes the current sensitive to frictional changes at the wafer surface. The largest changes occur when one material has been polished away, leaving a layer that has different polishing characteristics. Therefore, substantial changes in drive current are indicative of process endpoint.

SUMMARY

This chapter has provided a survey of sensor metrology and methods of monitoring semiconductor manufacturing processes. After identifying key measurement points in the process flow and differentiating between wafer state and equipment state measurements, a description of such measurement techniques ensued. Measuring key process and equipment state variables enables operators and engineers to ascertain product quality. However, conclusions regarding quality can be drawn only after this measurement data have been collected and analyzed. Methods for data analysis involve the application of various statistical tools. The fundamental concepts that support these tools are the subject of the next chapter.

PROBLEMS

3.1. A thin film of silicon dioxide covers a silicon wafer. Plane lightwaves of variable wavelengths are incident normal to the film. When one views the reflected wave, it is noted that complete destructive interference occurs at 600 nm and constructive interference occurs at 700 nm. Calculate the thickness of the SiO_2 film.

3.2. The correction factor for sheet resistance when thick materials are being measured with a Four-point probe is shown in Figure P3.2. Equation (3.24) must be multiplied by this factor to obtain accurate R_S values from $I-V$ measurements. Given that a uniformly doped silicon layer with a thickness equal to the probe spacings is measured and $V/I = 45$, compute R_S.

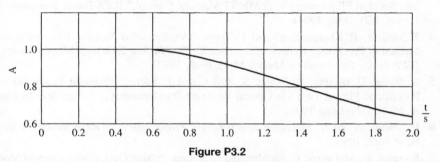

Figure P3.2

3.3. Describe four techniques for measuring the linewidth of patterned features on a substrate. Why is accurate linewidth measurement more difficult on wafer surfaces than on masks?

3.4. A 200-mm-diameter silicon wafer contains chips that are 0.25 cm^2. The wafer is initially clean and is then exposed to room air containing 1000 particles/ft^3 of diameter 0.5 μm and larger. On average, how long will it take to deposit one particle per chip, assuming a laminar air flow of 30 m/min?

3.5. Explain why a thermal conductivity gauge will not work in an ultrahigh vacuum.

3.6. The major source of uncertainty in pyrometry is uncertainty in emissivity. Planck's radiation law gives the spectral radiant exitance as a function of wavelength and temperature ($M_\lambda(\lambda, T)$) as

$$M_\lambda(T) = \varepsilon(\lambda)\frac{c_1}{\lambda^5(e^{c_2/\lambda T} - 1)}$$

where $\varepsilon(\lambda)$ is the wavelength-dependent emissivity of the emitting body and c_1 and c_2 are the first and second radiation constants (given by 3.7142 ×

10^{-16} W-m^2 and 1.4388×10^{-2} m·K, respectively). If the wafer temperature is $1000°C$, what wavelength is most desirable to minimize the effect of this uncertainty?

3.7. An ion implanter has a beam current of 30 mA. The wafer holder can accommodate thirty 100-mm-diameter wafers. For a 5-min implant at a 130 keV implant energy, compute the dose received by the wafers.

REFERENCES

1. D. Halliday and R. Resnick, *Physics*, NY: Wiley, New York, 1978.

2. J. Pope, R. Woodburn, J. Watkins, R. Lachenbruch, and G. Viloria, "Manufacturing Integration of Real-Time Laser Interferometry to Isotropically Etch Silicon Oxide Films for Contacts and Vias," *Proc. SPIE Conf. Microelectronic Processing*, Vol. 2091, 1993, pp. 185–196.

3. S. Maung, S. Banerjee, D. Draheim, S. Henck, and S. Butler, "Integration of In-Situ Spectral Ellipsometry with MMST Machine Control," *IEEE Trans. Semiconduct. Manuf.* **7**(2), (May 1994).

4. T. Vincent, P. Khargonekar, and F. Terry, "An Extended Kalman Filtering-Based Method of Processing Reflectometry Data for Fast *In-Situ* Etch Rate Measurements," *IEEE Trans. Semiconduct. Manuf.* **10**(1), (Feb. 1997).

5. K. Wong, D. Boning, H. Sawin, S. Butler, and E. Sachs, "Endpoint Prediction for Polysilicon Plasma Etch via Optical Emission Interferometry," *J. Vac. Sci. Technol. A.* **15**(3), (May/June 1997).

6. H. Tompkins and W. McGahan, *Spectroscopic Ellipsometry and Reflectometry*, Wiley, New York, 1999.

7. F. Yang, W. McGahan, C. Mohler, and L. Booms, "Using Optical Metrology to Monitor Low-K Dielectric Thin Films," *Micro* **31–38** (May 2000).

8. J. McGilp, D. Weaire, and C. Patterson (eds), *Epioptics*, Springer-Verlag, New York, 1995.

9. W. Press, B. Flannery, S. Teukolsky, and W. Vetterling, *Numerical Recipes in C*, Cambridge Univ. Press, Cambridge, MA, 1988.

10. B. Stutzman, H. Huang, and F. Terry, "Two-Channel Spectroscopic Reflectometry for *In-Situ* Monitoring of Blanket and Patterned Structures During Reactive Ion Etching," *J. Vac. Sci. Technol. B.* **18**(6), (Nov./Dec. 2000).

11. R. Jaeger, *Introduction to Microelectronic Fabrication*, Addison-Wesley, Reading, MA, 1993.

12. S. Wolf and R. Tauber, *Silicon Processing for the VLSI Era*, Lattice Press, Sunset Beach, CA, 2000.

13. A. Landzberg, *Microelectronics Manufacturing Diagnostics Handbook*, Van Nostrand Reinhold, New York, 1993.

14. C. Raymond, "Angle-Resolved Scatterometry for Semiconductor Manufacturing," *Microlithogry. World* (winter 2000).

15. J. Pineda de Gyvez and D. Pradhan, *Integrated Circuit Manufacturability*, IEEE Press, Piscataway, NJ, 1999.

16. S. Campbell, *The Science and Engineering of Microelectronic Fabrication*, Oxford Univ. Press, New York, 2001.

17. D. Manos and D. Flamm, *Plasma Etching: An Introduction*, Academic Press, San Diego, CA, 1989.

18. D. Stokes and G. May, "Real-Time Control of Reactive Ion Etching Using Neural Networks," *IEEE Trans. Semiconduct. Manuf.* **13**(4), 469–480 (Nov. 2000).

19. P. Raynaud, T. Amilis, and Y. Segui, "Infrared Absorption Analysis of Organosilicon/Oxygen Plasmas in a Microwave Multipolar Plasma Excited by Distributed Electron Cyclotron Resonance," *Appl. Surf. Sci.* **138–139**, 285–291 (1999).

20. C. Almgren, "The Role of RF Measurements in Plasma Etching," *Semicond. Intl.* 99–104 (Aug. 1997).

4

STATISTICAL FUNDAMENTALS

OBJECTIVES

- Explain, in general terms, the issues surrounding process variability.
- Introduce the statistical fundamentals necessary for analyzing semiconductor manufacturing processes.
- Describe and differentiate between discrete and continuous probability distributions.
- Discuss the concepts of sampling, estimation, statistical significance, confidence intervals, and hypothesis testing.

INTRODUCTION

In Chapter 3, various monitoring tools used to generate data necessary for process control were presented. For high-volume semiconductor manufacturing, such testing and inspection methods are essential for producing high-quality ICs. The term "quality" here refers to the fitness of a product for its designated use. In this sense, quality requires conformance of all products to a set of specifications and the reduction of any variability in the manufacturing process. A key metric for process quality is product *yield*, which is discussed in Chapter 5. Maintaining quality involves the use of *statistical process control* (SPC), which is the subject of Chapter 6. Since product variability is often described in statistical

Fundamentals of Semiconductor Manufacturing and Process Control,
By Gary S. May and Costas J. Spanos
Copyright © 2006 John Wiley & Sons, Inc.

terms, statistical methods necessarily play a central role in quality control and yield improvement efforts. Therefore, this chapter provides a concise review of some basic statistical fundamentals, along with appropriate examples from the semiconductor manufacturing domain.

In terms of semiconductor manufacturing processes, the most relevant aspect of quality is *quality of conformance*, or how well manufactured products conform to the specifications and tolerances required by their design and intended use. Every semiconductor device or circuit possesses a number of elements that collectively describe its fitness for use. These elements are referred to as *quality characteristics*.

Perhaps the major barrier to perfecting quality in a manufacturing environment is *variability*. Variability is inherent in every product—no two products are ever identical. For example, the dimensions of two thin films used for interconnect will vary according to the precise conditions and equipment used to deposit and pattern the films. Small variations might have negligible impact on the final product, but large variations can lead to final products that are unacceptable. *Quality improvement* may be defined as the reduction of such variability in processes and products.

Statistics allow engineers to make decisions about a process or population based on the analysis of a sample from that population. For example, two well-known statistics are the *sample average* and *sample variance*. Suppose that $x_1, x_2, \ldots, x_n$ are observations in a sample of size n. The statistic used to estimate the mean value (μ) of this population based on the sample is the sample average (x), which is given by

$$\overline{x} = \frac{x_1 + x_2 + \cdots + x_n}{n} = \frac{1}{n} \sum_{i=1}^{n} x_i \qquad (4.1)$$

The variance (σ^2), or spread, in a dataset is a statistic that can be estimated by the sample variance (s^2):

$$s^2 = \frac{1}{n-1} \sum_{i=1}^{n} (x_i - \overline{x})^2 \qquad (4.2)$$

The square root of the sample variance is known as the *sample standard deviation*. Generally, the larger the variance, the greater the spread in the sample data.

Statistical methods provide the principal means by which products are sampled, tested, and evaluated in a manufacturing environment. In the remainder of the chapter, various statistical methodologies are introduced as tools for use in quality control and improvement.

4.1. PROBABILITY DISTRIBUTIONS

A *probability distribution* is a mathematical model that relates the value of a random variable to its probability of occurrence. There are two types of probability distributions: discrete and continuous. *Discrete* distributions are used to describe random variables that can take on only certain specific values, such as the number of defects on a semiconductor wafer. On the other hand, when the random variable can have any value on a continuous scale (such as linewidth in a sample population of interconnect), the probability distribution is *continuous*. Examples of discrete and continuous probability distributions are shown in Figure 4.1.

4.1.1. Discrete Distributions

The discrete distribution is characterized by a series of vertical lines whose height represents the probability (Figure 4.1a). The probability that a random variable x is equal to a specific value x_i is given by

$$P\{x = x_i\} = p(x_i) \tag{4.3}$$

Two examples of discrete probability distributions that arise frequently in manufacturing applications are the binomial distribution and the Poisson distribution.

4.1.1.1. *Hypergeometric*

Let N represent the size of a finite population of items. Suppose that D of these items (where $D \leq N$) fall into a specific class of interest, such as the number of defective items in the population. If a random sample of n items is selected from the population without replacement, then the number of items in the sample that belong in the class of interest (x) is a random variable that follows the *hypergeometric* distribution. The probability of selecting x items belonging to the class is given by

$$P(x) = \frac{\binom{D}{x} \binom{N-D}{n-x}}{\binom{N}{n}} \tag{4.4}$$

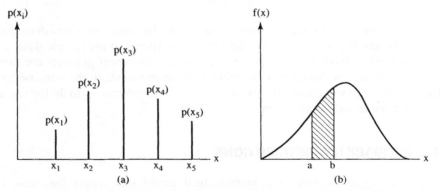

Figure 4.1. (a) Discrete and (b) continuous probability distributions [1].

where

$$\binom{a}{b} = \frac{a!}{b!(a-b)!}.$$

The mean (μ) and variance (σ^2) of the binomial distribution are

$$\mu = \frac{nD}{N} \tag{4.5}$$

$$\sigma^2 = \frac{nD}{N}\left(1 - \frac{D}{N}\right)\left(\frac{N-n}{N-1}\right) \tag{4.6}$$

The hypergeometric distribution is an appropriate model for encountering defective samples when selecting a random sample of n items without replacement from a population N of these items, of which D are nonconforming or defective. In semiconductor manufacturing, this is analogous to selecting a sample of n dies from a lot of wafers containing N total dies, D of which are known to be defective.

Example 4.1. Suppose that a lot of wafers contains 100 dies, 5 of which are known to be defective. If 10 of these dies are selected at random for inspection, what is the probability of finding less than two defective dies in the sample?

Solution: Here, $N = 100$, $n = 10$, and $D = 5$. To find the probability of less than two defective dies, we apply Eq. (4.4) as follows:

$$P(x < 2) = P(x \leq 1) = P(0) + P(1)$$

$$= \frac{\binom{5}{0}\binom{95}{10}}{\binom{100}{10}} + \frac{\binom{5}{1}\binom{95}{9}}{\binom{100}{10}} = 0.923$$

Therefore, the probability of finding less than two defective dies is 92.3%.

4.1.1.2. Binomial

Suppose that a process consists of n independent trials. Each trial has two possible outcomes: "success" or "failure." Trials with these characteristics are called *Bernoulli* trials. Let p be the probability of success for any given trial (thus, $0 < p < 1$). If p is constant, then the probability of achieving x successes in n trials is

$$P(x) = \binom{n}{x} p^x (1-p)^{n-x} \quad x = 0, 1, \ldots, n \tag{4.7}$$

The mean (μ) and variance (σ^2) of the binomial distribution are

$$\mu = np \tag{4.8}$$

$$\sigma^2 = np(1-p) \tag{4.9}$$

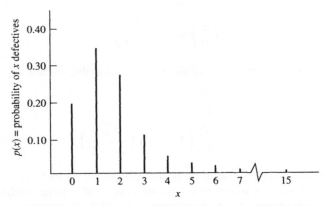

Figure 4.2. Binomial distribution with $p = 0.10$ and $n = 15$ [1].

The binomial model is used for sampling from an infinite population, and p represents the fraction of defective or nonconforming parts in that population. In this situation, x is the number of nonconforming parts identified in a random sample of n items. A typically shaped binomial distribution corresponding to $p = 0.10$ and $n = 15$ is shown in Figure 4.2.

Example 4.2. Suppose that a wire bonding process has an average of 1% defective bonds. If an inspector selects a random sample of 100 bonds, what is the probability of more than two of the bonds being defective?

Solution: In this case, $n = 100$ and $p = 0.01$. To find the probability of greater than two defective bonds, we apply Eq. (4.7) as follows:

$$P(x) = \binom{100}{x} (0.01)^x (0.99)^{100-x} \qquad x = 0, 1, \ldots, 100$$

Note that

$$P(x > 2) = 1 - P(x \le 2) = P(0) + P(1) + P(2)$$

$$= \sum_{x=0}^{2} \binom{100}{x} (0.01)^x (0.99)^{100-x}$$

$$= (0.99)^{100} + 100(0.01)^1(0.99)^{99} + 4950(0.01)^2(0.99)^{98} \cong 0.92$$

Therefore, the probability of finding more than two defective bonds is $1 - 0.92 = 0.08$ (8%).

An important random variable used in statistical process control is the *sample fraction nonconforming* ($\hat{p}$), which is

$$\hat{p} = \frac{x}{n} \tag{4.10}$$

This variable is the ratio of defective items to sample size. The probability distribution for $\hat{p}$ is derived from the binomial, since

$$P(\hat{p} \le a) = P\left(\frac{x}{n} \le a\right) = P(x \le na) = \sum_{x=0}^{na} \binom{n}{x} p^x (1-p)^{n-x} \quad (4.11)$$

where $[na]$ = greatest integer less than or equal to na. It can be shown that the mean and variance of $\hat{p}$ are $\mu(\hat{p}) = p$ and $\sigma^2(\hat{p}) = [p(1-p)]/n$, respectively.

4.1.1.3. Poisson

Another important discrete distribution is the Poisson distribution, which is characterized by the expression

$$P(x) = \frac{e^{-\lambda}\lambda^x}{x!} \quad (4.12)$$

where x is an integer and λ is a constant > 0. The mean and variance of the Poisson distribution are

$$\mu = \lambda \quad (4.13)$$

$$\sigma^2 = \lambda \quad (4.14)$$

respectively. The Poisson distribution is used to model the number of defects that occur in a single product. To illustrate, consider the following example.

Example 4.3. Suppose that the number of wire bonding defects that occur has a Poisson distribution with $\lambda = 4$. What is the probability that a randomly selected package will have two or fewer defects?

Solution: Applying (4.12) gives $P\{x \le 2\} = \sum_{x=0}^{2}(e^{-4}4^x)/x! = 0.238$.

The Poisson distribution corresponding to $\lambda = 4$ is shown in Figure 4.3. The Poisson distribution is known for its skewed shape (i.e., the long "tail" to

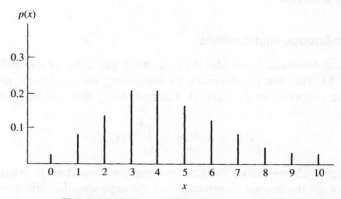

Figure 4.3. Poisson distribution with $\lambda = 4$ [1].

the right). As λ becomes larger, the shape of the distribution becomes more symmetric.

The Poisson distribution can be derived as a limiting form of the binomial distribution. In a binomial distribution with parameters n and p, as n approaches infinity and p approaches zero in such a way that $\lambda = np$ is a constant, then the Poisson distribution results.

4.1.1.4. Pascal

Like the binomial distribution, the Pascal distribution is based on a series of Bernoulli trials. For a sequence of independent trials, each with a probability of success (or failure) given by p, let x denote the trial in which the rth success (or failure) occurs. Under these circumstances, x is a Pascal random variable with the following probability distribution

$$P(x) = \binom{x-1}{r-1} p^r (1-p)^{x-r} \tag{4.15}$$

where $r \geq 1$ is an integer and $x \geq r$. The mean and variance of the Pascal distribution are

$$\mu = \frac{r}{p} \tag{4.16}$$

$$\sigma^2 = \frac{r(1-p)}{p^2} \tag{4.17}$$

respectively.

There are two special cases of the Pascal distribution that are of interest in semiconductor manufacturing applications. The first is when $r > 0$ and not necessarily an integer. The resulting distribution in this case is called the *negative binomial* distribution, which is particularly useful in modeling IC yield (see Chapter 5). The second special case occurs when $r = 1$, which results in the *geometric* distribution. This is the distribution of the number of Bernoulli trials until the first success.

4.1.2. Continuous Distributions

A continuous distribution provides the probability that x lies in a specific interval (Figure 4.1b). This can be computed by integrating the continuous distribution between the endpoints of the interval. The probability that x is between a and b is given by

$$P\{a \leq x \leq b\} = \int_a^b f(x)\, dx \tag{4.18}$$

Two examples of continuous distributions that are important in statistical process control are the normal distribution and the exponential distribution. Each is described in more detail below.

4.1.2.1. Normal

The normal distribution is undoubtedly the most important and best known probability distribution in applied statistics. The *probability density function* for a normally distributed random variable x is given by

$$f(x) = \frac{1}{\sigma\sqrt{2\pi}} \exp\left[-\frac{1}{2}\left(\frac{x-\mu}{\sigma}\right)^2\right] \tag{4.19}$$

The notation $x \sim N(\mu, \sigma^2)$ is often used to imply that x is normally distributed with mean μ and variance σ^2. The normal distribution has a symmetric bell shape, as shown in Figure 4.4.

A useful graphic to interpret the value of the standard deviation of the normal distribution is shown in Figure 4.5. This figure shows that 68.26% of the area under a normal curve lies in the interval $\mu \pm 1\sigma$, 95.46% of the area lies in the interval $\mu \pm 2\sigma$, and 99.73% of the area lies in the interval $\mu \pm 3\sigma$.

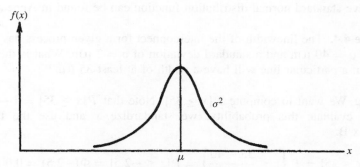

Figure 4.4. The normal distribution [1].

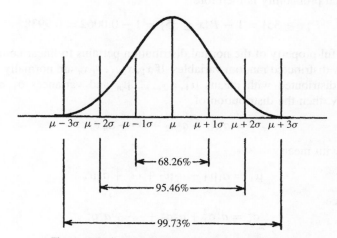

Figure 4.5. Areas under the normal distribution [1].

The *cumulative normal distribution* is defined as the probability that x is less than or equal to some value a, or

$$P(x \leq a) = F(a) = \int_{-\infty}^{a} f(x)\,dx \tag{4.20}$$

Unfortunately, this integral cannot be evaluated in closed form. Instead, the following change of variables is used:

$$z = \frac{x - \mu}{\sigma} \tag{4.21}$$

This allows the integral in Eq. (4.20) to be evaluated independently of μ and σ^2. In other words

$$P(x \leq a) = P\left\{z \leq \frac{a - \mu}{\sigma}\right\} = \Phi\left(\frac{a - \mu}{\sigma}\right) \tag{4.22}$$

where Φ is the cumulative distribution function of the *standard normal distribution* (i.e., the normal distribution with $\mu = 0$ and $\sigma = 1$). A table of values for the cumulative standard normal distribution function can be found in Appendix B.

Example 4.4. The linewidth of the interconnect for a given process has a mean value of $\mu = 40$ µm and a standard deviation of $\sigma = 2$ µm. What is the probability that a particular line will have a width of at least 35 µm?

Solution: We want to compute $P\{x \geq 35\}$. Note that $P\{x \geq 35\} = 1 - P\{x \leq 35\}$. To evaluate this probability, we standardize x and use the table in Appendix B.

$$P\{x \leq 35\} = P\left\{z \leq \frac{35 - 40}{z}\right\} = P\{z \leq -2.5\} = \Phi(-2.5) = 0.0062$$

The required probability is therefore

$$P\{x \geq 35\} = 1 - P\{x \leq 35\} = 1 - 0.0062 = 0.9938$$

One useful property of the normal distribution pertains to linear combinations of normally distributed random variables. If $x_1, x_2, \ldots, x_n$ are normally and independently distributed with means $\mu_1, \mu_2, \ldots, \mu_n$ and variances $\sigma_1^2, \sigma_2^2 \ldots, \sigma_n^2$, respectively, then the distribution of

$$y = a_1 x_1 + a_2 x_2 + \cdots + a_n x_n$$

is normal with mean

$$\mu_y = a_1 \mu_1 + a_2 \mu_2 + \cdots + a_n \mu_n \tag{4.23}$$

and variance

$$\sigma_y^2 = a_1^2 \sigma_1^2 + a_2^2 \sigma_2^2 + \cdots + a_n^2 \sigma_n^2 \tag{4.24}$$

where $a_1, a_2, \ldots, a_n$ are constants.

4.1.2.2. Exponential

The exponential distribution is widely used in reliability engineering as a model for the time to failure of a component or system. The probability density function for a random variable x that has this distribution is

$$f(x) = \lambda e^{-\lambda x} \tag{4.25}$$

where $\lambda > 0$ is a constant. A graph of the density function appears in Figure 4.6. The mean and variance of the exponential distribution are

$$\mu = \frac{1}{\lambda} \tag{4.26}$$

$$\sigma^2 = \frac{1}{\lambda^2} \tag{4.27}$$

respectively. The cumulative exponential distribution function is

$$F(a) = P(x \le a) = \int_0^a \lambda e^{-\lambda t}\, dt = 1 - e^{-\lambda a} \quad a \ge 0 \tag{4.28}$$

The parameter λ is used to model the *failure rate* of a system, and the mean of the distribution $(1/\lambda)$ is called the *mean time to failure.*

Example 4.5. An electronic component has a useful lifetime that is described by an exponential distribution with a failure rate of 10^{-4} per hour (i.e., $\lambda = 10^{-4}$). What is the probability that this component will fail before its expected life?

Solution: We want to compute $P\{x \le 1/\lambda\}$. We evaluate this probability as follows:

$$P\left\{x \le \frac{1}{\lambda}\right\} = \int_0^{1/\lambda} \lambda e^{-\lambda t}\, dt = 1 - e^{-1} = 0.6321$$

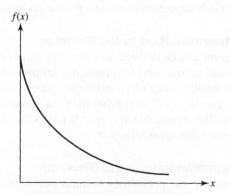

Figure 4.6. The exponential distribution.

There is an important relationship between the exponential and Poisson distributions. If the Poisson distribution is assumed to model the number of occurrences of a failure in the interval $(0, t]$, then applying Eq. (4.12) gives

$$P(x) = \frac{e^{-\lambda t}(\lambda t)^x}{x!} \tag{4.29}$$

If $x = 0$, there are no failures in the interval $(0, t]$, and, $P(0) = e^{-\lambda t}$. This may also be regarded as the probability that the first failure occurs *after* time t, or

$$P\{y > t\} = P(0) = e^{-\lambda t} \tag{4.30}$$

where y is a random variable representing the time interval until the first failure. Since

$$F(t) = P\{y \le t\} = 1 - e^{-\lambda t} \tag{4.31}$$

and

$$f(y) = dF(y)/dy \tag{4.32}$$

we can conclude that

$$f(y) = \lambda e^{-\lambda y} \tag{4.33}$$

is the distribution of the interval to the first failure. Note that Eq. (4.33) is just the exponential distribution with parameter λ. Therefore, if the number of failures has a Poisson distribution with parameter λ, then the *interval between* failures is exponential with parameter λ.

4.1.3. Useful Approximations

For certain process control applications, approximating one probability distribution with another can significantly simplify the analysis. This approach is particularly useful in situations when the original distribution is complex or not well tabulated. Two such approximations are presented in the following.

4.1.3.1. Poisson Approximation to the Binomial
The Poisson distribution can be derived as a limiting form of the binomial distribution when p approaches zero and n approaches infinity with $\lambda = np$ constant. This implies that for small values of p and large values of n, the Poisson distribution with $\lambda = np$ can be used to approximate the binomial distribution. This approximation is usually reasonable for $p < 0.1$, but the larger the n and the smaller the p, the better the approximation.

4.1.3.2. Normal Approximation to the Binomial
The binomial distribution was previously defined as a sum of n Bernoulli trials, each with an associated probability of success p. If n is large, then the central

limit theorem may be used to justify a normal approximation to the binomial distribution with mean np and variance $np(1 - p)$. In other words

$$P(x = a) = \binom{n}{a} p^a (1 - p)^{n-a} = \frac{1}{\sqrt{2\pi np(1 - p)}} e^{-(1/2)[(a-np)^2/np(1-p)]}$$

(4.34)

Since the binomial distribution is discrete and the normal distribution is continuous, the following continuity correction is commonly applied to this approximation

$$P(x = a) \cong \Phi\left(\frac{a + \frac{1}{2} - np}{\sqrt{np(1 - p)}}\right) - \Phi\left(\frac{a - \frac{1}{2} - np}{\sqrt{np(1 - p)}}\right)$$

(4.35)

where Φ is the standard normal cumulative distribution function. Probability intervals are evaluated similarly. In other words

$$P(a \leq x \leq b) \cong \Phi\left(\frac{b + \frac{1}{2} - np}{\sqrt{np(1 - p)}}\right) - \Phi\left(\frac{a - \frac{1}{2} - np}{\sqrt{np(1 - p)}}\right)$$

(4.36)

This approximation is satisfactory for $p \approx \frac{1}{2}$ and $n > 10$. For larger values of p, larger values of n are required. In general, the approximation is inadequate for $p < 1/(n + 1)$ or $p > n/(n + 1)$, or for values of the random variable outside the interval $np \pm 3\sqrt{np(1 - p)}$.

Since the binomial distribution can be approximated by the normal, and since the binomial and Poisson distributions are closely related, the Poisson distribution can also be approximated by the normal. The normal approximation to the Poisson distribution with $\mu = \lambda$ and $\sigma^2 = \lambda$ is satisfactory for $\lambda \geq 15$.

4.2. SAMPLING FROM A NORMAL DISTRIBUTION

Statistics allow inferences to be made or conclusions to be drawn about a population based on a *sample* chosen from that population. *Random sampling* refers to any method of sample selection that lacks systematic direction or bias. A random sample of size n consists of observations $x_1, x_2, \ldots, x_n$ selected so that the observations x_i are *independently and identically distributed* (IID). In other words, random sampling allows every sample an equal likelihood of being selected. If it can be further assumed that the samples come from a normal distribution, then it is said that the samples are IIDN.

Statistical inference procedures use quantities such as the sample mean $(\overline{x})$ and sample variance (s^2) to draw conclusions about the central tendency and dispersion, respectively, of a population based on a sample. If the probability distribution from which a sample was taken is known, then the distribution of statistics such as $\overline{x}$ and s^2 can be determined from the sample data. For example, suppose that a random variable x is normally distributed with mean μ and variance σ^2. If $x_1, x_2, \ldots, x_n$ is a random sample of size n from this population, then the distribution of $(\bar{x})$ is $N[\mu, (\sigma^2/n)]$, which follows directly from Eqs. (4.23) and (4.24). In general, the probability distribution of a statistic is called the *sampling distribution*.

4.2.1. Chi-Square Distribution

An important sampling distribution that originates from the normal distribution is the *chi-square* (χ^2) distribution. If $x_1, x_2, \ldots, x_n$ are normally distributed random variables with mean zero and variance one, then the random variable

$$\chi_n^2 = \chi_1^2 + \chi_2^2 + \cdots + \chi_n^2$$

is distributed as chi-square with n *degrees of freedom*. The probability density function of χ^2 is

$$f\left(\chi^2\right) = \frac{1}{2^{n/2}\Gamma(n/2)} \left(\chi^2\right)^{(n/2)-1} e^{-\chi^2/2} \qquad (4.37)$$

where Γ is the gamma function. If a random sample of size n is collected from a $N(\mu, \sigma^2)$ distribution, and this sample yields a sample variance of s^2, it can be shown that

$$\frac{(n-1)s^2}{\sigma^2} \approx \chi_{n-1}^2 \qquad (4.38)$$

that is, the sampling distribution of $(n-1)s^2/\sigma^2$ is χ_{n-1}^2. The chi-square distribution is used to make inferences about the variance of a normal distribution. A few chi-square distributions are shown in Figure 4.7. A table of values for the cumulative chi-square distribution function is given in Appendix C.

4.2.2. *t* Distribution

The *t* distribution is another useful sampling distribution based on the normal distribution. If x and χ_k^2 are standard normal and chi-square random variables,

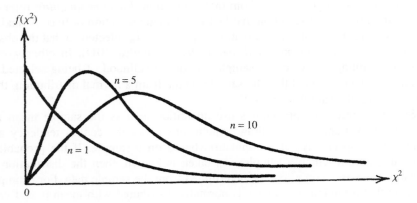

Figure 4.7. Several χ^2 distributions.

then the random variable

$$t_k \equiv \frac{x}{\sqrt{\chi_k^2/k}}$$

is distributed as t with k degrees of freedom. The probability density function of t is

$$f(t) = \frac{\Gamma[(k+1)/2]}{\sqrt{k\pi}(k/2)} \left(\frac{t^2}{k}+1\right)^{-(k+1)/2} \tag{4.39}$$

For a random sample of size n collected from a $N(\mu, \sigma^2)$ distribution with a sample mean $\bar{x}$ and a sample variance of s^2, it can be shown that

$$\frac{\bar{x}-\mu}{s/\sqrt{n}} \sim t_{n-1} \tag{4.40}$$

The t distribution is used to make inferences about the mean of a normal distribution. A few t distributions are shown in Figure 4.8. Note that as $k \to \infty$, the t distribution becomes the standard normal distribution. A table of values for the cumulative t distribution function is given in Appendix D.

4.2.3. *F* Distribution

The last sampling distribution to be considered that is based on the chi-square distribution is the F distribution. If χ_u^2 and χ_v^2 are chi-square random variables with u and v degrees of freedom, then the ratio

$$F_{u,v} \equiv \frac{\chi_u^2/u}{\chi_v^2/v}$$

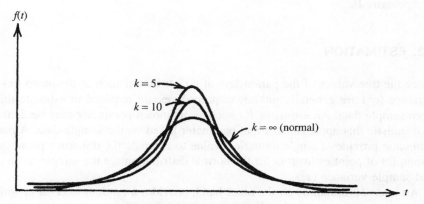

Figure 4.8. Several t distributions.

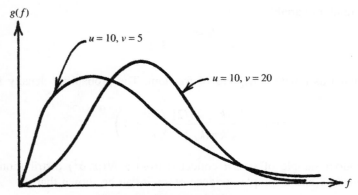

Figure 4.9. Several F distributions.

is distributed as F with u and v degrees of freedom. The probability density function of F is

$$g(F) = \frac{\Gamma\left(\dfrac{u+v}{2}\right)\left(\dfrac{u}{v}\right)^{u/2}}{\Gamma\left(\dfrac{u}{2}\right)\Gamma\left(\dfrac{v}{2}\right)} \; \frac{F^{u/2-1}}{\left[\left(\dfrac{u}{2}\right)F + 1\right]^{(u+v)/2}} \qquad (4.41)$$

Consider two independent normal processes, $x_1 \sim N(\mu_1, \sigma_1^2)$ and $x_2 \sim N(\mu_2, \sigma_2^2)$. If random samples of sizes n_1 and n_2 yield sample variances s_1^2 and s_2^2, respectively, then it can be shown that

$$\frac{s_1^2/\sigma_1^2}{s_2^2/\sigma_2^2} \sim F_{n_1-1, n_2-1} \qquad (4.42)$$

The F distribution can thus be used to make inferences in comparing the variances of two normal distributions. A few F distributions are shown in Figure 4.9. A table of values for the cumulative F distribution function is given in Appendix E.

4.3. ESTIMATION

Since the true values of the parameters of a distribution such as the mean (μ) or variance (σ^2) are generally unknown, procedures are required to estimate them from sample data. An estimator for such an unknown parameter may be defined as a statistic that approximates that parameter based on the sample data. A *point estimator* provides a single numerical value to estimate the unknown parameter. Examples of point estimators for the normal distribution are the sample mean ($\bar{x}$) and sample variance (s^2).

An *interval estimator*, on the other hand, provides a random interval in which the true value of the parameter being estimated falls with some probability. These intervals are called *confidence intervals*. A summary of some of the more useful confidence intervals for the normal distribution follows.

4.3.1. Confidence Interval for the Mean with Known Variance

Suppose that a sample of n independent observations $x_1, x_2, \ldots, x_n$ on a random variable x is taken. If $(\bar{x})$ is computed from the sample, then a $100(1 - \alpha)\%$ confidence interval on the mean μ of this population is defined as

$$\bar{x} - z_{(\alpha/2)}\frac{\sigma}{\sqrt{n}} \leq \mu \leq \bar{x} + z_{(\alpha/2)}\frac{\sigma}{\sqrt{n}} \tag{4.43}$$

where $z_{\alpha/2}$ is the value of the $N(0, 1)$ distribution such that $P\{z \geq z_{\alpha/2}\} = \alpha/2$.

4.3.2. Confidence Interval for the Mean with Unknown Variance

Suppose that a sample of n independent observations $x_1, x_2, \ldots, x_n$ on a normally distributed random variable x is taken. If $\bar{x}$ and s^2 are computed from the sample, then a $100(1 - \alpha)\%$ confidence interval on the mean μ of this population is defined as

$$\bar{x} - t_{(\alpha/2),n-1}\frac{s}{\sqrt{n}} \leq \mu \leq \bar{x} + t_{(\alpha/2),n-1}\frac{s}{\sqrt{n}} \tag{4.44}$$

where $t_{(\alpha/2),n-1}$ is the value of the t distribution with $n - 1$ degrees of freedom such that $P\{t_{n-1} \geq t_{(\alpha/2),n-1}\} = \alpha/2$.

Example 4.6. Suppose that the linewidth of $n = 16$ with supposedly identical interconnect traces is measured. The sample mean and sample standard deviation for these measurements are $\bar{x} = 49.86$ μm and $s = 1.66$ μm, respectively. What is the 95% confidence interval on this estimate of the mean?

Solution: Since $t_{0.025,15} = 2.132$, the 95% confidence interval on m can be found from Eq. (4.44) as follows:

$$49.86 - (2.132)1.66/\sqrt{16} \leq \mu \leq 49.86 + (2.132)1.66/\sqrt{16}$$

$$49.98 \leq \mu \leq 50.74$$

Thus, the estimate of the mean linewidth is 49.86 ± 0.88 μm with 95% confidence.

4.3.3. Confidence Interval for Variance

Suppose that a sample of n IIDN observations $x_1, x_2, \ldots, x_n$ on a random variable x is taken. If s^2 is computed from the sample, then a $100(1 - \alpha)\%$ confidence interval on the variance σ^2 of this population is defined as

$$\frac{(n - 1)s^2}{\chi^2_{(\alpha/2),n-1}} \leq \sigma^2 \leq \frac{(n - 1)s^2}{\chi^2_{1-(\alpha/2),n-1}} \tag{4.45}$$

where $\chi^2_{(\alpha/2),n-1}$ is the value of the χ^2 distribution with $n - 1$ degrees of freedom such that $P\{\chi^2_{n-1} \geq \chi^2_{\alpha/2,n-1}\} = \alpha/2$.

Example 4.7. For the dataset in Example 4.6, what is the 95% confidence interval on the estimate of the variance?

Solution: Since $\chi^2_{0.025,15} = 27.49$, $\chi^2_{0.975,15} = 6.27$, and $s^2 = 2.76$, the 95% confidence interval on σ^2 can be found from Eq. (4.45) as follows:

$$\frac{(15)(2.76)}{27.49} \leq \sigma^2 \leq \frac{(15)(2.76)}{6.27}$$

$$1.51 \leq \sigma^2 \leq 6.60$$

4.3.4. Confidence Interval for the Difference between Two Means, Known Variance

Consider two normal random variables from two different populations: x_1 with mean μ_1 and variance σ_1^2, and x_2 with mean μ_2 and variance σ_2^2. Suppose that samples of n_1 observations $x_{11}, x_{12}, \ldots, x_{1n_1}$ on random variable x_1 and n_2 observations $x_{21}, x_{22}, \ldots, x_{2n_2}$ on random variable x_2 are taken. If $\bar{x}_1$ and $\bar{x}_2$ are computed from the two samples and the variances are known, then a $100(1 - \alpha)\%$ confidence interval on the difference between the means of these two populations is defined as follows:

$$\left\{ \bar{x}_1 - \bar{x}_2 - z_{(\alpha/2)}\sqrt{\frac{\sigma_1^2}{n_1} + \frac{\sigma_2^2}{n_2}} \right\} \leq (\mu_1 - \mu_2) \leq \left\{ \bar{x}_1 - \bar{x}_2 + z_{(\alpha/2)}\sqrt{\frac{\sigma_1^2}{n_1} + \frac{\sigma_2^2}{n_2}} \right\}$$

$$(4.46)$$

4.3.5. Confidence Interval for the Difference between Two Means, Unknown Variances

Consider two normal random variables from two different populations: x_1 with mean μ_1 and variance σ_1^2, and x_2 with mean μ_2 and variance σ_2^2. Suppose that samples of n_1 observations $x_{11}, x_{12}, \ldots, x_{1n_1}$ on random variable x_1 and n_2 observations $x_{21}, x_{22}, \ldots, x_{2n_2}$ on random variable x_2 are taken. Assume that the means and variances are unknown, but the variances are equal; that is, $\sigma_1^2 = \sigma_2^2 = \sigma^2$.

If $\bar{x}_1, \bar{x}_2, s_1^2$, and s_2^2 are computed from the two samples, then a *pooled estimate of the common variance* of the two populations is

$$s_p^2 = \frac{(n_1 - 1)s_1^2 + (n_2 - 1)s_2^2}{n_1 + n_2 - 2}$$

$$(4.47)$$

Under these conditions, a $100(1 - \alpha)\%$ confidence interval on $\mu_1 - \mu_2$ is defined as

$$\left\{ \overline{x}_1 - \overline{x}_2 - t_{(\alpha/2),\nu} s_p \sqrt{\frac{1}{n_1} + \frac{1}{n_2}} \right\} \leq (\mu_1 - \mu_2)$$

$$\leq \left\{ \overline{x}_1 - \overline{x}_2 + t_{(\alpha/2),\nu} s_p \sqrt{\frac{1}{n_1} + \frac{1}{n_2}} \right\} \qquad (4.48)$$

where $\nu = n_1 + n_2 - 2$.

Example 4.8. The average contact pad size for two different ICs is to be compared. $n_1 = n_2 = 10$ pads are selected at random, and their IIDN side dimensions are measured. For the first IC, $\overline{x}_1 = 90.70$ µm and $s_1^2 = 1.34$ µm²; for the second IC, $\overline{x}_2 = 90.80$ µm and $s_2^2 = 1.07$ µm². What is the 99% confidence interval for the difference in pad size for the two ICs?

Solution: Assuming that the variances for pad size on each IC are the same, the pooled estimate of the common variance is found from Eq. (4.47) as follows:

$$s_p^2 = \frac{(n_1 - 1)s_1^2 + (n_2 - 1)s_2^2}{n_1 + n_2 - 2} = \frac{(9)1.34 + (9)1.07}{10 + 10 - 2} = 1.21$$

The 99% confidence interval on $\mu_1 - \mu_2$ can be found from Eq. (4.48) as follows:

$$\left\{ \overline{x}_1 - \overline{x}_2 - t_{0.005,18} s_p \sqrt{\frac{1}{n_1} + \frac{1}{n_2}} \right\}$$

$$\leq (\mu_1 - \mu_2) \leq \left\{ \overline{x}_1 - \overline{x}_2 + t_{0.005,18} s_p \sqrt{\frac{1}{n_1} + \frac{1}{n_2}} \right\}$$

$$\left\{ 90.70 - 90.80 - (2.878)(1.1) \sqrt{\frac{1}{10} + \frac{1}{10}} \right\} \leq (\mu_1 - \mu_2)$$

$$\leq \left\{ 90.70 - 90.80 + (2.878)(1.1) \sqrt{\frac{1}{10} + \frac{1}{10}} \right\}$$

$$- 1.51 \leq \mu_1 - \mu_2 \leq 1.31$$

4.3.6. Confidence Interval for the Ratio of Two Variances

Consider two normal random variables from two different populations: x_1 with mean μ_1 and variance σ_1^2, and x_2 with mean μ_2 and variance σ_2^2. Suppose that samples of n_1 observations $x_{11}, x_{22}, \ldots, x_{1n_1}$ on random variable x_1 and n_2 observations $x_{21}, x_{22}, \ldots, x_{2n_2}$ on random variable x_2 are taken. If $\bar{x}_1, \bar{x}_2, s_1^2, s_2^2$ are computed from the two samples, then a $100(1 - \alpha)\%$ confidence interval on σ_1^2/σ_2^2 is defined as

$$\frac{s_1^2}{s_2^2} F_{1-(\alpha/2), v_2, v_1} \leq \frac{\sigma_1^2}{\sigma_2^2} \leq \frac{s_1^2}{s_2^2} F_{(\alpha/2), v_2, v_1} \tag{4.49}$$

where $v_1 = n_1 - 1$, $v_2 = n_2 - 1$, and $F_{\alpha/2, u, v}$ is the value of the F distribution with u and v degrees of freedom such that $P\{F_{u,v} \geq F_{\alpha/2, u, v}\} = \alpha/2$.

Example 4.9. Consider the dataset in Example 4.8. What is the 95% confidence interval for the ratio of the variances of contact pad size for the two ICs?

Solution: From Appendix E, $F_{0.025, 9, 9} = 4.03$ and $F_{0.975, 9, 9} = 0.248$. Using Eq. (4.49), the required confidence interval is

$$\frac{1.34}{1.07}(0.248) \leq \frac{\sigma_1^2}{\sigma_2^2} \leq \frac{1.34}{107}(4.03)$$

$$0.31 \leq \frac{\sigma_1^2}{\sigma_2^2} \leq 5.05$$

4.4. HYPOTHESIS TESTING

A *statistical hypothesis* is a statement about the values about the parameters of a probability distribution. A *hypothesis test* is an evaluation of the validity of the hypothesis according to some criterion. Hypotheses are expressed in the following manner:

$$H_0: \mu = \mu_0$$
$$H_1: \mu \neq \mu_0 \tag{4.50}$$

where μ is the unknown mean of the distribution and μ_0 is a hypothesized value of μ. The statement $H_0: \mu = \mu_0$ is called the *null hypothesis*, and $H_1: \mu \neq \mu_0$ is called the *alternative hypothesis*. Hypothesis testing procedures form the basis for many of the statistical process control techniques described in Chapter 6. To perform a hypothesis test, select a random sample from a population, compute an appropriate test statistic, and then either accept or reject the null hypothesis H_0.

Two types of error may result when performing such a test. If the null hypothesis is rejected when it is actually true, then a *type I error* has occurred. On the other hand, if the null hypothesis is accepted when it is actually false, this is called a *type II error*. The probabilities for each of these errors are denoted as follows:

$$\alpha = P(\text{type I error}) = P(\text{reject } H_0 | H_0 \text{ is true})$$
$$\beta = P(\text{type II error}) = P(\text{accept } H_0 | H_0 \text{ is false})$$

For statistical process control applications, α is considered the probability of a *false alarm* and β is the probability of a *missed alarm*. The statistical power of a test is defined as follows:

$$\text{Power} = 1 - \beta = P(\text{reject } H_0 | H_0 \text{ is false})$$

The power, therefore, represents the probability of correctly rejecting H_0. The basic procedure required for hypothesis testing involves specifying a desired value of α, and then designing a test that produces a small value of β. A few common test scenarios are illustrated in the following sections.

4.4.1. Tests on Means with Known Variance

Let x be a normally distributed random variable with unknown mean μ and known variance σ^2. Suppose that the hypothesis that the mean is equal to some constant value μ_0 must be tested. This hypothesis is described by Eq. (4.50). The procedure to perform the test requires taking a random sample of n independent observations and computing the following test statistic:

$$z_0 = \frac{\bar{x} - \mu_0}{\sigma/\sqrt{n}} \tag{4.51}$$

The null hypothesis H_0 is rejected if $|z_0| > z_{\alpha/2}$, where $z_{\alpha/2}$ is the value of the standard normal distribution such that $P\{z \geq z_{\alpha/2}\} = \alpha/2$. In some cases, it may be necessary to test the hypothesis that the mean is larger than μ_0. Under these circumstances, the *one-sided alternative hypothesis* is H_1: $\mu > \mu_0$, and H_0 is rejected only if $z_0 > z_\alpha$. To test the hypothesis that the mean is smaller than μ_0, the one-sided alternative hypothesis is H_1: $\mu < \mu_0$, and H_0 is rejected if $z_0 < -z_\alpha$.

Suppose now that there are two populations with unknown means (μ_1 and μ_2) that must be compared. Assume that the two populations have known variances σ_1^2 and σ_2^2. To compare the two means, test the following hypothesis:

$$H_0: \ \mu_1 = \mu_2$$
$$H_1: \ \mu_1 \neq \mu_2 \tag{4.52}$$

To perform this test, n_1 and n_2 sample observations from each population are collected and then the test statistic

$$z_0 = \frac{\bar{x}_1 - \bar{x}_2}{\sqrt{\dfrac{\sigma_1^2}{n_1} + \dfrac{\sigma_2^2}{n_2}}} \tag{4.53}$$

is computed. H_0 is rejected if $|z_0| > z_{\alpha/2}$. The one-sided tests are similar to those described above.

Example 4.10. Suppose that it must be determined whether the mean thickness of a film exceeds 175 Å. The standard deviation of this thickness is known to be

10 Å. A random sample of 25 locations on a wafer yields an average thickness of $\bar{x} = 182$ Å.

Solution: The following hypothesis test is of interest:

$$H_0: \mu = 175$$
$$H_1: \mu > 175$$

The value of the test statistic is

$$z_0 = \frac{\bar{x} - \mu_0}{\sigma/\sqrt{n}} = \frac{182 - 175}{10/\sqrt{25}} = 3.50$$

If a type I error of $\alpha = 0.05$ is specified, then, from Appendix B, $z_\alpha = z_{0.05} = 1.645$. Therefore, H_0 is rejected, and the mean thickness does exceed 175 Å.

4.4.2. Tests on Means with Unknown Variance

Let x be a normal random variable with unknown mean μ and unknown variance σ^2. Suppose that the hypothesis that the mean is equal to some constant value μ_0 must be tested. Since the variance is unknown, it must be estimated by the sample variance s^2. The procedure to perform the test then requires taking a random sample of n observations and computing the following test statistic:

$$t_0 = \frac{\bar{x} - \mu_0}{s/\sqrt{n}} \qquad (4.54)$$

H_0 is rejected if $|t_0| > t_{\alpha/2, n-1}$, where $t_{\alpha/2, n-1}$ is the value of the t distribution with $n - 1$ degrees of freedom such that $P\{t \geq t_{\alpha/2, n-1}\} = \alpha/2$. In some cases, the hypothesis that the mean is larger than μ_0 must be tested. Under these circumstances, the one-sided alternative hypothesis is $H_1: \mu > \mu_0$, and H_0 is rejected only if $t_0 > t_{\alpha, n-1}$. To test the hypothesis that the mean is smaller than μ_0, the one-sided alternative hypothesis is $H_1: \mu < \mu_0$, and H_0 is rejected if $t_0 < -t_{\alpha, n-1}$.

Suppose now that there are two normal populations with unknown means (μ_1 and μ_2) that must be compared. Assume that the two populations have unknown variances σ_1^2 and σ_2^2. To compare the two means, the hypothesis given by Eq. (4.52) is tested. The test procedure depends on whether the two variances can reasonably be assumed to be equal. If they are equal, and n_1 and n_2 sample observations are collected from each population, then a "pooled" estimate of the common variance of the two populations is given by Eq. (4.47). The appropriate test statistic is then

$$t_0 = \frac{\bar{x}_1 - \bar{x}_2}{s_p\sqrt{\dfrac{1}{n_1} + \dfrac{1}{n_2}}} \qquad (4.55)$$

H_0 is rejected if $|t_0| > t_{\alpha/2,n_1+n_2-2}$. The one-sided tests are similar to those described above. If the variances are *not equal*, then the appropriate test statistic is

$$t_0 = \frac{\bar{x}_1 - \bar{x}_2}{\sqrt{\dfrac{s_1^2}{n_1} + \dfrac{s_2^2}{n_2}}} \tag{4.56}$$

and the number of degrees of freedom for t_0 are

$$v = \frac{\left(\dfrac{s_1^2}{n_1} + \dfrac{s_2^2}{n_2}\right)^2}{\dfrac{(s_1^2/n_1)^2}{n_1+1} + \dfrac{(s_2^2/n_2)}{n_2+1}} - 2 \tag{4.57}$$

Once again, H_0 is rejected if $|t_0| > t_{\alpha/2,v}$, and the one-sided tests are similar to those described above.

Example 4.11. Consider the data in Example 4.8. Suppose that the hypothesis that the mean pad size for the first IC is equal to the mean pad size for the second IC must be tested, or

$$H_0: \mu_1 = \mu_2$$

$$H_1: \mu_1 \neq \mu_2$$

Solution: Assuming, $\sigma_1^2 = \sigma_2^2$, which is reasonable if the ICs have undergone the same manufacturing process, then $s_p = 1.10$. The test statistic is then

$$t_0 = \frac{\bar{x}_1 - \bar{x}_2}{s_p\sqrt{\dfrac{1}{n_1} + \dfrac{1}{n_2}}} = -0.20$$

If a type I error of $\alpha = 0.01$ is specified, then, from Appendix D, $t_{0.005,18} = 2.878$. Since $|t_0| < t_{(\alpha/2),n-1}$, H_0 must be accepted, and there is no strong evidence that the two means are different.

4.4.3. Tests on Variance

Suppose that we want to test the hypothesis that the variance of a normal distribution is equal to some constant value σ_0^2. The hypotheses are expressed as follows:

$$H_0: \sigma^2 = \sigma_0^2$$

$$H_1: \sigma^2 \neq \sigma_0^2 \tag{4.58}$$

The appropriate test statistic is

$$\chi_0^2 = \frac{(n-1)s^2}{\sigma_0^2} \tag{4.59}$$

where s^2 is the sample variance computed from a random sample of n observations. Hypothesis H_0 is rejected if $\chi_0^2 > \chi_{(\alpha/2),n-1}^2$ or if $\chi_0^2 < \chi_{1-(\alpha/2),n-1}^2$, where $\chi_{(\alpha/2),n-1}^2$ and $\chi_{1-(\alpha/2),n-1}^2$ are the upper $\alpha/2$ and lower $1 - (\alpha/2)$ percentage points of the χ^2 distribution with $n-1$ degrees of freedom. For the one-sided alternative hypothesis H_1: $\sigma^2 > \sigma_0^2$, we reject H_0 if $\chi_0^2 > \chi_{\alpha,n-1}^2$. To test the hypothesis that the variance is smaller than σ_0^2, the one-sided alternative hypothesis is H_1: $\sigma^2 < \sigma_0^2$, and we reject H_0 if $\chi_0^2 < \chi_{1-\alpha,n-1}^2$.

Now consider two normal populations with variances σ_1^2 and σ_2^2. To compare these populations, n_1 and n_2 sample observations from each are collected, and the hypothesis

$$H_0: \sigma_1^2 = \sigma_2^2$$
$$H_1: \sigma_1^2 \neq \sigma_2^2 \tag{4.60}$$

is tested. The test statistic is

$$F_0 = \frac{s_1^2}{s_2^2} \tag{4.61}$$

Hypothesis H_0 is rejected if $F_0 > F_{(\alpha/2),n_1-1,n_2-1}$ or if $F_0 < F_{1-(\alpha/2),n_1-1,n_2-1}$, where $F_{(\alpha/2),n_1-1,n_2-1}$ and $F_{1-(\alpha/2),n_1-1,n_2-1}$ are the upper $\alpha/2$ and lower $1 - (\alpha/2)$ percentage points of the F distribution with $n_1 - 1$ and $n_2 - 1$ degrees of freedom. For the one-sided alternative hypothesis H_1: $\sigma_1^2 > \sigma_2^2$, H_0 is rejected if $F_0 > F_{\alpha,n_1-1,n_2-1}$. For the one-sided alternative hypothesis H_1: $\sigma_1^2 < \sigma_2^2$, H_0 is rejected if $F_0 > F_{\alpha,n_2-1,n_1-1}$.

Example 4.12. Consider once again the data in Example 4.8. Suppose that the hypothesis that the variances of the pad sizes are equal is to be tested, or

$$H_0: \sigma_1^2 = \sigma_2^2$$
$$H_1: \sigma_1^2 \neq \sigma_2^2$$

Solution: Given that $s_1^2 = 1.34$ and $s_2{}^2 = 1.07$, the test statistic is

$$F_0 = \frac{s_1^2}{s_2^2} = 1.25$$

If a type I error of $\alpha = 0.05$ is specified, then, from Appendix E, $F_{0.025,9,9} = 4.03$. Since $F_0 < F_{\alpha/2,n_1-1,n_2-1}$, H_0 is accepted, and there is no strong evidence that the variances are different.

SUMMARY

This chapter provided an introduction to the concept of process variability and a brief survey of the statistical tools used to analyze semiconductor manufacturing processes. An understanding of these statistical fundamentals is essential for describing, analyzing, modeling, and controlling these processes, all of which are the subject of subsequent chapters.

PROBLEMS

4.1. A random sample of 50 dies is collected from each lot in a given processes. Calculate the probability that we will find less than three defective dies in this sample if the yield of the process is 98%.

4.2. An IC manufacturing process is subject to defects that obey a Poisson distribution with a mean of four defects per wafer.

    (a) Assuming that a single defect will destroy a wafer, calculate the functional yield of the process.

    (b) Suppose that we can add extra redundant dies to account for the defects. If one redundant die is needed to replace exactly one defective die, how many dies are required to ensure a yield of at least 50%?

4.3. Suppose the concentration of particles produced in an etching operation on any given day is normally distributed with a mean of 15.08 particles/ft^3 and a standard deviation of 0.05 particles/ft^3. The specifications on the process call for a concentration of $15.00 +/- 0.1$ particles/ft^3. What fraction of etching systems conform to specifications?

4.4. The time to failure of printed circuit boards is modeled by the following exponential distribution probability density function:

$$f(t) = 0.125e^{-0.125t} \quad \text{for} \quad t > 0$$

where t is the time in years. What percentage of the circuit boards will fail within one year?

4.5. A new process has been developed for spin coating photoresist. Ten wafers have been tested with the new process, and the results of thickness measurements (in μm) are shown below and are assumed to be IIDN. Find a 99% confidence interval on the mean photoresist thickness.

13.3946	13.4002
13.3987	13.3957
13.3902	13.4015
13.4001	13.3918
13.3965	13.3925

4.6. Suppose that we are interested in calibrating a chemical vapor deposition furnace. The furnace will be shut down for repairs if significant difference is found between the thermocouples that are measuring the deposition temperature at the two ends of the furnace tube. The following temperatures have been measured during several test runs:

Thermocouple 1 ($^\circ$C)	Thermocouple 2 ($^\circ$C)
606.5	604.0
605.0	604.5
605.5	605.5
605.5	605.7
606.2	605.5
606.5	605.2
603.7	606.0
607.7	606.5
607.7	607.7
604.2	604.2

(a) Using the appropriate hypothesis test, determine whether we can be 95% confident that these temperatures are the same at both ends of the tube.

(b) Find the 90% confidence interval for the ratio of the two variances $(\sigma_{T1}^2/\sigma_{T2}^2)$.

REFERENCE

1. D. Montgomery, *Introduction to Statistical Quality Control*, Wiley, New York, 1993.

5

YIELD MODELING

OBJECTIVES

- Provide a general definition of yield.
- Differentiate between functional and parametric yield.
- Introduce various yield models and simulators.
- Address financial aspects of yield.

INTRODUCTION

The primary objective of any semiconductor manufacturing operation is to produce outputs that meet required performance specifications. However, the variability inherent in manufacturing processes can lead to deformations or nonconformities in semiconductor products. Such process disturbances often result in faults, or unintentional changes in the performance or conformance of the finished integrated circuits. The presence of such faults is quantified by the *yield*.

Yield is in many ways the most important financial factor in producing ICs. This is because yield is inversely proportional to the total manufacturing cost—the higher the yield, the lower the cost.

Fundamentals of Semiconductor Manufacturing and Process Control,
By Gary S. May and Costas J. Spanos
Copyright © 2006 John Wiley & Sons, Inc.

147

5.1. DEFINITIONS OF YIELD COMPONENTS

Yield can be defined in many different ways. The first, and perhaps most basic definition, is that of *manufacturing yield*. This figure simply measures the proportion of successfully fabricated products compared to the number that have started the process. This definition applies to integrated circuits, which are batch-fabricated on semiconductor wafers, as well as to printed circuit boards, which are processed as individual parts.

Wafers that for one reason or another get scrapped along the way contribute to *wafer yield* losses. These losses can occur as a result of equipment malfunctions, wafer transport problems, or other difficulties. Clearly, identifying and removing problematic wafers as early as possible is an important objective, as it preserves processing resources. While factories implement early tests for that purpose, frequently wafers have to be rejected near the end, when they fail the various electrical or "probe" tests that are performed to confirm the overall electrical properties along the way, or the final electrical test done on the various devices and simple circuit structures. Further refining these definitions, production engineers distinguish the following three manufacturing yield components:

Wafer yield—the percentage of wafers that make it to final probing

Probe testing yield—the percentage of wafers that make it through the probe testing steps

Final testing yield—the percentage of wafers that make it through the final electrical testing step

Once a wafer has been successfully completed to the point that the product die can be electrically tested, then the figure of interest is the *design yield*, or *die yield*. There are two basic *die yield* components:

Functional yield (also known as "hard" or "catastrophic" yield)—the proportion of fully functional ICs

Parametric yield (also known as "soft" yield)—the overall performance achieved by the functional ICs

The one that can be determined first is the functional yield, which is usually limited by processing defects (such as particles), or artifacts that in general destroy the functionality of a circuit. These artifacts might cause short circuits, open circuits, or other types of "binary" failures. *Functional yield* is typically measured with high but finite precision, by running a series of functionality tests before individual ICs are diced from the wafer. These functionality tests are designed to balance the test coverage and the testing cost, and the overall objective is to avoid packaging, or worse, shipping nonfunctional ICs. Functional yield depends not only on process and material cleanliness but also on IC design practices. The issue of understanding, modeling, and improving functional yield will be discussed in some depth in this chapter.

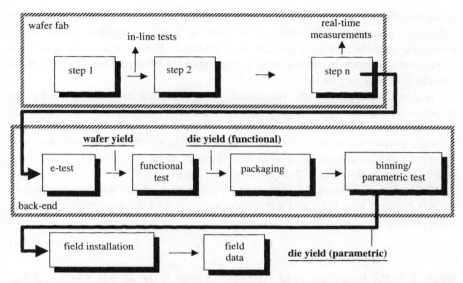

Figure 5.1. Manufacturing process flow from the perspective of yield monitoring and control.

It is usually after the individual ICs have been diced from the wafer that the latter yield component, the *parametric yield,* is determined. Performance might be quantified by various metrics, such as speed of execution or power consumption. The natural variability of the process, as well as the non-catastrophic impact of some types of defects, will lead to a statistical spread of the various device parameters, and this spread will in turn result in a spread of IC performances. During this last stage of testing, IC products are typically separated into various performance "bins" and parametric variation determines the percentage of ICs that end up into each bin. The issues that determine parametric spread relate to process control practices, as well as process and material variations. The impact of these variations on the parametric yield can be further controlled by the appropriate design practices.

All yield components are subject to intense scrutiny, by means of material and process studies and IC failure analysis. Once the *assignable causes* of yield loss have been eliminated, the emphasis shifts to understanding and quantifying the systematic causes, leading to a body of work focusing on yield modeling and simulation. The concept of *design for manufacturability* then comes into play, in an attempt to mitigate the impact of these causes by means of appropriate circuit and process design consideration. Figure 5.1 outlines the overall process flow, from the perspective of yield monitoring and control.

5.2. FUNCTIONAL YIELD MODELS

The development of models to estimate the functional yield of microelectronic circuits and packages is fundamental to manufacturing. A model that provides accurate estimates of manufacturing yield can help predict product cost, determine optimum equipment utilization, or be used as a metric against which actual measured manufacturing yields can be evaluated. Yield models are also used to support decisions involving new technologies and the identification of problematic products or processes.

As mentioned previously, functional yield is significantly impacted by the presence of *defects*. Defects can result from many random sources, including contamination from equipment, processes, or handling; mask imperfections; and airborne particles. Physically, these defects include shorts and opens (short and open circuits), misalignment, photoresist splatters and flakes, pinholes, scratches, and crystallographic flaws. This is illustrated by Figure 5.2.

Yield models are usually presented as a function of the average number of defects per unit area (D_0) and the *critical area* (A_c) of the electronic system. In other words,

$$Y = f(A_c, D_0) \tag{5.1}$$

where Y is the functional yield. The *critical area* is the area in which a defect occurring has a high probability of resulting in a fault. For example, if the particles in Figure 5.2 (which repeats Figure 2.8) are conductive, only particle 3 has fallen into an area in which it causes a short between the two metal lines that it bridges. The relationship between the yield, defect density, and critical area is complex. It depends on the circuit geometry, the density of photolithographic patterns, the number of photolithography steps used in the manufacturing process, and other factors. A few of the more prevalent models that attempt to quantify this relationship are described in the following sections.

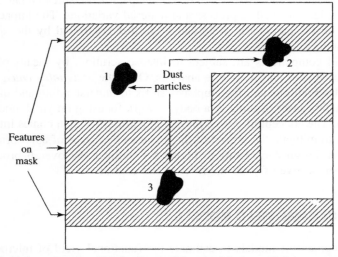

Figure 5.2. Various ways in which foreign particles can interfere with interconnect patterns.

5.2.1. Poisson Model

The Poisson yield model requires that defects be considered as perfect points that are spatially uncorrelated and uniformly distributed (with a defect density D_0) across a substrate. The Poisson model further requires that each defect result in a fault. J. Piñeda de Gyvez provides an excellent derivation of this model [2]. Let C be the number of circuits on a substrate (the number of ICs, modules, etc.), and let M be the number of possible defect types. Under these conditions, there are C^M unique ways in which the M defects can be distributed on the C circuits. For example, if there are three circuits (C1, C2, and C3) and three defect types (e.g., M1 = metal open, M2 = metal short, and M3 = metal 1–metal 2 short), then there are

$$C^M = 3^3 = 27 \tag{5.2}$$

possible ways in which these three defects can be distributed over three chips. These combinations are illustrated in Table 5.1.

If one circuit is removed (i.e., is found to contain no defects), the number of ways to distribute the M defects among the remaining circuits is

$$(C - 1)^M \tag{5.3}$$

Thus, the probability that a circuit will contain zero defects of any type is

$$\frac{(C - 1)^M}{C^M} = \left(1 - \frac{1}{C}\right)^M \tag{5.4}$$

Substituting $M = CA_cD_0$, the yield is the number of circuits with zero defects, or

$$Y = \lim_{C \to \infty} \left(1 - \frac{1}{C}\right)^{CA_cD} = \exp(-A_cD_0) \tag{5.5}$$

Table 5.1. Table of unique fault combinations.

	C1	C2	C3		C1	C2	C3
1	M1 M2 M3			15	M3		M2 M1
2		M1 M2 M3		16		M1 M2	M3
3			M1 M2 M3	17		M1 M3	M2
4	M1 M2	M3		18		M2 M3	M1
5	M1 M3	M2		19		M1	M2 M3
6	M2 M3	M1		20		M2	M1 M3
7	M1 M2		M3	21		M3	M2 M1
8	M1 M3		M2	22	M1	M2	M3
9	M2 M3		M1	23	M1	M3	M2
10	M1	M2 M3		24	M2	M1	M3
11	M2	M1 M3		25	M2	M3	M1
12	M3	M2 M1		26	M3	M1	M2
13	M1		M2 M3	27	M3	M2	M1
14	M2		M1 M3				

For N circuits to have zero defects, this becomes

$$Y = \exp(-A_c D_0)^N = \exp(-N A_c D_0) \tag{5.6}$$

The same result can be obtained using Poisson statistics directly. Poisson statistics represent an approximation of the Maxwell–Boltzmann (or *binomial*) distribution when large sample sizes are used. Recall the Poisson probability distribution given by Eq. (4.12). If x is the number of faults per circuit and $\lambda = N A_c D_0$ is the fault density, the yield is defined at $x = 0$, or

$$Y = P(x = 0) = \exp(-N A_c D_0) \tag{5.7}$$

We thus achieve an equivalent expression to that given by [Eq. (5.6).]

The Poisson model is simple and relatively easy to derive. It provides a reasonably good estimate of yield when the critical area is small. However, if D_0 is calculated based on small-area circuits, using the same D_0 for large-area yield computations results in a yield estimate that is overly pessimistic compared to actual measured data.

5.2.2. Murphy's Yield Integral

B. T. Murphy first proposed that the value of the defect density (D) should not be constant [1]. Instead, he reasoned that D must be summed over all circuits and substrates using a normalized probability density function $f(D)$. The yield can then be calculated using the integral

$$Y = \int_0^\infty e^{-A_c D} f(D) dD \tag{5.8}$$

Various forms of $f(D)$ form the basis for the differences between many analytical yield models. The Poisson model described in the previous section assumes that $f(D)$ is a delta function, that is

$$f(D) = \delta(D - D_0) \tag{5.9}$$

where D_0 is the average defect density as before (see Figure 5.3a). Using this density function, the yield is determined from Eq. (5.8) as

$$Y_{\text{Poisson}} = \int_0^\infty e^{-A_c D} f(D) dD = \exp(-A_c D_0) \tag{5.10}$$

as shown before.

Murphy initially investigated a uniform density function as shown in Figure 5.3b. The evaluation of the yield integral for the uniform density function gives

$$Y_{\text{uniform}} = \frac{1 - e^{-2D_0 A_c}}{2 D_0 A_c} \tag{5.11}$$

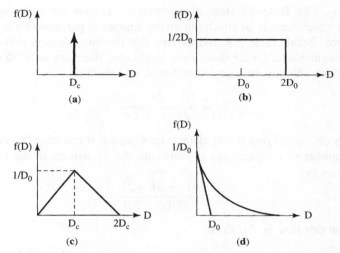

Figure 5.3. (a) Probability density function (pdf) for the Poisson model; (b) pdf for the uniform Murphy model; (c) pdf for the triangular Murphy model; (d) pdf for the exponential Seeds model [2].

Murphy later believed that a Gaussian distribution would be a better reflection of the true defect density distribution than the delta function. However, since evaluating the yield integral with a Gaussian function substituted for $f(D)$ would not have resulted in a closed-form solution, he approximated it using the triangular function in Figure 5.3c. This function results in the yield expression

$$Y_{\text{triangular}} = \left(\frac{1 - e^{-D_0 A_c}}{D_0 A_c} \right)^2 \tag{5.12}$$

The triangular Murphy yield model is widely used today in industry to determine the effect of manufacturing process defect density.

R. B. Seeds was the first to verify Murphy's predictions [3]. However, Seeds theorized that high yields were caused by a large population of low defect densities (which are not high enough to cause faults) and a small proportion of high defect densities (i.e., high enough to cause faults). He therefore proposed the exponential density function given by

$$f(D) = \frac{1}{D_0} \exp\left(\frac{-D}{D_0} \right) \tag{5.13}$$

and shown in Figure 5.3d. This function implies that the probability of observing a low defect density is significantly higher than that of observing a high defect density. Substituting this exponential function in the Murphy integral and integrating yields

$$Y_{\text{exponential}} = \frac{1}{1 + D_0 A_c} \tag{5.14}$$

It should be noted that the Seeds model may also be derived in an alternate manner using Bose–Einstein statistics. This was accomplished independently by Price [4]. The Bose–Einstein distribution is relevant for indistinguishable particles in which there is no constraint on the number of particles that can occupy a given state. Recall Table 5.1, and assume that the three defects (M1–M3) are now indistinguishable. Under these new conditions, there are only 10 combinations of defects that are uniquely identifiable, and there are

$$Z_1 = \frac{(C + M - 1)!}{M!(C - 1)!} \tag{5.15}$$

unique ways of identifying the M defects on C chips. If one chip has no defects, then the number of unique ways to distribute the M defects in the rest of the $(C - 1)$ chips is

$$Z_2 = \frac{(C + M - 2)!}{M!(C - 2)!} \tag{5.16}$$

The yield in this case is Z_2/Z_1, or

$$Y = \left[\frac{(C - 1)!}{(C - 2)!}\right]\left[\frac{(C + M - 2)!}{(C + M - 1)!}\right] = \frac{C - 1}{C + M - 1} = \frac{\left(1 - \dfrac{1}{C}\right)}{\left(1 + \dfrac{M}{C} + \dfrac{1}{C}\right)} \tag{5.17}$$

If we now substitute $M = CA_cD_0$, taking the limit as C tends to infinity gives

$$Y = \lim_{C \to \infty} \frac{\left(1 + \dfrac{1}{C}\right)}{\left(1 + \dfrac{M}{C} + \dfrac{1}{C}\right)} = \frac{1}{1 + A_cD_0} \tag{5.18}$$

which is the same as the model given in Eq. (5.14).

Although the Seeds model is simple, its yield predictions for large-area chips are too optimistic. This is because the assumption of indistinguishable defects is seldom valid for IC fabrication processes, where defects are often visually distinguishable from one another. Therefore, this model has not been widely used in industry.

5.2.3. Negative Binomial Model

Okabe et al. recognized the physical nature of defect distributions and proposed the gamma probability density function [5]. C. H. Stapper has likewise written several papers on the development and applications of yield models using the gamma density function [6]. The gamma probability density function is given by

$$f(D) = [\Gamma(\alpha)\beta^\alpha]^{-1}D^{\alpha-1}e^{-D/\beta} \tag{5.19}$$

where α and β are two parameters of the distribution and $\Gamma(\alpha)$ is the gamma function. The shape of $\Gamma(\alpha)$ is shown for several values of α in Figure 5.4. In

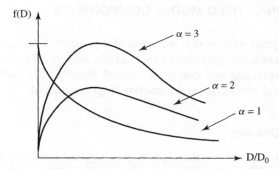

Figure 5.4. Probability density function for the gamma distribution.

this distribution, the average defect density is $D_0 = \alpha\beta$, and the variance of D_0 is $\alpha\beta^2$. The yield model derived by substituting Eq. (5.19) into Murphy's integral is

$$Y_{\text{gamma}} = \left(1 + \frac{A_c D_0}{\alpha}\right)^{-\alpha} \qquad (5.20)$$

This model is commonly referred to as the *negative binomial* model. The parameter α is generally called the "cluster" parameter since it increases with decreasing variance in the distribution of defects.

If α is high, that means that the variability of defect density is low (little clustering). Under these conditions, the gamma density function approaches a delta function, and then the negative binomial model reduces to the Poisson model. Mathematically, this means

$$Y = \lim_{\alpha \to \infty} \left(1 + \frac{A_c D_0}{\alpha}\right)^{-\alpha} = \exp(-A_c D_0) \qquad (5.21)$$

If α is low, on the other hand, the variability of defect density across the substrate is significant (much clustering), and the gamma model reduces to the Seeds exponential model, or

$$Y = \lim_{\alpha \to 0} \left(1 + \frac{A_c D_0}{\alpha}\right)^{-\alpha} = \frac{1}{1 + A_c D_0} \qquad (5.22)$$

The parameter α must be determined empirically. Methods for doing so that involve particle counting using laser reflectometry exist [7], but several authors have found that values of $\alpha = 2$ provide a good approximation for a variety of logic and memory circuits [8]. Therefore, if the critical area and defect density are known (or can be accurately measured), the negative binomial model is an excellent general-purpose yield predictor that can be used for a variety of electronics manufacturing processes.

5.3. FUNCTIONAL YIELD MODEL COMPONENTS

The functional yield models that we have discussed thus far have been defined in terms of independent parameters such as D_0 and A_c. These parameters are statistically independent and can be measured directly. The following sections describe these and other critical parameters in greater detail.

5.3.1. Defect Density

Defect density is clearly a critical parameter in yield modeling and production yield planning. A "defect" is an unintended pattern on the wafer surface. It can consist of either extra material or missing material. In order to properly describe defect density, a few other terms must be defined:

Contamination—any foreign material on a wafer surface or embedded in a thin film. Sources of contamination include human skin, dirt, dust, or particles resulting from an oxidized gas, residual chemicals, or sputtering.

Defect—any alteration in the desired physical pattern intended to be printed. Typical defects include metal stringers, open and short circuits, notches, splotches, bridges, or hillocks.

Fault—an electrical circuit failure caused by a defect.

On the basis of these definitions, we observe that contamination is a random physical event that may or may not lead to a defect. Similarly, a defect may or may not result in a fault. The correlation between contamination, defects, and faults is weak. Mapping contamination to defects or defects to faults is difficult and time-consuming.

A physical interpretation of defect density should incorporate the size distribution of defects, as well as the probability that a defect will cause a failure. Typically, defects smaller than the minimum feature size will not cause failures. However, if a defect of a particular size causes a fault, then a larger defect at the same location will also cause a fault. An example of the effect of defects of different sizes at the same location is shown in Figure 5.5. The two adjacent metal lines in this figure will be shorted (short-circuited) by a defect of greater size than the spacing between them.

In general, the defect density is defined mathematically as the area under the defect size distribution curve for specific size limits. For a mature manufacturing process, defect density has been shown experimentally to follow an inverse power-law relationship with respect to size [9]. In other words

$$D(x) = \frac{N}{x^p} \qquad (5.23)$$

where x is the defect diameter (assuming spherical defects), N is a technology-dependent parameter, and p must be determined empirically. This power law also

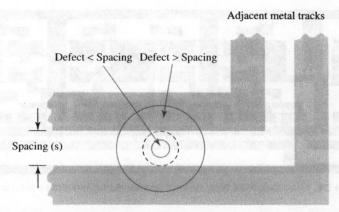

Figure 5.5. Illustration of the effect of defect size distribution on critical area [2].

assumes that defects are located randomly across the wafer surface. The average defect density may thus be determined by integrating this expression, or

$$D_0 = N \int_{x_0}^{\infty} D(x)dx = N \int_{x_0}^{\infty} \frac{1}{x^p} dx = \frac{N}{1-p}(1 - x_0^{1-p}) \qquad (5.24)$$

where x_0 is the minimum defect diameter, which is usually the minimum feature size for a given technology. Neither the defect density nor the critical area can be determined without the defect size distribution.

A simpler method of extracting the defect density involves using a particular yield model to solve for D_0 mathematically. For example, using the negative binomial model for a single chip gives

$$D_0 = \frac{\alpha(\sqrt[\alpha]{1/Y} - 1)}{A_c} \qquad (5.25)$$

This approach works best for similar products fabricated using the same mature technology with chip areas within a factor of 2–3, but should be applied cautiously otherwise. The most useful aspect of this approach is in using the calculated value of D_0 as a metric of manufacturing process performance.

5.3.2. Critical Area

The concept of critical area is used to account for the fact that not all parts of a chip layout are equally likely to fail because of the presence of defects. This allows greater accuracy when calculating the defect sensitivity of a chip layout. Consider Figure 5.6, in which the dark areas represent the first metal layer for a given circuit. The crosshatched area represents the sensitive regions at the minimum spacing for this technology. The critical area is a measure of such sensitive regions for the entire chip.

Figure 5.6. Subcircuit metal layer in which shaded region indicates critical area [2].

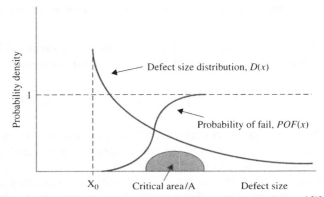

Figure 5.7. Graphical representation of the critical area integral [2].

Critical area is defined mathematically by the following relationship:

$$A_c = A \int_{x_0}^{\infty} \text{PoF}(x)D(x)dx \qquad (5.26)$$

where A is the chip area, x_0 is the minimum defect size, $D(x)$ is the defect size distribution, and PoF(x) is the probability of failure, which is a strong function of the defect size. A graphical interpretation of this relationship is shown in Figure 5.7. Several methods have been reported to determine critical area, and in each case, the calculation of PoF is crucial. More detail on these techniques is provided in Section 5.4.

5.3.3. Global Yield Loss

The yield models discussed thus far have focused solely on yield loss due to the presence of *local* defects. However, *global defects* can also be present. Global yield loss is usually spatially correlated and often manifests itself as a consequence of variability in electrical parameters (such as transistor threshold voltage) caused by process fluctuations (such as temperature or film thickness

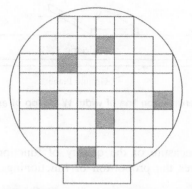

Figure 5.8. Example of a wafer map [2].

variations). Global yield loss, which is quantified *by parametric yield* models (see Section 5.4), is often identifiable as an anomalous spatial pattern on a wafer, such as an annual ring or cluster of failing chips. Yield models can take global defects into account by incorporating a factor Y_0. For example, the negative binomial model that includes global yield loss effects is

$$Y = Y_0 \left(1 + \frac{A_c D_0}{\alpha} \right)^{-\alpha} \tag{5.27}$$

The Y_0 factor is not related to the defect density or the critical area.

Although it is difficult to determine global yield loss analytically, spatial analysis techniques can be used to evaluate whether the measured yield loss is consistent with this model. Spatial analysis typically requires wafer maps generated from automated test equipment. In these maps, failing chips are categorized by the similarity of failures (e.g., function fail, speed fail). An example of a typical wafer map appears in Figure 5.8.

One way to estimate the value of Y_0 is to use a "windowing" technique in which individual chips are grouped together into windows of increasing size. The effective yield of each window size is then plotted against the effective chip size. The y intercept of this plot is the yield with an area of zero. Thus, Y_0 is equal to one minus this intercept. If the intercept is at 100% yield, there is no global yield loss.

5.4. PARAMETRIC YIELD

Even in a defect-free manufacturing environment, random processing variations can lead to varying levels of system performance. These variations result from global defects that cause the fluctuation of numerous physical and environmental parameters (linewidths, film thicknesses, ambient humidity, etc.), which in turn manifest themselves as variations in final system performance (such as speed or noise level). These performance variations lead to "soft" faults and are characterized by the parametric yield of the manufacturing process.

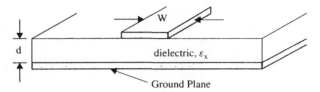

Figure 5.9. A microstrip transmission line of width W on top of an insulating dielectric with thickness d.

Parametric yield is a measure of the quality of functioning systems, whereas *functional yield* measures the proportion of functioning units produced by the manufacturing process.

A common method used to evaluate parametric yield is *Monte Carlo simulation*. In the Monte Carlo approach, a large number of pseudorandom sets of values for circuit or system parameters are generated according to an assumed probability distribution (usually the normal distribution) based on sample means and standard deviations extracted from measured data. For each set of parameters, a simulation is performed to obtain information about the predicted behavior of a circuit or system, and the overall performance distribution is then extracted from the set of simulation results.

To illustrate the Monte Carlo technique, consider as a performance metric the characteristic impedance (Z_0) of a microstrip transmission line of width W on top of an insulating dielectric with thickness d (refer to Figure 5.9). Under the condition that $W/d \ll 1$, it can be shown that

$$Z_0 = \frac{60}{\sqrt{\varepsilon_e}} \ln \left(\frac{8d}{W} + \frac{W}{4d} \right) \tag{5.28}$$

where ε_e is the effective dielectric constant of the insulator, which is given by

$$\varepsilon_e = \frac{\varepsilon_r + 1}{2} + \frac{\varepsilon_r - 1}{2\sqrt{1 + \dfrac{12d}{W}}} \tag{5.29}$$

where ε_r is the relative permittivity of the insulator [10]. From these equations, it is clear that Z_0 is a function of the physical dimensions, d and W, or $Z_0 = f(d, W)$. Both of these dimensions are subject to manufacturing process variations. They can thus be characterized as varying according to normal distributions with means μ_d and μ_W and standard deviations σ_d and σ_W, respectively (see Figure 5.10a).

Using the Monte Carlo approach, we can estimate the parametric yield of microstrips produced by a given manufacturing process within a certain range of characteristic impedances by computing the value of Z_0 for every possible combination of d and W. The result of these computations is a final performance distribution such as the one shown in Figure 5.10b. This probability density function can then be used to compute the proportion of microstrips having a given

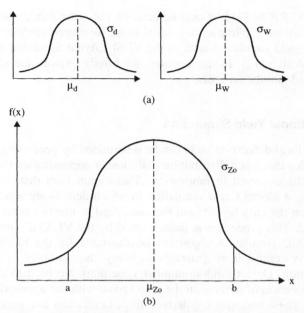

Figure 5.10. (a) Normal probability density functions for *W* and *d*; (b) overall pdf for characteristic impedance.

range of impedances. For example, if we wanted to compute the percentage of microstrips manufactured that would have the a value of Z_0 between two limits *a* and *b*, we would evaluate the integral

$$\text{Yield(microstrips with } a < Z_0 < b) = \int_a^b f(x)dx \qquad (5.30)$$

Thus, once the overall distribution of a given output metric is known, it is possible to estimate the fraction of manufactured parts with any range of performance. Estimation of parametric yield is useful for system designers since it helps identify the limits of the manufacturing process to facilitate and encourage *design for manufacturability*.

5.5. YIELD SIMULATION

It is highly desirable for IC manufacturers to be able to predict yield loss prior to circuit fabrication. This enables corrective action to be taken before production starts and can prevent misprocessing. Yield simulation software tools are the primary means for facilitating yield prediction.

Local and global defects are the two basic sources of yield loss. The effects of global defects, which result in parametric yield loss, have been modeled in statistical process simulators such as the FABRication of Integrated Circuits

Simulator (FABRICS) [11].[1] Local defects, on the other hand, which can cause catastrophic failures that impact functional yield, have been modeled using Monte Carlo–based yield simulators such as the VLSI LAyout Simulator for Integrated Circuits (VLASIC) [12]. In this section, we briefly explore the capabilities of these two yield simulation tools.

5.5.1. Functional Yield Simulation

The effect of local defects on yield can be determined by generating a population of chip samples that has a distribution that closely approximates the distribution of circuit faults observed in fabrication. This circuit fault distribution may be obtained using a Monte Carlo simulation in which defects are repeatedly generated, placed on the chip layout, and then analyzed to identify what circuit faults have occurred. This procedure is implemented by the VLASIC simulator.

The VLASIC simulation algorithm is illustrated by the block diagram in Figure 5.11. A control loop generates as many chip samples as desired for a given simulation. Defect random-number generators are used to determine the number and location of defects on the chip layout with the appropriate statistical distributions. These statistics are derived from fabrication line measurements.

Once the defects have been placed on the layout, fault analysis is used to determine what, if any, circuit faults have occurred. The resulting faults are then filtered so that those faults that do not affect functional yield are ignored. The output is a chip sample containing the list of faults that have occurred during simulated fabrication. When simulation is complete, the list of unique chip faults

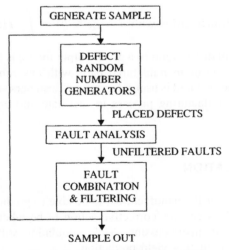

Figure 5.11. VLASIC main loop [12].

[1]The FABRICS parametric yield simulator was developed by Maly and Strojwas of Carnegie Mellon University in 1982.

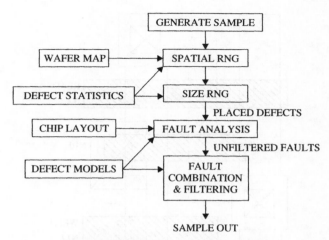

Figure 5.12. VLASIC system structure [12].

and their frequency of occurrence is passed to postprocessors designed to predict yield, optimize design rules, generate test vectors, or evaluate process sensitivity.

A detailed view of the VLASIC system structure is shown in Figure 5.12. Defects in VLASIC have both size and spatial distributions. A wafer map is used to place defects on a wafer. The output of the defect random-number generators is a list of defect types, locations within the chips, and defect diameters. The defect statistics are similar to those described in Section 5.3.

For each defect, the fault analysis phase calls a series of procedures to examine the layout geometry in the immediate vicinity of the defects to determine whether any circuit faults have occurred. A separate procedure is used for each fault type (shorts, opens, etc.). Since the defects and layout features are represented as polygons, the analysis procedures manipulate layout geometry using general-purpose polygon operations. The resulting output of fault analysis is a list of unfiltered circuit faults caused by the defects. The unfiltered faults then pass through a filtering–and combination phase in which faults that do not cause a change in DC circuit operation are ignored, and some faults are combined together to form a composite fault, respectively.

Fault analysis and filtering operations both depend on input from defect models. These models describe a fabrication process as a number of patterned layers in which defects are represented as modifications to the layout of each layer (i.e., extra or missing material). The models also specify the circuit faults that can be caused by each defect type, which layers are affected by the defect, and the manner in which layers are electrically connected. After filtering, the resulting output is a list of circuit faults that have occurred during simulated fabrication. Each fault is specified by its type, size, location, type of defect that caused it, location of the fault in the circuit graph, and number of times the fault occurred. The fault list is then ready for postprocessing.

To illustrate the use of VLASIC, consider the simulation of a simplified chip containing only a single three-transistor dynamic RAM cell (see Figure 5.13).

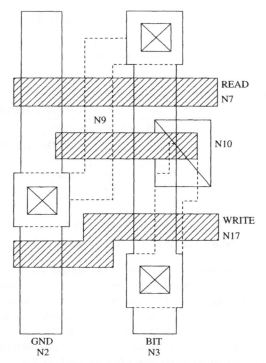

Figure 5.13. Three-transistor DRAM cell [12].

Suppose that the chip is placed at 100 locations on the wafer as shown in Figure 5.14. The process conditions used in the simulation are given in Table 5.2. Note that relatively high defect densities are used in order to obtain an average of 2.5 defects per sample. The α parameter for the negative binomial model is also high, indicating very little spatial clustering of defects.

The result of simulating 1000 chip samples is shown in Figure 5.15. This output represents a list of unique chip fabrication outcomes. Each unique outcome has a frequency count and a list of fault groups (i.e., sets of faults caused by a single defect). For each fault, the fault type, defect type that caused it, defect location, defect diameter, and fault description are provided. For the single instance where several defects of the same type have caused the same fault to occur (i.e., several oxide pinholes shorting the same nets together), defect size and defect location are meaningless, since only the values for the first defect causing the fault are recorded.

The simulated fabrication of the 1000 samples results in 25 unique chip faults, the distribution of which appears in Table 5.3. Despite an average of 2.5 defects per sample, only 5.9% of the simulated chips had a circuit fault. This result is typical of yield simulations. To explore reasons for this, note that only 3.3% of the DRAM cell area contains a gate oxide. Thus, a gate oxide pinhole defect has only a 1 in 30 chance of causing a gate-to-channel short (circuit). Consider also the case of extra metal defects, recalling that extra material defects must have a

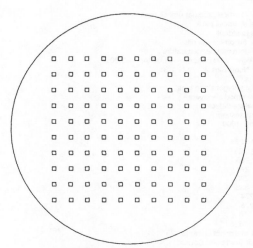

Figure 5.14. Wafer map.

Table 5.2. DRAM cell process conditions.

Defect density	
Extra/missing metal	20,000 cm^{-2}
Extra/missing polysilicon	20,000 cm^{-2}
Extra/missing active	20,000 cm^{-2}
First-level pinholes	20,000 cm^{-2}
Gate oxide pinholes	20,000 cm^{-2}
Design rules	
Metal width/space	6 μm
Metal contact width	4 μm
Polysilicon width/space	4 μm
Active width/space	4 μm
Polysilicon/active space	2 μm
Diameter of peak density	
Extra/missing metal	2 μm
Extra/missing polysilicon	2 μm
Extra/missing active	2 μm
Maximum defect diameter	18 μm
Between-lot alpha	100
Between-wafer alpha	100
Wafers per lot	1
Radial distribution	None
Minimum line spacing	0
Minimum linewidth	0

minimum diameter to cause a circuit fault. In this simulation, the combination of the peak defect size of 2 μm (using a defect size distribution similar to that shown in Figure 5.6) and minimum metal width/space of 6 μm leads to only a 1 in 18 chance for a defect of this type causing a short.

```
SAM> vlasic −t omcellproc.dat omcell.pack
Reading wirelist omcell.pack
Writing faults to stdout
Parsing wafer file omcellwaf.cif
Initializing random number generators
left: −27.00 right: 3.00 bottom: −51.00 top: −3.00
NumXBins: 1 NumYBins 2
Allocating bins
Putting wirelist polygons into bins
Generating intermediate layers
Place and analyze defects
Results of fault testing:
Sample Count: 1000
Trial Count:
type POSM1: 234
type NEGM1: 237
type POSP: 283
type NEGP: 271
type POSD: 298
type NEGD: 335
type PIN1: 274
type PIN2: 276
type PING: 318
Total Trials:  2526
Total number of distinct chiplists:  25
Distinct Circuit Faults and Counts:

941 NIL

9 SHORT PIN1 X −10 Y −37 Diam 0 N3 N17

6 SHORT PIN1 X −25 Y −15 Diam 0 N2 N7

5 NEWVIA PING X −15 Y −16 Diam 0 NCHAN N7

5 SHORT PIN1 X −9 Y −17 Diam 0 N3 N7

4 SHORT PIN1 X −5 Y −21 Diam 0 N3 N10

4 NEWVIA PING X −13 Y −25 Diam 0 NCHAN N10

3 SHORT PIN1 X −20 Y −25 Diam 0 N2 N10

3 NEWVIA PING X −3 Y −37 Diam 0 NCHAN N17

3 SHORT PIN1 X −20 Y −37 Diam 0 N2 N17

2 SHORT POSP X 1 Y −27 Diam 16.09 N10 N17

2 SHORT POSP X −4 Y −12 Diam 7.43 N3 N7

1 SHORT POSM1 X −16 Y −14 Diam 17.19 N2 N3

1 SHORT POSP X −10 Y −43 Diam 9.04 N3 N17

1 SHORT PIN1 X −20 Y −37 Diam 0 N2 N17
  SHORT PIN1 X −25 Y −15 Diam 0 N2 N7

1 SHORT PIN1 X −10 Y −37 Diam 0 N3 N17
  SHORT PIN1 X −5 Y −21 Diam 0 N3 N10

1 NEWVIA PING X −15 Y −16 Diam 0 NCHAN N7
  SHORT PIN1 X −5 Y −21 Diam 0 N3 N10
  SHORT PIN1 X −20 Y −25 Diam 0 N2 N10

1 OPEN NEGP X −21 Y −41 Diam 12.25 N17/1 LEFT NP N17/2 Tran DO G

1 NEWVIA PING X −3 Y −37 Diam 0 NCHAN N17
  OPEN NEGM1 X −7 Y −43 Diam 7.58 N3/1 Tran DO SD N3/2 BOTTOM NM1 TOP NM1

1 OPEN NEGP X −7 Y −23 Diam 4.89 N10/1 Tran DO SD N10/2 Tran D1 G

1 OPEND NEGD X −14 Y −21 Diam 16.04 Tran D2
    OPEND NEGD X −14 Y −21 Diam 16.04 Tran D1

1 NEWGD POSP X −10 Y −19 Diam 8.43 CVTMULTI Tran D1 D2 SD: N2/0 N3/0 N9/0 G: N7/0 N10/0
    SHORT POSP X −10 Y −19 Diam 8.43 N7 N10

1 OPEND NEGD X −5 Y −35 Diam 4.91 Tran DO

1 SHORTD NEGP X −17 Y −15 Diam 8.10 Tran D2
    OPEN NEGP X −17 Y −15 Diam 8.10 N7/1 Tran D2 G N7/2 LEFT NP

1 OPEN POSP X −24 Y −28 Diam 18.00 N2/1 Tran D1 SD LEFT ND N2/2 LEFT NM1 BOTTOM NM1
    SHORT POSP X −24 Y −28 Diam 18.00 N2 N10 N17
```

Figure 5.15. VLASIC DRAM example [12].

Another noteworthy aspect of this simulation that is typical of all yield simulations is that chips with a single simple fault are much more common than those with multiple fault groups. This is directly attributable to the fact that single defects are more common than multiple defects. Complex multiple fault groups are rare because large extra or missing material defects must be present to cause

Table 5.3. DRAM chip sample distribution.

94.1%	No faults
4.2%	One oxide pinhole short
0.6%	One extra material short
0.2%	Two oxide pinhole shorts
0.2%	One missing material open
0.1%	Three oxide pinhole shorts
0.1%	Two-open device
0.1%	One-open device
0.1%	One oxide pinhole short and one missing material open
0.1%	One new gate device and one extra material short
0.1%	One shorted device and one missing material open
0.1%	One extra material open and one extra material short

them. Of the four fault groups that occurred in this example, the smallest defect causing one was 8.1 μm in diameter. Only about 1 in 33 defects is this large.

5.5.2. Parametric Yield Simulation

The FABRICS parametric yield simulator embodies an approach to modeling the IC fabrication process that accounts for the statistical fluctuations that occur during manufacturing. This simulator is capable of generating values for the parameters of IC circuit elements (resistances, capacitances, transconductances, etc.), as well as estimates of inline measurements typically made during fabrication (junction depths, sheet resistances, oxide thicknesses, etc.). These quantities are described statistically as random variables characterized by a joint probability density function.

FABRICS accounts for the dependence of IC elements on both layout and process parameters. Each process step is modeled individually, with its outcome dependent on a set of control parameters, a set of process disturbances, and the outcome of the previous process step (see Figure 5.16). To formally describe

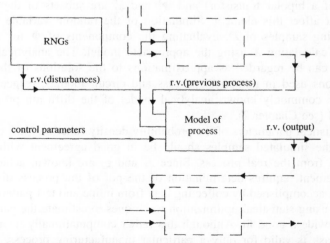

Figure 5.16. Model of a single process step (rv = "random varaible") [11].

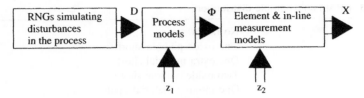

Figure 5.17. Basic FABRICS structure [11].

the FABRICS simulation procedure, let X be a vector of random variables that denotes the parameters of the IC elements or values of inline measurements. In addition, let the vector z_1 represent the process control parameters (temperatures, times, gas flows, etc.), and let the vector z_2 represent the layout dimensions. Finally, let D be a vector of random variables representing uncontrollable process disturbances. These disturbances are simulated in FABRICS as appropriately defined random-number generators (RNGs).

A basic flowchart describing the structure of the *FABRICS* simulator is shown in Figure 5.17. *FABRICS* uses analytical models for each manufacturing process step and circuit element of the form:

$$\mathbf{\Phi}_j = g_j(\hat{\mathbf{\Phi}}^j, \mathbf{z}_1^j, \mathbf{D}^j) \qquad j = 1, \ldots, m \qquad (5.31)$$

where $\mathbf{\Phi}_j$ is a component of the m-dimensional vector $\mathbf{\Phi}$ of physical parameters which describes the outcome of a given step (i.e., oxide thicknesses, doping profile parameters, misalignment, etc.), $\hat{\mathbf{\Phi}}^j$ is a vector of physical parameters obtained from previous steps, and z_1^j and $\mathbf{D}^j$ are vectors containing those components of z_1 and D affecting the jth physical parameter. Models of the IC circuit elements are of the form

$$\mathbf{X}_i = h_i(\hat{\mathbf{\Phi}}^i, \mathbf{z}_2^i) \qquad i = 1, \ldots, n \qquad (5.32)$$

where X_i is an electrical parameter associated with a given circuit element (such as the β of a bipolar transistor) and $\hat{\mathbf{\Phi}}^i$ and z_2^j are subsets of the vectors $\mathbf{\Phi}$ and z_2 that affect this element. Simulation of the random variable X consists of generating samples of D, evaluating the components of $\mathbf{\Phi}$ for subsequent steps, and calculating X using the appropriate model. The analytical functions g_j and h_i can be regarded as approximations to the solutions of the differential equations used in numerical process and circuit models, respectively. For example, a commonly known analytical model of the diffusion process is the *erfc* model (see Chapter 2).

The statistical parameters of the probability density function (pdf) of X resulting from the simulated samples should be in good agreement with measured parameters from the real process. Since z_1 and z_2 are known, achieving such good agreement requires determination of the pdf of the process disturbances, f_D. This is accomplished by collecting data from inline and test pattern measurements and using statistical optimization techniques to estimate the parameters of f_D that provide a good fit. Although this rather computationally intensive identification task is valid for only a particular manufacturing process, the results can be used to simulate the manufacture for a variety of ICs, irrespective of the IC layout. Therefore, once f_D is known, the simulator may be used instead of the actual fabrication process to optimize layout or fine-tune process control parameters.

The random variable of process disturbances affecting one chip, D_i, is simulated with a RNG that generates data with a mean equal to μD_i and a standard deviation of σD_i. Assuming that μD_i and σD_i change randomly from one chip to another, each disturbance must be simulated by a two-level RNG that accounts for local (within-chip) and global (chip-to-chip) variations. This approach is illustrated in Figure 5.18, which shows a two-level structure consisting of three RNGs. RNG1 simulates a disturbance within a chip by generating a normally distributed random variable D_i. RNG2 and RNG3 provide RNG1 with μD_i and σD_i for the chip, respectively. The inputs to RNG2 are the mean of means (μ_μ) and standard deviation of means (σ_μ) of the chips in the wafer. Similarly, inputs to RNG3 are the mean of standard deviations (μ_σ) and standard deviation of standard deviations (σ_σ) of the chips in the wafer.

A more detailed data flow diagram for FABRICS is shown in Figure 5.19. Data entered into the simulator include process parameters, IC layout dimensions, and control parameters used to activate the RNGs and models in the correct sequence.

To illustrate the operation of FABRICS, consider the production of the MC1530 operational amplifier (shown schematically in Figure 5.20) as a typical bipolar manufacturing process. Suppose that we want to ascertain the effect modifying the surface concentration of phosphorus during the predeposition of the emitter layer of the transistors in this circuit. Since the modification of the surface concentration will result in a change in the sheet resistance of the emitter layer (R_{SE}), we will use FABRICS to examine the relationship between the parametric yield and R_{SE}.

Using FABRICS Monte Carlo simulation in conjunction with a circuit simulator [such as SPICE (a simulation program with integrated circuit emphasis)], the yield of the amplifier for six different phosphorus surface concentrations is

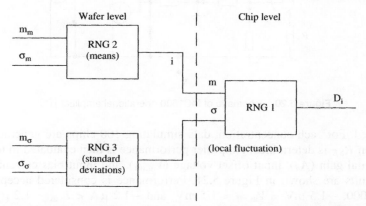

Figure 5.18. Illustration of two-level RNG architecture that simulates within-chip and chip-to-chip process variations [11].

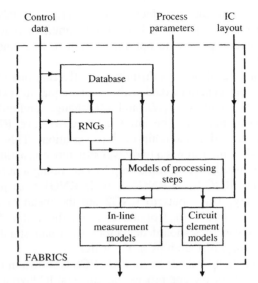

Figure 5.19. Detailed FABRICS data flow [11].

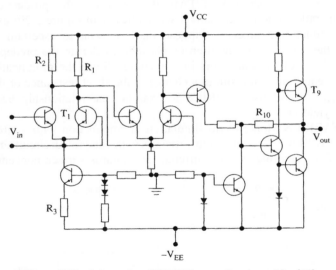

Figure 5.20. Schematic of MC1530 operational amplifier [11].

computed. For each concentration, data simulating 100 chips are generated, and the mean R_{SE} is determined. Amplifier performance is then evaluated in terms of differential gain (A_d), input offset voltage ($V_{in,off}$), and input bias current (I_{bias}). The results are shown in Figure 5.21. Performance is considered acceptable if $A_d > 8000$, -1.5 mV $< V_{in,off} < 1.5$ mV, and -1.2 μA $< I_{bias} < 1.2$ μA. The best yield is obtained when R_{SE} is near 4 Ω/square, and cannot be increased very much without a significant drop in yield.

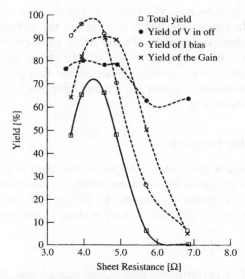

Figure 5.21. Yield of MC1530 operational amplifier versus emitter sheet resistance [11].

FABRICS is a powerful tool for computing parametric yield that can easily be tuned to the random variations of a real manufacturing process. It can be used for bipolar, MOS, or any other process technology, so long as appropriate data and analytical physical models are available.

5.6. DESIGN CENTERING

Yield simulation tools provide a mechanism for yield optimization and quality enhancement through accounting for manufacturing variations in IC components during the design phase. The objective of such efforts is to minimize circuit performance sensitivity with respect to potential component and parameter fluctuations. The objective of *design centering* is to maximize yield by identifying an optimal set of design parameters (x_{opt}) such that yield is optimized.

This concept is illustrated graphically for a simple two-parameter design space in Figure 5.22. In this figure, the region labeled *A* represents the region of acceptable circuit performance. The oval centered at the coordinates of the design

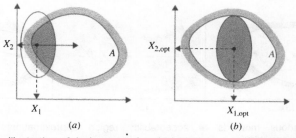

Figure 5.22. Illustration of design centering: (a) initial low yield; (b) optimized yield [2].

parameters x_1 and x_2 represents the area over which these two parameters may vary during manufacturing. The yield is therefore the shaded area represented by the overlap of these two regions. Figure 5.22a corresponds to a situation in which a poor choice of x_1 and x_2 results in low initial yield. In contrast, by centering the design with an optimal choice in the two parameters ($x_{1,opt}$ and $x_{2,opt}$), yield is maximized. Another term for the process of design centering is *design for manufacturability*.

5.6.1. Acceptability Regions

The *acceptability region* is defined as the part of the space of performance parameters in which all constraints imposed on circuit performance are fulfilled. For the two-dimensional example represented by Figure 5.22, the area A represents the acceptability region. In general, A is an m-dimensional hypersurface defined by the inequality

$$S_j^L \leq y_j \leq S_j^U \qquad j = 1, \dots, m \tag{5.33}$$

where y_j is one of m performance parameters and S_j^L and S_j^U are the (usually designer-defined) lower and upper bounds imposed on these parameters.

Since complicated relationships between performance parameters and bounds can be defined, acceptability regions can also be very complicated, including nonconvex regions or internal unacceptabability regions ("holes"). In order to determine whether a given point in the circuit parameter space belongs to A or its complement, an indicator function $I(x)$ is used, where

$$I(x) = \begin{cases} 1 & x \in A \\ 0 & x \notin A \end{cases} \tag{5.34}$$

The points for which $I(x) = 1$ are called successful, or "pass" points, and those for which $I(x) = 0$ are called "fail" points.

Except for some simple cases, the shape of the acceptability region in the performance space is unknown and nearly impossible to define completely. However, for yield optimization purposes, either implicit or explicit knowledge of A and its boundaries is required. Therefore, it is necessary to approximate the shape of A. Several methods are available to do so [2]. A few of these are illustrated in Figure 5.23.

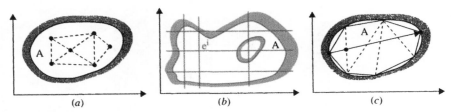

(a)   *(b)*   *(c)*

Figure 5.23. Various methods of acceptability region approximation: (a) point-based; (b) ODOS; (c) simplicial [2].

For example, in the *point-based* approximation shown in Figure 5.23a, subsequent approximations to A are generated using Monte Carlo simulation [13]. After each new point is generated, it is determined whether it belongs to A_{i-1} (i.e., the latest approximation to A). If it does, the sampled point is considered successful. If it does not belong to A_{i-1} and the next circuit simulation reveals that it belongs to A, the polyhedron is expanded to include the new point. In the method of approximation called *one-dimensional orthogonal search* (ODOS) shown in Figure 5.23b, line segments passing through the points e_i are randomly sampled in the performance space parallel to the coordinate axes and used for the approximation of A [14]. ODOS is very efficient for large linear circuits, since the intersections with A can be *directly* found from analytical formulas. The *simplicial approximation* is based on approximating the boundary of A in the performance space by a polyhedron [15]. The boundary of A is assumed to be convex (see Figure 5.23c). The simplicial approximation is obtained by locating points on the boundary of A by a systematic expansion of the polyhedron. The search for the next vertex is always performed in the direction passing through the center of the largest face of the polyhedron already existing and perpendicular to that face.

5.6.2. Parametric Yield Optimization

After estimation of the acceptability region, yield optimization is the objective of design centering techniques. In so doing, design centering attempts to inscribe the largest hypersphere of input parameter variation (also called the *norm body*; see Figure 5.22) into the approximation of the acceptability region A. The center of the largest norm body is taken as the optimal vector of input parameters x. Consider, for example, a simplicial approximation to the acceptability region. Under these conditions, several yield optimization schemes have been proposed. The most typical is as follows.

After a nominal point x belonging to the acceptability region is identified, line searches via circuit simulation are performed from that point to obtain some points located on the boundary of A. Several simulations may be required to find one boundary point. To form a polyhedron in an n-dimensional space, at least $(n + 1)$ boundary points need to be found. Once the first polyhedron approximation to A is obtained, the largest possible norm body is inscribed into it (using linear programming techniques), and its center is assumed as the first approximation to the center of A. The next steps involve improvements to the current approximation, $\tilde{A}$, by expanding the simplex. The center of the largest polyhedron face is found, and a line search is performed from the center along the line passing through it in a direction orthogonal to the face considered, to obtain another vertex point on the boundary of A. The polyhedron is then inflated to include the new point generated. This process is repeated until no further improvement is obtained.

Intuitively, this method should improve yield but, because of the approximation used, will not necessarily maximize it. For example, the simplicial approximation will not be accurate if A is nonconvex, and it will fail if A is not simply connected, since some parts of the approximation A will be outside the actual acceptability region. Notably, the computational cost of obtaining the simplicial approximation quickly increases for high-dimensional performance spaces. Thus, this method is most suitable for problems with a small number of designable parameters.

5.7. PROCESS INTRODUCTION AND TIME-TO-YIELD

While the final or "steady state" yield is important, one does not achieve the final yield instantaneously. Indeed, extensive field studies show that new processes arrive in the manufacturing line with limited initial yield [16]. As shown graphically in Figure 5.24, the initial process introduction period is followed by a period of intense learning where the various key yield detractors are identified and removed. The length of this "rapid learning" period is of paramount importance, as it often limits the amount of time it takes to bring a new product to market. Time-to-market is critical, if one is to capture significant market share and avoid the rapid price erosion that follows the introduction of high-end IC products. The final period, in which yield levels off and approaches a maximum, is one characterized by small gains due to the removal of the last few yield detractors. During this period, the investment of further effort and expense for marginal returns is questionable.

The discussion above underscores the importance of a "dynamic" study of yield. The objective of studying the metric known as *time-to-yield* is to identify key methods, tools, and actions that can accelerate the initial learning period following the introduction of a new process. As one might suspect, many factors affect time-to-yield, and some do so in ways that are not easily quantifiable. For example, it has been shown that time-to-yield can be accelerated by simply accelerating the processing cycle, as this allows for more *'work-in-progress'* (WIP) turns and more rapid acquisition of the required process understanding. Another factor that accelerates time-to-yield appears to be the systematic (and

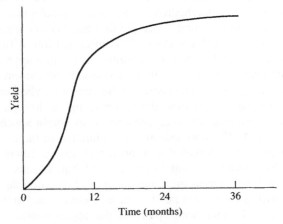

Figure 5.24. Yield learning curve.

often automated) collection of data, especially if this is done in the context of a well-structured quality control program.

While these aspects are difficult to quantify, one can draw interesting conclusions from carefully organized field studies. One such study was done in the context of the Competitive Semiconductor Manufacturing program, run at the University of California at Berkeley in 1994. This program involved studying a large number of IC production facilities, and it included detailed questionnaires, as well as site visits by multidisciplinary groups of experts. In this study, researchers recorded yield–time data for a variety of facilities, covering products ranging from memories to high-end logic ICs. While the main objective of the study was to capture the main reasons behind achieving high yield numbers, the data also offered a rare opportunity to examine the time-to-yield figure on a qualitative basis. Some of the yield data are shown in Figure 5.25.

Graphs like Figure 5.24 plot a "normalized" total yield figure that is appropriately adjusted for die size, minimum feature size, and other properties. Using several such datasets for various IC technologies, the authors of the study created and calibrated an empirical model of the following form

$$W_j = \alpha_o j + \alpha_{1j}(\text{die Size}) + \alpha_{2j} \log(\text{process age}) \qquad (5.35)$$

where j is the index defining the various semiconductor manufacturing facilities participating in the study, (die size) is in cm^2, and (process age) is the time in months from the oldest to the most recent die yield data point. This model implies that the yield increases logarithmically as processes mature, and the coefficient α_{2j} is a quantitative figure of merit that captures the unique ability of each facility to rapidly improve the yield of a new process. Table 5.4 attempts to capture the impact of various factors on the α_{2j} "yield learning coefficient." In this table, because of the limited sample size, the facilities are divided into three

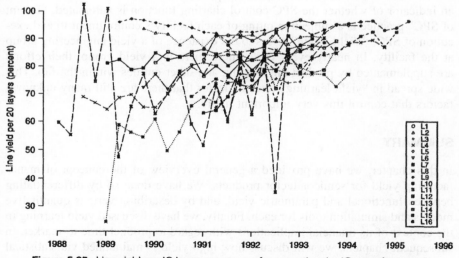

Figure 5.25. Line yield per IC layer versus year for several major IC manufacturers.

Table 5.4. Impact of various factors on yield learning.

SPC AUTOMATION			EXTENT OF SPC USE		
	Yes	No		High	Med	Low
High	3	0	High	1	2	0
Med	6	0	Med	2	3	1
Low	2	4	Low	1	3	2

PAPERLESS WAFER TRACKING			EXTENT OF CAM AUTOMATION		
	Yes	No		Full	Semi	Manual
High	1	2	High	1	2	0
Med	3	3	Med	3	2	1
Low	1	5	Low	1	1	4

YIELD MODEL IN USE			YIELD GROUP PRESENT IN FACTORY		
	Homegrown	Away		Yes	No
High	2	1	High	3	0
Med	1	5	Med	3	3
Low	2	4	Low	4	2

GEOGRAPHIC FAB REGION		
	USA	Asia	Europe
High	3	0	0
Med	1	4	1
Low	4	2	0

categories relating the yield learning speed (low, medium, high). Facilities with $\alpha_2 > 1.00$ received a high yield improvement rating, and facilities with $\alpha_2 < 0.30$ were given a low improvement rating. The number of facilities falling in each category is also noted below.

Here, the term "yield model" refers the specific formula used by that organization to predict the yield of the process as a function of a measure of defect density. "Paperless" is an indicator of the extent of computer-aided manufacturing in the facility. Only three of the facilities under study were fully paperless (i.e., had no lot travelers or run cards accompanying production lots). "SPC automation" is an indicator of whether the SPC control charting function is automated. "Extent of SPC" practice is a subjective rating of each facility's commitment to and execution of SPC. "Yield group" refers to the existence of a yield engineering group at the facility. In nearly every case where there is a yield group, their efforts are supplemented by product engineering and other entities within the fab. This wide spread in "yield learning" rates indicates that there are still many unknown factors that control this very important figure.

SUMMARY

In this chapter, we have provided a general overview of the concept of manufacturing yield for semiconductor products. We have done so by differentiating between functional and parametric yield, and by describing various quantitative models and simulation tools for each. Finally, we have discussed yield learning in the context of its financial implications with regard to product time-to-market. In subsequent chapters, we will discuss how high yield is maintained via statistical process control techniques.

PROBLEMS

5.1. Assuming a Poisson model, calculate the maximum defect density allowable on 100,000 NMOS transistors in order to achieve a functional yield of 95%. Assume that the gate of each device is 10 μm wide and 1 μm long.

5.2. Use Murphy's yield integral to derive Eqs. (5.11), (5.12), and (5.14).

5.3. Suppose that the probability density function of the defect density for a given IC manufacturing process is given by

$$f(D) = -100D + 10$$

$$0 \leq D \leq 0.1$$

If D is measured in cm^{-2} and the critical area for this IC is 100 cm^2, what functional yield can we expect for the process over the range of defect densities from 0.05 to 0.1 cm^{-2}?

5.4. Consider the effect of defects on IC interconnect. Figure P5.4 illustrates the impact of defect size on critical area for circular defects of diameter x. The area in which the center of such defects must fall to cause a failure increases linearly as a function of defect size. It can be expressed as

$$A_c(x) = L(x + w - 2R)$$

$$R \leq x \leq \infty$$

where L is the interconnect length and R is the allowable gap in the interconnect line. Suppose that the normalized probability density function of defect sizes is given by

$$g(x) = \frac{X_U^2 X_L^2}{x^3(X_U^2 - X_L^2)}$$

where X_L and X_U are the lower and upper limits of the range of defect sizes, respectively.

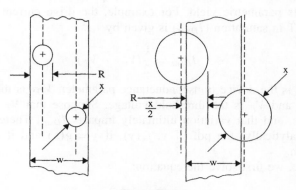

Figure P5.4

(a) Given $R \geq X_L$, find an expression for the average critical area (A_{av}) by evaluating the integral

$$A_{av} = \int_{X_L}^{X_U} A_c(x)g(x)dx$$

(b) Show that as the upper limit on defect size approaches infinity, then

$$A_{av} = \frac{LX_L^2 w}{2R^2}$$

5.5. Assume that 10,000 units of a product with area 0.5 cm^2 and 200 chips per wafer are to be produced in three manufacturing areas, with a D_0 of 0.9, 1.1, and 1.3 cm^{-2}, respectively. How many wafers need to be ordered? Assume a negative binomial model with $\alpha = 2$, and also assume that these steps will be followed by assembly and test. The combined assembly and test steps have a yield of 95%.

5.6. A new product with a critical area of 0.45 cm^2 is to be produced using a technology with a defect density of 0.5 cm^{-2}. Three similar products are already being produced using this technology, and their critical areas and yield data appear in Table P5.6. Analyze the data and calculate the short-term and long-term yield expectations using the Poisson model.

Table P5.6

A_C(cm^2)	Measured Yield (%)
0.1	81
0.2	78
0.4	70

5.7. Suppose we are given the joint distribution of several parameters which vary in an IC manufacturing process, and we would like to evaluate the impact of these variations on the overall performance of the IC by evaluating its parametric yield. For example, the drive current in mA of a MOSFET in saturation (I_{Dsat}) is given by

$$I_{Dsat} = \frac{k}{2}(V_{GS} - V_T)^2$$

where k is the device transconductance parameter, V_{GS} is the gate-source voltage, and V_T is the threshold voltage. Suppose that V_T is subject to variation, and that variation ultimately impacts I_{Dsat}. There is a way to find (analytically) the pdf of y, $f_y(y)$, if $y = g(x)$ and if the pdf of x is known.

To do so, we first solve the equation:

$$y = g(x)$$

for x in terms of y. If $x_1, x_2, \ldots, x_n$ are the real roots of this expression, then

$$f_y(y) = \frac{f_x(x_1)}{|g'(x_1)|} + \cdots + \frac{f_x(x_n)}{|g'(x_n)|}$$

where $g'(x) = \dfrac{dg(x)}{dx}$

(a) Find the analytical expression for the pdf for the drive current, if V_T is uniformly distributed between 0.3 and 0.8 V, and if all other parameters are deterministic.

(b) If $k = 1$ mA/V^2, determine the parametric yield for a large population of transistors that achieve drive currents between 1.5 and 2.0 mA, if V_{GS} is 2.5 V.

5.8. An oncoming 150-mm wafer has 10 randomly spaced point defects on it. The chip size is 1.0 cm^2, and the final target yield is 75%. If there are eight additional processing levels, what is the maximum number of defects that we can afford to accumulate on each level? (Use the Poisson yield model.)

5.9. A manufacturing facility has a yield that is controlled purely by random defects. The density of these random defects depends on the design rule used. More specifically, for a 1-μm design rule, the density is 0.5/cm^2, while for a 0.5-μm design rule, the density is 2.0/cm^2. (Use the Poisson yield model.)

(a) A given product takes 1.0 cm^2. Further, 90% of this area is using 1 μm design rules, while the rest 10% is using the 0.5 μm design rules. Estimate the yield of this product.

(b) This product can be redesigned (shrunk) to take only 0.5 cm^2, but now 50% of the chip is using the 0.5 μm design rules. Estimate the yield of the redesigned product.

(c) What would be the ratio of good die per wafer of the redesigned product to that of the original product?

5.10. Suppose that you use 200-mm wafers, and also assume that you can get functional dies only within the inner 190-mm diameter (outer 5-mm margin is full of defects). On the one product that you have run so far, a chip with area 5 × 5 mm, the yield is 80%.

(a) Using the simple Poisson model, find the defect density (in the good area of the wafer) and plot the yield as a function of S, where S is the square root of the area of the die in production. Plot the total and the good die per wafer as a function of S on the same graph.

(b) Repeat the calculations and plots in (a) using the negative binomial model ($\alpha = 1.5$).

(c) Suppose that an alternative explanation for the data were that some fraction f of the wafer were perfect and the rest were totally dead.

This is the "black–white" model that assumes a perfect deterministic clustering of defects. What is f? Plot the "good die" per wafer for this model on the same graph as in (a)–(b).

(d) What defect density reduction would you have to achieve to yield 50% of the available die at $S = 15$ mm according to models (a), (b), and (c)?

REFERENCES

1. B. Murphy, "Cost-Size Optima of Monolithic Integrated Circuits," *Proc. IEEE* **52**(12), 1537–1545 (Dec. 1964).

2. J. Piñeda de Gyvez and D. Pradhan, *Integrated Circuit Manufacturability*, IEEE Press, 1999.

3. R. Seeds, "Yield and Cost Analysis of Bipolar LSI," *IEEE Intl. Electron Devices Meet.*, Washington, DC, Oct., 1967.

4. J. Price, "A New Look at Yield of Integrated Circuits," *Proc. IEEE* (Lett.) **58**, 1290–1291 (Aug. 1970).

5. T. Okabe, Nagata, and Shimada, "Analysis of Yield of Integrated Circuits and a New Expression for the Yield," in *Defect and Fault Tolerance in VLSI Systems*, C. Stapper, ed., Vol. 2, Plenum Press, 1990, pp. 47–61.

6. C. Stapper, "Fact and Fiction in Yield Modeling," *Microelectron. J.* **210**(1–2), 129–151 (May 1989).

7. J. Cunningham, "The Use and Evaluation of Yield Models in Integrated Circuit Manufacturing," *IEEE Trans. Semiconduct. Manuf.* **3**(2), 60–71 (May 1990).

8. C. Stapper and R. Rossner, "A Simplified Method for Modeling VLSI Yields," *Solid State Electron.* **25**(6), 655–657 (July 1973).

9. C. Stapper, "Modeling Defects in Integrated Circuit Photolithographic Patterns," *IBM J. Res. Devel.* **28**(4), 461–475 (July 1984).

10. W. Brown, *Advanced Electronic Packaging*, IEEE Press, Piscataway, NJ, 1999.

11. W. Maly and A. Strojwas, "Statistical Simulation of the IC Manufacturing Process," *IEEE Trans. CAD ICs Syst.* **1**(3) (July 1982).

12. D. Walker, *Yield Simulation for Integrated Circuits*, Kluwer, Boston, 1987.

13. S. Director and G. Hatchel, "A Point Basis for Statistical Design," *Proc. IEEE Intl. Symp. Circuits and Systems*, New York, May, 1978.

14. J. Ogrodzki, L. Opalski, and M. Styblinski, "Acceptability Regions for a Class of Linear Networks," *Proc. IEEE Intl. Symp. Circuits and Systems*, Houston, May 1980.

15. S. Director and G. Hatchel, "The Simplical Approximation Approach to Design Centering," *IEEE Trans. Circuits Syst.* **CAS-24**(7) (July 1977).

16. S. Cunningham, C. J. Spanos, and K. Voros, "Semiconductor Yield Improvement: Results and Best Practices", *IEEE Trans. Semiconduct. Manuf.* **8**(2), 103–109 (May 1995).

6

STATISTICAL PROCESS CONTROL

OBJECTIVES

- Provide an overview of statistical process control (SPC) techniques.
- Define and describe various types of control charts.
- Differentiate between control charts for attributes and control charts for variables.
- Introduce a few advanced SPC concepts.

INTRODUCTION

Manufacturing processes must be stable, repeatable, and of high quality to yield products with acceptable performance. This implies that all individuals involved in manufacturing a product (including operators, engineers, and management) must continuously seek to improve manufacturing process output and reduce variability. Variability reduction is accomplished in a large part by strict process control. The application of process control in manufacturing continues to expand in the semiconductor industry. In this chapter we will focus on statistical process control techniques a as means to achieve high-quality products.

Statistical process control (SPC) refers to a powerful collection of problem-solving tools used to achieve process stability and reduce variability. Perhaps the primary and most technically sophisticated of these tools is the control chart. The

Fundamentals of Semiconductor Manufacturing and Process Control,
By Gary S. May and Costas J. Spanos
Copyright © 2006 John Wiley & Sons, Inc.

control chart was developed by Dr. Walter Shewhart of Bell Telephone Laboratories in the 1920s. For this reason, control charts are also often referred to as *Shewhart control charts*.

6.1. CONTROL CHART BASICS

A control chart is used to detect the occurrence of shifts in process performance so that investigation and corrective action may be undertaken to bring an incorrectly behaving manufacturing process back under control. A typical control chart is shown in Figure 6.1. This chart is a graphical display of a quality characteristic that has been measured from a sample versus the sample number or time. The chart consists of: (1) a *centerline*, which represents the average value of the characteristic corresponding to an in-control state; (2) an *upper control limit* (UCL); and (3) a *lower control limit* (LCL). The control limits are selected such that if the process is under statistical control, nearly all the sample points will plot between them. Points that plot outside the control limits are interpreted as evidence that the process is out of control.

There is a close connection between control charts and the concept of hypothesis testing, which was discussed in Chapter 4. Essentially, the control chart represents a continuous series of tests of the hypothesis that the process is under control. A point that plots within the control limits is equivalent to accepting the hypothesis of statistical control, and a point outside the limits is equivalent to rejecting this hypothesis. We can think of the probability of a type I error (a "false alarm") as the probability of concluding that the process is out of statistical control when it really is under control, and of the probability of a type II error (a "missed alarm") as the probability of concluding that the process is under control when it really is not.

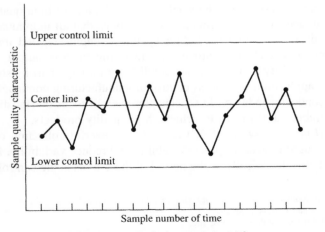

Figure 6.1. Typical control chart [1].

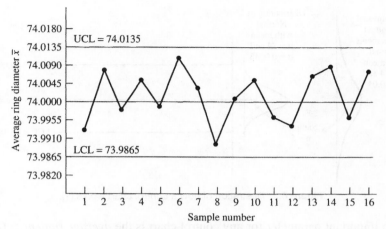

Figure 6.2. $\bar{x}$ chart for via diameter [1].

To illustrate, consider an example pertaining to the formation of vias in a dielectric layer. Suppose that this process can be controlled at mean via diameter of 74 μm, and the standard deviation of the diameter is 0.01 μm. A control chart for via diameter is shown in Figure 6.2. For every product wafer, a sample of five via diameters are measured, and that sample average, $\bar{x}$, is plotted on the chart. Note that all the points fall within the control limits, indicating that the via formation process is under statistical control.

Let's examine how the control limits in this example were determined. For a sample size of $n = 5$ vias, the standard deviation of the sample average is

$$\frac{\sigma}{\sqrt{n}} = \frac{0.01}{\sqrt{5}} = 0.0045 \ \mu m \tag{6.1}$$

If we assume that x is normally distributed, we would expect $100(1 - \alpha)\%$ of the sample mean diameters to fall within $74 + z_{\alpha/2}(0.0045)$ and $74 - z_{\alpha/2}(0.0045)$. If the constant $z_{\alpha/2}$ is selected to be 3, the upper and lower control limits become

$$UCL = 74 + 3(0.0045) = 74.0135 \ \mu m$$

$$LCL = 74 - 3(0.0045) = 73.9865 \ \mu m$$

These are typically called "3-sigma" (3σ) control charts, where "sigma" refers to the standard deviation of the sample average computed in Eq. (6.1). Note that the selection of the control limits is equivalent to testing the hypothesis

$$H_0: \ \mu = 74$$

$$H_1: \ \mu \neq 74$$

where σ = 0.01 is known. Essentially, the control chart just tests this hypothesis repeatedly for each sample. This is illustrated graphically in Figure 6.3.

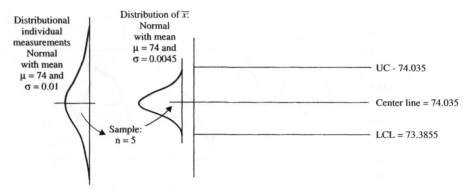

Figure 6.3. Illustration of how a control chart works [1].

An important parameter for any control chart is the *average runlength* (ARL), which is defined as the average number of samples taken before the control limits are exceeded. Mathematically, the ARL is $1/P$ (*a sample point plots out of control*). Thus, if the process is *in control*, the ARL is

$$ARL = 1/\alpha \qquad (6.2)$$

where α is the probability of a type I error. If the process is *out of control*, then the ARL is

$$ARL = \frac{1}{1 - \beta} \qquad (6.3)$$

where β is the probability of a type II error.

6.2. PATTERNS IN CONTROL CHARTS

A control chart may indicate an out-of control condition when a point plots beyond the control limits or when a sequence of points exhibit nonrandom behavior. For example, consider the charts shown in Figure 6.4. The pattern in Figure 6.4a is called a "trend" (or "run"). Although most of the points in this chart are within the control limits, they are not indicative of statistical control because their pattern is very nonrandom. A pattern of several consecutive points on the same side of the centerline is also called a "run." A run of several points has a very low probability of occurrence in a truly random sample.

Other types of patterns may also indicate an out-of-control state. For example, the chart in Figure 6.4b exhibits cyclic (or periodic) behavior, even though all the points are within the control limits. This type of pattern might result from operator fatigue, raw-materials depletion, or other periodic problems. Several other special patterns in control charts might be suspicious, including

- *Mixtures*—points from two or more source distributions
- *Shifts*—abrupt changes
- *Stratification*—charts that exhibit unusually small variability

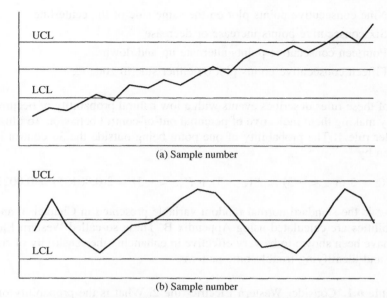

Figure 6.4. Examples of patterns in control charts: (a) trend; (b) cyclic.

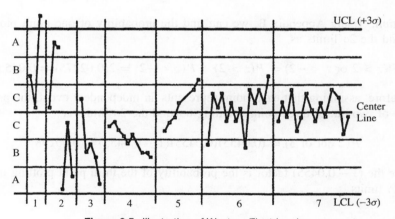

Figure 6.5. Illustration of Western Electric rules.

In order to detect patterns such as these, special rules must be applied. The Western Electric *Statistical Quality Control Handbook* [3] provides a set of rules for detecting nonrandom patterns in control charts. Referring to Figure 6.5, these rules state that a process is out of control if either:

1. Any single point plots beyond the 3σ control limits.
2. Two out of three consecutive points plot beyond the 2σ warning limits (zone A).
3. Four out of five consecutive points plot beyond 1σ (zone B).

4. Nine consecutive points plot on the same side of the centerline.

5. Six consecutive points increase or decrease.

6. Fourteen consecutive points alternate up and down.

7. Fifteen consecutive points plot on either side in zone C.

Each of these rules describes events with a low natural probability of occurrence, thereby making them indicative of potential out-of-control behavior. To illustrate, consider rule 1. The probability of one point being outside the 3σ control limits is given by

$$P(z > 3 \text{ or } z < -3) = P(z > 3) + P(z < -3) = 2(0.00135) = 0.0027$$

where z is the standard normal random variable presented in Chapter 4, and the probabilities are calculated using Appendix B. These so-called Western Electric rules have been shown to be very effective in enhancing the sensitivity of control charts and identifying troublesome patterns.

Example 6.1. Consider Western Electric rule 2. What is the probability of two out of three consecutive points plotting beyond the 2σ warning limits?

Solution: Using Appendix B, we can find the probability of one point plotting beyond the 2σ limits as

$$P(z > 2 \text{ or } z < -2) = P(z > 2) + P(z < -2) = 2(0.02275) = 0.0455$$

Therefore, assuming that each point represents an independent event, the probability of two out of three points plotting beyond the 2σ limits is

$$P(2 \text{ out of } 3) = (0.0455)(0.0455)(1 - 0.0455) = 0.00198$$

where the $(1 - 0.0455)$ factor is the probability of the third point plotting inside the 2σ limits.

6.3. CONTROL CHARTS FOR ATTRIBUTES

Some quality characteristics cannot be easily represented numerically. For example, we may be concerned with whether a contact is defective. In this case, the contact is classified as either "defective" or "nondefective" (or equivalently, "conforming" or "nonconforming"), and there is no numerical value associated with its quality. Quality characteristics of this type are referred to as *attributes*. In this section, three commonly used control charts for attributes are presented: (1) the fraction nonconforming chart (*p chart*), (2) the defect chart (*c chart*), and (3) the defect density chart (*u chart*).

6.3.1. Control Chart for Fraction Nonconforming

The fraction nonconforming is defined as the number of nonconforming items in a population divided by the total number of items in the population. The control chart for fraction nonconforming is called the *p chart*, which is based on the binomial distribution (see Section 4.1.1.1). Suppose that the probability that any product in a manufacturing process will not conform is p. If each unit is produced independently, and a random sample of n products yields D units that are nonconforming, then D has a binomial distribution. In other words

$$P(D = x) = \binom{n}{x} p^x (1 - p)^{n-x} \qquad x = 0, 1, \ldots n \qquad (6.4)$$

The sample fraction nonconforming ($\hat{p}$) is defined as

$$\hat{p} = \frac{D}{n} \qquad (6.5)$$

As noted in Section 4.1.1.1, the mean and variance of $\hat{p}$ are $\mu_{\hat{p}} = p$ and $\sigma_{\hat{p}}^2 = p(1 - p)/n$, respectively. On the basis of these relationships, we can set up the centerline and $\pm 3\sigma$ control limits for the p chart as follows:

$$UCL = p + 3\sqrt{\frac{p(1 - p)}{n}}$$

$$\text{Centerline} = p \qquad (6.6)$$

$$LCL = p - \sqrt{\frac{p(1 - p)}{n}}$$

This above implementation of the p chart assumes that p is known (or given). If p is not known, it must be computed from the observed data. The usual procedure is to select m preliminary samples, each of size n. If there are D_i nonconforming units in the ith sample, then the fraction nonconforming is

$$\hat{p}_i = \frac{D_i}{n} \qquad i = 1, 2, \ldots, m \qquad (6.7)$$

and the average of the individual fractions nonconforming is

$$\overline{p} = \frac{1}{mn} \sum_{i=1}^{mn} D_i = \frac{1}{m} \sum_{i=1}^{mn} \hat{p}_i \qquad (6.8)$$

The centerline and control limits for the p chart under these conditions are

$$UCL = \overline{p} + 3\sqrt{\frac{\overline{p}(1 - \overline{p})}{n}}$$

$$\text{Centerline} = \overline{p} \qquad (6.9)$$

$$LCL = \overline{p} - 3\sqrt{\frac{\overline{p}(1 - \overline{p})}{n}}$$

Example 6.2. Consider a wire bonding operation. Suppose that 30 samples of size $n = 50$ have been collected from 30 chips. Given a total of 347 defective bonds found, set up the $\pm 3\sigma$ p chart for this process.

Solution: Using Eq. (6.8), we have

$$\overline{p} = \frac{1}{mn} \sum_{i=1}^{m} D_i = \frac{347}{(30)(50)} = 0.2313$$

This is the centerline for the p chart. The upper and lower control limits can be found from Eq. (6.9) as

$$\text{UCL} = \overline{p} + 3\sqrt{\frac{\overline{p}(1 - \overline{p})}{n}} = 0.4102$$

$$\text{LCL} = \overline{p} - 3\sqrt{\frac{\overline{p}(1 - \overline{p})}{n}} = 0.0524$$

It should be pointed out that the limits defined by Eq. (6.9) are actually just *trial* control limits. They permit the determination of whether the process was in control when the m samples were collected. To test the hypothesis that the process was in fact under control during this period, the sample fraction nonconforming from each sample on the chart must be plotted and analyzed. If all points are inside the control limits and no systematic trends are evident, then it may be concluded that the process was indeed under control, and the trial limits are reasonable.

If, on the other hand, one or more of the $\hat{p}_i$ statistics plots out of control when compared to the trial control limits, then the hypothesis of past control must be rejected, and the trial limits are no longer valid. It then becomes necessary to revise the trial control limits by first examining each out-of-control point in an effort to identify an assignable cause. If a cause can be found, the point in question is discarded and the control limits are recalculated using the remaining points. The remaining points are then reexamined, and this process is repeated until all points plot in control, at which point the trial limits may be adopted as valid.

6.3.1.1. Chart Design

Constructing a p chart requires that the sample size, frequency of sampling, and width of the control limits all be specified. Obviously, the sample size and sampling frequency are interrelated. Assuming 100% inspection for a given production rate, selecting a sampling frequency fixes the sample size.

Various rules have been suggested for the choice of sample size (n). If p is very small, n must be sufficiently large that we have a high probability of finding at least one nonconforming unit in a sample in order for the p chart to be effective. Otherwise, the control limits might end up being so narrow that the presence of only a single nonconforming unit in a sample might indicate an out-of-control condition. For example, if $p = 0.01$ and $n = 8$, then the 3σ upper control limit is

$$\text{UCL} = p + \sqrt{\frac{p(1-p)}{n}} = 0.1155$$

With only one nonconforming unit, $\hat{p} = 0.125$, and the process appears to be out of control.

To avoid this problem, Duncan has suggested that the sample size be large enough to ensure an approximately 50% chance of detecting a process shift of some specified amount [2]. For example, let $p = 0.01$, and suppose that we want the probability of detecting a shift from $p = 0.01$ to $p = 0.05$ to be 0.5. Assuming that the normal approximation to the binomial distribution applies, this implies that n must be selected such that the UCL exactly coincides with the fraction nonconforming in the out-of-control state. In general, if δ is the magnitude of this process shift, then n is given by

$$\delta = k\sqrt{\frac{p(1-p)}{n}} \tag{6.10}$$

In our example, $\delta = 0.05 - 0.01 = 0.04$, and if 3σ limits are used (i.e., $k = 3$), then

$$n = \left(\frac{k}{\delta}\right)^2 p(1-p) = \left(\frac{3}{0.04}\right)^2 (0.01)(0.99) = 56$$

If the in-control value of the fraction nonconforming is small, it is also desirable to choose n large enough so that the p chart will have a positive lower control limit. This will allow us to detect samples that have an unusually small number of nonconforming items. In other words, we want

$$\text{LCL} = p - k\sqrt{\frac{p(1-p)}{n}} > 0 \tag{6.11}$$

or

$$n > \frac{(1-p)}{p}k^2 \tag{6.12}$$

Note that this is not always practical. If we want the chart in our example to have a positive LCL, this will require that $n \geq 891$.

6.3.1.2. Variable Sample Size

In some applications, the sample size for the fraction nonconforming control chart is not fixed. In these cases, there are several approaches to constructing the p chart. The first, and probably the simplest, approach is to determine control limits according to the specific size of each sample. In other words, if the ith sample is of size n_i, the upper and lower 3σ control limits are placed at $p \pm 3\sqrt{p(1-p)/n_i}$. However, this results in control limits that vary for each sample, as shown in Figure 6.6.

The approach described above is somewhat unappealing. A second approach is to use the *average* sample size to compute the control limits. This assumes that the

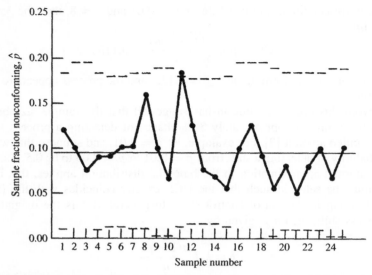

Figure 6.6. Example of control chart for fraction nonconforming with variable sample size [1].

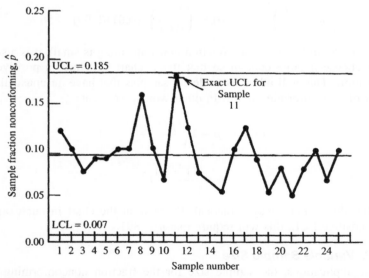

Figure 6.7. Control chart for fraction nonconforming based on average sample size [1].

sample sizes will not differ appreciably over the duration of the chart. The result is a set of control limits that are approximate, but constant, and therefore more satisfying and easier to interpret. Applying this approach to the same dataset used in Figure 6.6 results in the chart shown in Figure 6.7. Care must be exercised in the interpretation of points near the approximate control limits, however. Notice that $\hat{p}$ for sample 11 in Figure 6.7 is close to the upper control limit, but appears to be in control. When compared the exact limits used in Figure 6.6, though, this point appears to be out of control. Similarly, points outside the approximate limits may indeed be inside their exact limits.

Using the second approach, care must also be taken in analyzing patterns such as those indicated in the Western Electric rules. Since the sample size actually changes from run to run, such analyses are practically meaningless. A solution to this problem is to use a "standardized" control chart where all points are plotted using standard deviation units. This type of chart has a centerline at zero and upper and lower control limits at ± 3, respectively, for 3σ control. The variable plotted on the chart is

$$Z_i = \frac{\hat{p}_i - p}{\sqrt{\dfrac{p(1 - p)}{n_i}}} \tag{6.13}$$

where p (or $\bar{p}$) is the process nonconforming in the in-control state. The standardized chart for the same dataset as in Figures 6.6 and 6.7 is shown in Figure 6.8. Tests for patterns can be safely applied to this chart since the relative changes from one point to another are all expressed in the same units.

6.3.1.3. Operating Characteristic and Average Runlength

The operating characteristic (OC) curve of a control chart is a graph of the probability of incorrectly accepting the hypothesis of statistical control (i.e., a type II error) versus the fraction nonconforming. The OC provides a measure of the sensitivity of the chart to a given process shift. In the case of the p chart, the OC provides the graphical display of its ability to detect a shift from the nominal

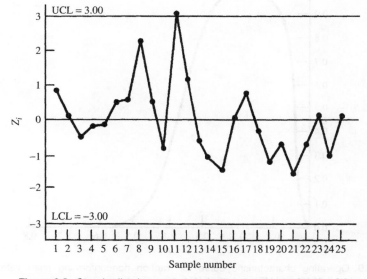

Figure 6.8. Standardized control chart for fraction nonconforming [1].

value of $\bar{p}$ to some new value. The probability of a type II error for this chart is given by

$$\beta = P\{\hat{p} < \text{UCL}|p\} - P\{\hat{p} \le \text{LCL}|p\} \qquad (6.14)$$
$$= P\{D < n\text{UCL}|p\} - P\{D \le n\text{LCL}|p\}$$

where D is a binomial random variable with parameters n and p. The probability defined by Eq. (6.14) can be obtained from the cumulative binomial distribution. A typical OC curve for the fraction nonconforming chart is shown in Figure 6.9.

The OC curve may also be used to compute the average runlength (ARL) for the fraction nonconforming chart. Recall that the ARL is given by Eqs. (6.2) and (6.3). From the OC in Figure 6.9, for $p = 0.2$, the process is in control, and the probability that a point plots within the control limits is 0.9973. The in-control ARL is therefore

$$\text{ARL} = \frac{1}{\alpha} = \frac{1}{0.0027} = 370$$

This implies that if the process is in control, there will be a "false alarm" about every 370 samples. Suppose that the process shifts out of control to $p = 0.3$. From Figure 6.9, a value of $p = 0.3$ corresponds to $\beta = 0.8594$. The out-of-control ARL is then

$$\text{ARL} = \frac{1}{1 - \beta} = \frac{1}{1 - 0.8594} = 7$$

This means that it will take seven samples, on average, for the p chart to detect this shift.

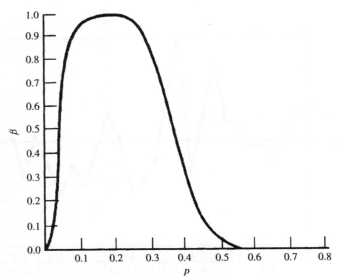

Figure 6.9. Operating characteristic curve for fraction nonconforming chart with $n = 50$, $\bar{p} = 0.2$, LCL $= 0.0303$, and UCL $= 0.3697$ [1].

6.3.2. Control Chart for Defects

When a specification is not satisfied in a product, a defect or nonconformity may result. In many cases, it is preferable to directly control the actual number of defects rather than the fraction nonconforming. In such cases, it is possible to develop control charts for either the total number of defects or the defect density. These charts assume that the presence of defects in samples of constant size is appropriately modeled by the Poisson distribution; that is

$$P(x) = \frac{e^{-c}c^x}{x!} \qquad (6.15)$$

where x is the number of defects and $c > 0$ is the parameter of the Poisson distribution. Since c is both the mean and variance of the Poisson distribution, the control chart for defects (c chart) with 3σ limits is given by

$$UCL = c + 3\sqrt{c}$$
$$\text{Centerline} = c \qquad (6.16)$$
$$LCL = c - 3\sqrt{c}$$

assuming that c is known. (*Note*: If these calculations yield a negative value for the LCL, the standard practice is to set the LCL $= 0$.) If c is not known, it may be estimated from an observed average number of defects in a sample ($\bar{c}$). In this case, the control chart becomes

$$UCL = \bar{c} + 3\sqrt{\bar{c}}$$
$$\text{Centerline} = \bar{c} \qquad (6.17)$$
$$LCL = \bar{c} - 3\sqrt{\bar{c}}$$

Example 6.3. Suppose that the inspection of 26 silicon wafers yields 516 defects. Set up a c chart for this situation.

Solution: We estimate $\bar{c}$ using

$$\bar{c} = \frac{516}{26} = 19.85$$

This is the centerline for the c chart. The upper and lower control limits can be found from Eq. (6.17) as

$$UCL = \bar{c} + 3\sqrt{\bar{c}} = 33.22$$
$$LCL = \bar{c} - 3\sqrt{\bar{c}} = 6.484$$

6.3.3. Control Chart for Defect Density

Suppose that we would like to set up a control chart for the *average* number of defects over a sample size of n products. If there were c total defects among the

n samples, then the average number of defects per sample is

$$u = \frac{c}{n} \tag{6.18}$$

The parameters of a 3σ defect density chart (u chart) are then given by

$$UCL = \bar{u} + 3\sqrt{\frac{\bar{u}}{n}}$$

$$Centerline = \bar{u} \tag{6.19}$$

$$LCL = \bar{u} - 3\sqrt{\frac{\bar{u}}{n}}$$

where $\bar{u}$ is the average number of defects over m groups of sample size n.

Example 6.4. Suppose that a manufacturer wants to establish a defect density chart. Twenty different samples of size $n = 5$ wafers are inspected, and a total of 193 defects are found. Set up the u chart for this situation.

Solution: We estimate u using

$$\bar{u} = \frac{u}{m} = \frac{c}{mn} = \frac{193}{(20)(5)} = 1.93$$

This is the centerline for the u chart. The upper and lower control limits can be found from Eq. (6.19) as

$$UCL = \bar{u} + 3\sqrt{\frac{\bar{u}}{n}} = 3.79$$

$$LCL = \bar{u} - 3\sqrt{\frac{\bar{u}}{n}} = 0.07$$

The operating characteristic (OC) curves for both the c and u charts are derived from the Poisson distribution. For the c chart, the OC represents the probability of type II error (β) as a function of the true mean number of defects. The expression for β is

$$\beta = P\{x < UCL|c\} - P\{x \leq LCL|c\} \tag{6.20}$$

where x is a Poisson random variable with parameter c. A typical OC for a c chart is shown in Figure 6.10.

For the u chart, the OC is generated from

$$\beta = P\{x < UCL|u\} - P\{x \leq LCL|u\}$$

$$= P\{c < nUCL|u\} - P\{c \leq nLCL|u\}$$

$$= P\{nLCL < c \leq nUCL|u\}$$

$$= \sum_{c=\langle nLCL \rangle}^{[nUCL]} \frac{e^{-nu}(nu)^c}{c!} \tag{6.21}$$

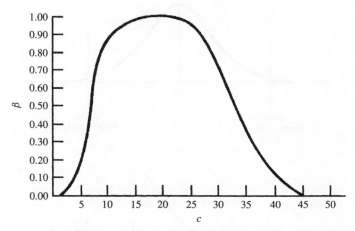

Figure 6.10. OC curve for fraction c chart with LCL = 6.48 and UCL = 33.22 [1].

where $\langle n\text{LCL}\rangle$ represents the smallest integer greater than or equal to $n\text{LCL}$ and $[n\text{UCL}]$ is the largest integer less than or equal to $n\text{UCL}$. These summation limits occur because the total number of defects observed must be an integer.

6.4. CONTROL CHARTS FOR VARIABLES

In many cases, quality characteristics are expressed as specific numerical measurements, rather than assessing the probability or presence of defects. For example, the thickness of an oxide layer is an important characteristic to be measured and controlled. Control charts for continuous variables such as this can provide more information regarding manufacturing process performance than attribute control charts like the p, c, and u charts.

When attempting to control continuous variables, it is important to control both the mean and the variance of the quality characteristic. This is true because shifts or drifts in either of these parameters can result in significant misprocessing. Consider a process represented by Figure 6.11. In Figure 6.11a, both the mean and the standard deviation are in control at their nominal values (μ_0 and σ_0). Under these conditions, most of the process output falls within the specification limits. However, in Figure 6.11b, the mean has shifted to a value $\mu_1 > \mu_0$, leading to a higher fraction of nonconforming product. Similarly, in Figure 6.11c, the standard deviation has shifted to a value $\sigma_1 > \sigma_0$, also resulting in more nonconforming products (even though the mean remains at its nominal value). Control of the mean is achieved using the $\bar{x}$ chart, and variance can be monitored using either the standard deviation (as in the s chart) or the range (as in the R chart). The x and R (or s) are among the most important and useful SPC tools.

6.4.1. Control Charts for $\bar{x}$ and R

We showed in Section 4.2.2 that if a quality characteristic is normally distributed with a known mean μ and standard deviation σ, then the sample mean ($\bar{x}$) for a

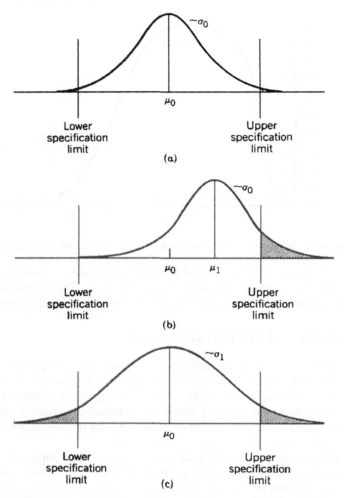

Figure 6.11. Illustration of the need to control both process mean and standard deviation: (a) nominal mean and standard deviation; (b) mean shifted to $\mu_1 > \mu_0$; (c) standard deviation shifted to $\sigma_1 > \sigma_0$ [1].

sample of size n is also normally distributed with mean μ and standard deviation $\sigma/\sqrt{n}$. Under these conditions, the probability that a sample mean will be between

$$\mu + z_{\alpha/2}\frac{\sigma}{\sqrt{n}} \tag{6.22}$$

and

$$\mu - z_{\alpha/2}\frac{\sigma}{\sqrt{n}} \tag{6.23}$$

is $1 - \alpha$. As a result, Eqs. (6.22) and (6.23) can be used as upper and lower control limits for a control chart for the sample mean. For 3σ control, we replace $z_{\alpha/2}$ by 3. This chart is called the $\bar{x}$ *chart*.

In practice, μ and σ rarely will be known. They must therefore be estimated from sample data. Suppose that m samples of size n are collected. If $\bar{x}_1, \bar{x}_2, \ldots, \bar{x}_m$ are the sample means, the best estimator for μ is the grand average $(\bar{\bar{x}})$, which is given by

$$\bar{\bar{x}} = \frac{\bar{x}_1 + \bar{x}_2 + \cdots + \bar{x}_m}{m} \tag{6.24}$$

Since $\bar{\bar{x}}$ estimates μ, $\bar{\bar{x}}$ is used as the centerline of the $\bar{x}$ chart.

To estimate σ, we can use the ranges of the m samples. The *range* (R) is defined as the difference between the maximum and minimum observation:

$$R = x_{\max} - x_{\min} \tag{6.25}$$

Another random variable $W = R/\sigma$ is called the *relative range*. The mean of W is a parameter called d_2, which is a function of the sample size n. (Values of d_2 for various sample sizes are given in Appendix F). Consequently, an estimator for σ is R/d_2. Let $R_1, R_2, \ldots, R_m$ be the ranges of the samples. The average range is then given by

$$\bar{R} = \frac{R_1 + R_2 + \cdots + R_m}{m} \tag{6.26}$$

and an estimate of σ is then

$$\hat{\sigma} = \frac{\bar{R}}{d_2} \tag{6.27}$$

If the sample size is small (i.e., $n < 10$), then the range is nearly as good an estimate of σ as the sample standard deviation (s).

If $\bar{\bar{x}}$ is used as an estimate of μ and $\bar{R}/d_2$ is used to estimate σ, then the parameters if the $\bar{x}$ chart are

$$\text{UCL} = \bar{\bar{x}} + \frac{3\bar{R}}{d_2\sqrt{n}}$$

$$\text{Centerline} = \bar{\bar{x}} \tag{6.28}$$

$$\text{LCL} = \bar{\bar{x}} - \frac{3\bar{R}}{d_2\sqrt{n}}$$

Note that the quantity $3/d_2\sqrt{n}$ is a constant that depends only on sample size. It is therefore possible to rewrite Eq. (6.28) as

$$\text{UCL} = \bar{\bar{x}} + A_2\bar{R}$$

$$\text{Centerline} = \bar{\bar{x}} \tag{6.29}$$

$$\text{LCL} = \bar{\bar{x}} - A_2\bar{R}$$

where the constant $A_2 = 3/d_2\sqrt{n}$ can be found tabulated for various sample sizes in Appendix F.

To control the range, the R chart is used. The centerline of the R chart is clearly $\overline{R}$, but to set up $\pm 3\sigma$ control limits for the R chart, we must first derive an estimate of the standard deviation of R ($\hat{\sigma}_R$). To do so, we again use the relative range. The standard deviation of W is d_3, which is a known function of n (see Appendix F). Since $R = W\sigma$, the true standard deviation of R is

$$\sigma_R = d_3\sigma \qquad (6.30)$$

Since σ is unknown, σ_R can be estimated from

$$\hat{\sigma}_R = d_3\frac{\overline{R}}{d_2} \qquad (6.31)$$

Therefore, the parameters of the R chart assuming 3σ control limits are

$$\text{UCL} = \overline{R} + 3d_3\frac{\overline{R}}{d_2}$$

$$\text{Centerline} = \overline{R} \qquad (6.32)$$

$$\text{LCL} = \overline{R} - 3d_3\frac{\overline{R}}{d_2}$$

If we let

$$D_3 = 1 - 3\frac{d_3}{d_2}$$

and

$$D_4 = 1 + 3\frac{d_3}{d_2}$$

then the parameters of the R chart may be defined as

$$\text{UCL} = \overline{R}D_4$$

$$\text{Centerline} = \overline{R} \qquad (6.33)$$

$$\text{LCL} = \overline{R}D_3$$

The constants D_3 and D_4 may also be found in Appendix F.

Example 6.5. Suppose that we want to establish an $\overline{x}$ chart to control linewidth for a lithography process. Twenty-five different samples of size $n = 5$ linewidths are measured. Suppose that the grand average for the 125 total lines measured is 74.001 μm and the average range for the 25 samples is 0.023 μm. What are the control limits for the $\overline{x}$ chart?

Solution: The value for d_2 for $n = 5$ (found in Appendix F) is 2.326. The upper and lower control limits for the $\overline{x}$ chart can therefore be found from

Eq. (6.28) as

$$UCL = \bar{\bar{x}} + \frac{3\bar{R}}{d_2\sqrt{n}} == 74.014 \ \mu m$$

$$LCL = \bar{\bar{x}} - \frac{3\bar{R}}{d_2\sqrt{n}} == 73.988 \ \mu m$$

6.4.1.1. Rational Subgroups

A fundamental idea in the use of control charts is the collection of sample data according to the *rational subgroup* concept. In general, this means that subgroups (i.e., samples of size n) should be selected so that if assignable causes for misprocessing are present, the chance for differences *between* subgroups will be maximized, whereas the chance for differences *within* a subgroup will be minimized. In other words, only random variation should be allowed within a subgroup.

The rational subgroup concept plays a particularly important role in the use of $\bar{x}$ and R control charts. The $\bar{x}$ chart monitors the average level of quality in a process, and the R chart measures the variability *within* a sample. In other words, the $\bar{x}$ chart monitors *between-sample* variabilty (variability in the process over time), and the R chart measures *within-sample* variability (instantaneous process variability for a given sample at a given time).

In semiconductor manufacturing, intuitive categories for rational subgroups include devices within a die, die within a wafer, or wafers in a lot. The following inequality represents the expected level of variation in these groupings:

(Within-die variation) < (within-wafer variation)

< (within-lot variation) < (lot-to-lot variation)

Care must be exercised when establishing such groupings. For example, grouping wafers within a quartz boat in a CVD furnace operation is inappropriate since reactant gas depletion effects down the length of the tube cause systematic variations in the deposition reaction [4].

Note from Eqs. (6.28) and (6.29) that the range is used to compute the control limits for the $\bar{x}$ chart. The range of a subgroup is used to estimate the standard deviation (σ) of that subgroup. This implies that the range across a lot, for example, should not be used to estimate the standard deviation between lots (i.e., the lot-to-lot variation); thus, *within-lot* statistics are different from *between-lot* statistics. The same is true for other rational subgroups. Since the within-lot variation is less than the between-lot variation, the wrong choice of the rational subgroup used to compute the range can bias the estimation and result in misleading interpretations of SPC data. Consider Figure 6.12, which shows two different $\bar{x}$ charts for monitoring linewidth in the same manufacturing process. In Figure 6.12a, the within-lot range has been used to compute the control limits, and the linewidth appears to be out of statistical control. However, when the between-lot range is used to compute the control limits (Figure 6.12b), there is apparently no problem.

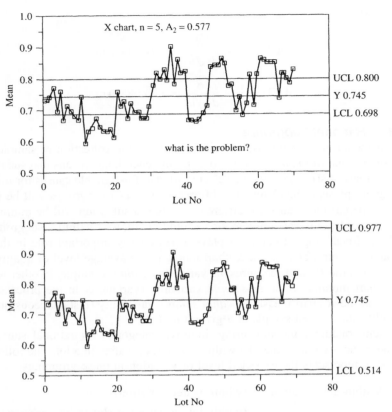

Figure 6.12. (a) $\bar{x}$ chart for linewidth control using within-lot range to compute control limits; (b) $\bar{x}$ chart for linewidth control using between-lot range to compute control limits.

6.4.1.2. Operating Characteristic and Average Runlength

Consider the operating characteristic (OC) curve for an $\bar{x}$ chart with a known standard deviation. If the process mean shifts from an in-control value (μ_0) to a new mean $\mu_1 = \mu_0 + k\sigma$, the probability of missing this shift on the next subsequent sample (i.e., the probability of type II error) is

$$\beta = P\{\text{LCL} \leq \hat{x} \leq \text{UCL} | \mu = \mu_0 + k\sigma\} \qquad (6.34)$$

Since $\bar{x} \sim N(\mu, \sigma^2/n)$, and the control limits are $\text{UCL} = \mu_0 + 3\sigma/\sqrt{n}$ and $\text{LCL} = \mu_0 - 3\sigma/\sqrt{n}$, Eq. (6.34) can be rewritten as

$$\beta = \Phi\left[\frac{\text{UCL} - (\mu_0 + k\sigma)}{\sigma/\sqrt{n}}\right] - \Phi\left[\frac{\text{LCL} - (\mu_0 + k\sigma)}{\sigma/\sqrt{n}}\right]$$

$$= \Phi\left[\frac{\mu_0 + 3\sigma/\sqrt{n} - (\mu_0 + k\sigma)}{\sigma/\sqrt{n}}\right] - \Phi\left[\frac{\mu_0 - 3\sigma/\sqrt{n} - (\mu_0 + k\sigma)}{\sigma/\sqrt{n}}\right]$$

$$= \Phi(3 - k\sqrt{n}) - \Phi(-3 - k\sqrt{n}) \qquad (6.35)$$

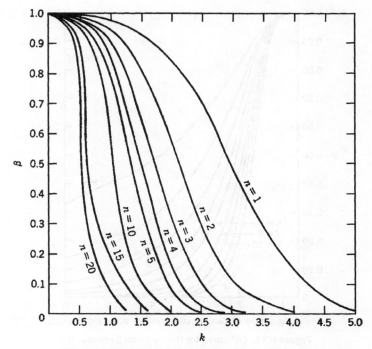

Figure 6.13. OC curve for $\bar{x}$-chart with 3-σ limits [1].

To construct the OC for the $\bar{x}$ chart, β is plotted versus k (the magnitude of the shift to be detected) for various sample sizes (see Figure 6.13). This figure shows that for small sample sizes ($n = 4$–6), the $\bar{x}$ chart is not particularly effective for detecting small shifts (i.e., shifts on the order of 1.5σ or less).

If the probability that a shift will be missed on the first sample after it occurs is β, then the probability that the shift will be detected in the first sample is $1 - \beta$. It then follows that the probability that the shift is detected on the second sample is $\beta(1 - \beta)$. Thus, the probability that a shift will be detected on the ith subsequent sample is

$$\beta^{i-1}(1 - \beta)$$

In general, the expected number of samples collected before the shift is detected is just the average runlength, so for the $\bar{x}$ chart, the ARL is

$$\mathrm{ARL} = \sum_{i=1}^{\infty} i\beta^{i-1}(1 - \beta) = \frac{1}{1 - \beta} \qquad (6.36)$$

This relationship suggests the advantage of using small sample sizes for the $\bar{x}$ chart. Even though small sample sizes result in a relatively high β, there is a good chance that a shift will be detected reasonably quickly in subsequent samples.

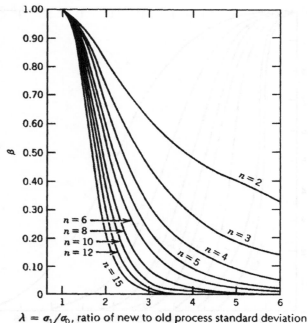

$\lambda = \sigma_1/\sigma_0$, ratio of new to old process standard deviation

Figure 6.14. OC curve for R-chart with 3σ limits [1].

To construct the OC for the R chart, the distribution of the relative range ($W = R/\sigma$) is used. Let the in-control value of the standard deviation be σ_0. The OC curve for the R chart then plots the probability of not detecting a shift to a new value (σ_1). Figure 6.14 shows the OC curve for b versus $\lambda = \sigma_1/\sigma_0$ for various values of n.

6.4.2. Control Charts for $\bar{x}$ and s

Although the range chart is quite popular, when the sample size is large (i.e., $n > 10$), it is desirable to estimate and control the standard deviation directly. This leads to control charts for $\bar{x}$ and s, where s is the sample standard deviation, which is computed using Eq. (4.2). Setting up these charts is similar to setting up $\bar{x}$ and R charts, except that for each sample, s is calculated rather than R.

The only caution that must be applied in this situation is that s cannot be used directly as the centerline of the s chart. This is due to the fact that s is *not* an unbiased estimator of s. (The term "unbiased" refers to the situation where the expected value of estimator is equal to the parameter being estimated.) Instead s actually estimates $c_4\sigma$, where c_4 is a statistical parameter that is dependent on the sample size (see Appendix F). In addition, the standard deviation of s is $\sigma\sqrt{1 - c_4^2}$. Using this information the control limits for the s chart can be set up as follows:

$$UCL = c_4\sigma + 3\sigma\sqrt{1 - c_4^2}$$

$$\text{Centerline} = c_4\sigma \tag{6.37}$$

$$LCL = c_4\sigma - 3\sigma\sqrt{1 - c_4^2}$$

It is customary to define two constants

$$B_5 = c_4 - 3\sqrt{1 - c_4^2}$$

$$B_6 = c_4 + 3\sqrt{1 - c_4^2} \tag{6.38}$$

As a result, the parameters of the s chart become

$$UCL = B_6\sigma$$

$$\text{Centerline} = c_4\sigma \tag{6.39}$$

$$LCL = B_5\sigma$$

If σ is unknown, then it must be estimated by analyzing past data. For m preliminary samples of size n, the average sample standard deviation is

$$\bar{s} = \frac{1}{m}\sum_{i=1}^{m} s_i \tag{6.40}$$

The statistic $\bar{s}/c_4$ is an unbiased estimator of σ. The parameters for the s chart then become

$$UCL = \bar{s} + 3\frac{\bar{s}}{c_4}\sqrt{1 - c_4^2}$$

$$\text{Centerline} = \bar{s} \tag{6.41}$$

$$LCL = \bar{s} - 3\frac{\bar{s}}{c_4}\sqrt{1 - c_4^2}$$

Once again, it is customary to define two constants:

$$B_3 = 1 - \frac{3}{c_4}\sqrt{1 - c_4^2}$$

$$B_4 = 1 + \frac{3}{c_4}\sqrt{1 - c_4^2} \tag{6.42}$$

Consequently, the parameters of the s chart become

$$UCL = B_4\bar{s}$$

$$\text{Centerline} = \bar{s} \tag{6.43}$$

$$LCL = B_3\bar{s}$$

Note that $B_4 = B_6/c_4$ and $B_3 = B_5/c_4$.

When $\bar{s}/c_4$ is used to estimate σ, the limits on the corresponding $\bar{x}$ chart may be defined as

$$\text{UCL} = \bar{\bar{x}} + \frac{3\bar{s}}{c_4\sqrt{n}}$$

$$\text{Centerline} = \bar{\bar{x}} \tag{6.44}$$

$$\text{LCL} = \bar{\bar{x}} - \frac{3\bar{s}}{c_4\sqrt{n}}$$

Let the constant $A_3 = 3/c_4\sqrt{n}$. It is therefore possible to rewrite Eq. (6.44) as

$$\text{UCL} = \bar{\bar{x}} + A_3\bar{s}$$

$$\text{Centerline} = \bar{\bar{x}} \tag{6.45}$$

$$\text{LCL} = \bar{\bar{x}} - A_3\bar{s}$$

Example 6.6. Consider the lithography process in Example 6.5. If $s = 0.009$ mm, what are the control limits for the s chart?

Solution: The value for c_4 for $n = 5$ (found in Appendix F) is 0.94. The upper and lower control limits can therefore be found from Eq. (6.41) as

$$\text{UCL} = \bar{s} + 3\frac{\bar{s}}{c_4}\sqrt{1 - c_4^2} = 0.019 \ \mu\text{m}$$

$$\text{LCL} = \bar{s} - 3\frac{\bar{s}}{c_4}\sqrt{1 - c_4^2} = 0 \ \mu\text{m}^1$$

6.4.3. Process Capability

Process capability quantifies what a process can accomplish when in control. Shewhart control charts are useful for estimating process capability. For example, suppose that the interconnect being defined by the lithography process described in Examples 6.5 and 6.6 must have a linewidth of 74.000 ± 0.05 μm. If these tolerances are not met, then some loss in product quality results. Tolerances such as this are called *specification limits*. Specification limits (SLs) differ from control limits in that they are externally imposed on the manufacturing process, whereas control limits are derived from the natural variability inherent in the process.

Control chart data can be used to investigate the capability of the process to produce linewidths according to the specification limits. Recall that our estimates for the process mean and standard deviation were

$$\bar{x} = 74.001 \ \mu\text{m}$$

$$\hat{\sigma} = \frac{R}{d_2} = 0.0099 \ \mu\text{m}$$

---

[1]Since the LCL is actually (slightly) negative in this case, we automatically set it to zero.

Assuming that the linewidth is normally distributed, we can estimate the fraction of nonconforming lines as

$$\hat{p} = P\{x < 73.95\} + P\{x > 74.05\}$$

$$= \Phi \left(\frac{73.95 - 74.001}{0.0099} \right) + 1 - \Phi \left(\frac{74.05 - 74.001}{0.0099} \right) \cong 0.00002$$

In other words, about 0.002% of the lines produced will be outside the specification limits. This means that the process is capable of achieving the specification limits 99.998% of the time. The remaining 0.002% of the lines will not meet the specifications no matter what steps are taken to improve the process.

Another way to express the process capability is in terms of the process capability ratio (PCR, or C_p). The PCR is defined as

$$C_p = \text{PCR} = \frac{\text{USL} - \text{LSL}}{6\sigma} \tag{6.46}$$

where USL and LSL are the upper and lower specification limits, respectively. Since σ is usually unknown, it frequently replaced by $\hat{\sigma} = R/d_2$. For the interconnect linewidth process, we can compute the PCR as

$$C_p = \text{PCR} = \frac{\text{USL} - \text{LSL}}{6\sigma} = \frac{74.05 - 73.95}{6(0.0099)} = 1.68$$

A PCR > 1 implies that the "natural" tolerance limits (NTLs) inherent in the process (as quantified by the $\pm 3\sigma$ control limits) are well inside the specification limits. This results in a relatively low number of nonconforming lines being produced. A common variation of the C_p parameter is C_{pk}, where

$$C_{pk} = \min \left\{ \left(\frac{\text{USL} - \mu}{3\sigma} \right), \left(\frac{\mu - \text{LSL}}{3\sigma} \right) \right\} \tag{6.47}$$

The C_{pk} parameter is a measure of the capability of the process to achieve control chart values that lie in the center of the specification range. This metric is useful when the specification limits are not symmetric about the centerline.

The PCR can also be interpreted using the quantity

$$P = \left(\frac{1}{\text{PCR}} \right) \times 100\% \tag{6.48}$$

This is just the percentage of the specification band that the process under consideration "uses up." For the interconnect linewidth example, we compute $P = 59.5\%$, which means that this process uses 59.5% of the specification band. Figure 6.15 illustrates the relationship between the PCR and the specification limits. In Figure 6.15a, the PCR is greater than one, which means that the process uses up much less that 100% of the tolerance band. In this case, few nonconforming products are produced. In Figure 6.15b, PCR = 1, which means that the process uses up all of the tolerance band. Finally, in Figure 6.15c, PCR < 1, and

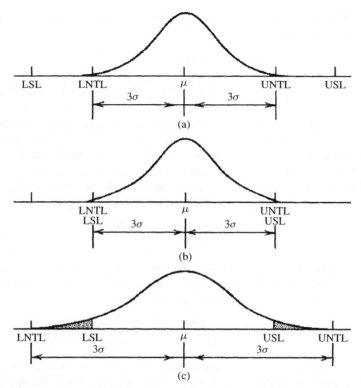

Figure 6.15. Illustration of relationship between specification limits, natural tolerance limits, and process capability ratio [1].

the process uses more than 100% of the tolerance band. In the latter case, a large number of nonconforming products will be produced.

6.4.4. Modified and Acceptance Charts

When $\bar{x}$ charts are used to control the fraction of nonconforming products, two important variations to standard SPC charts can be employed: the *modified* chart and the *acceptance* chart. Modified control limits are generally used when the natural tolerance limits of the process are smaller than the specification limits (i.e., PCR > 1). This occurs frequently in practice, particularly when a quality improvement program exists. In these situations, the modified control chart is designed to detect whether the true process mean (μ) is located such that the process yields a fraction nonconforming in excess of some specified value δ. Essentially, μ is allowed to vary over an interval $\mu_L \leq \mu \leq \mu_U$, where μ_L and μ_U represent lower and upper bounds on μ, respectively, that are consistent with producing a fraction nonconforming of at most δ. This scenario is represented graphically in Figure 6.16.

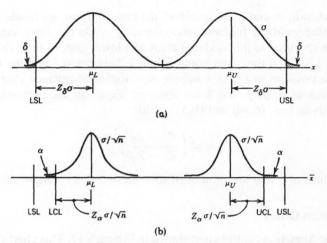

Figure 6.16. Control limits for modified control char: (a) distribution of process output; (b) distribution of the sample mean $\bar{x}$ [1].

To specify control limits for the modified chart (assuming a normally distributed process), for the fraction nonconforming to be less than δ, we must have

$$\mu_L = \text{LSL} + Z_\delta \sigma$$
$$\mu_U = \text{USL} - Z_\delta \sigma \tag{6.49}$$

where Z_δ is the upper $100(1-\delta)$ percentage point of the standard normal distribution. If the specified probability of type I error is α, the upper and lower control limits are then

$$\text{UCL} = \mu_U + \frac{Z_\alpha \sigma}{\sqrt{n}} = \text{USL} - \left(Z_\delta - \frac{Z_\alpha}{\sqrt{n}} \right) \sigma$$

$$\text{LCL} = \mu_L - \frac{Z_\alpha \sigma}{\sqrt{n}} = \text{LSL} + \left(Z_\delta - \frac{Z_\alpha}{\sqrt{n}} \right) \sigma \tag{6.50}$$

Note that using the modified chart is equivalent to testing the hypothesis that the process mean lies in the interval $\mu_L \leq \mu \leq \mu_U$.

Another approach to using the $\bar{x}$ chart to control the fraction nonconforming accounts for both the risk of rejecting a process operating at a satisfactory level (probability of type I error, or α) and the risk of accepting a process that is unsatisfactory (probability of type II error, or β). This second approach is called the *acceptance chart*. The design of this chart is based on a specified sample size (n) and a process fraction nonconforming (γ) that should be rejected with probability $1 - \beta$. In this case, the control limits for the acceptance chart are

$$\text{UCL} = \mu_U - \frac{Z_\beta \sigma}{\sqrt{n}} = \text{USL} - \left(Z_\gamma + \frac{Z_\beta}{\sqrt{n}} \right) \sigma$$

$$\text{LCL} = \mu_L + \frac{Z_\beta \sigma}{\sqrt{n}} = \text{LSL} + \left(Z_\gamma + \frac{Z_\beta}{\sqrt{n}} \right) \sigma \tag{6.51}$$

Note that when n, γ, and β are specified, the control limits are inside the μ_L and μ_U values that yield the fraction nonconforming γ. On the other hand, when n, δ, and α are specified in the modified chart, the lower control limit falls between μ_L and the LSL, and the upper control limit is between μ_U and the USL.

It is also possible to select a sample size for an acceptance chart such that desired values of α, β, γ, and δ are obtained. Equating the expressions for the control limits in Eqs. (6.50) and (6.51) yields

$$n = \left(\frac{Z_\alpha + Z_\beta}{Z_\delta - Z_\gamma}\right)^2 \tag{6.52}$$

Clearly, values of $\delta = \gamma$ are prohibited to achieve a finite sample size.

6.4.5. Cusum Chart

Consider the Shewhart control chart shown in Figure 6.17. This chart corresponds to a normally distributed process with a mean $\mu = 10$ and a standard deviation $\sigma = 1$. Note that all of the first 20 observations appear to be under statistical control. The last 10 observations in this chart were drawn from the same process after the mean has shifted to a new value $\mu = 11$. We can think of these latter observations as having been taken from the process after the mean has shifted out of statistical control by an amount 1σ. However, none of the last 10 points plots outside the control limits, so there is no strong evidence that the process is truly out of control. Even applying the *Western Electric rules*, the Shewhart chart has failed to detect the mean shift.

The reason for this failure is the relatively small magnitude of the shift. Shewhart charts are generally effective for detecting shifts on the order of 1.5–2σ or larger. For smaller shifts the "cumulative sum" (or *cusum*) control chart is preferred. The cusum chart incorporates historical information from a sequence of

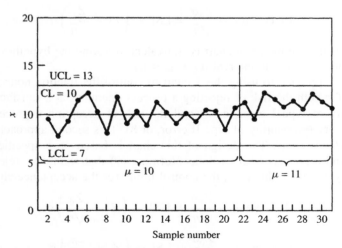

Figure 6.17. Shewhart chart for before and after a mean shift from $\mu = 10$ to $\mu = 11$ [1].

samples by plotting the cumulative sums of the sample deviations from a target value. If samples of size $n \geq 1$ are collected and μ_0 is the target for the process mean, the cusum chart is formed by plotting the quantity

$$C_i = \sum_{j=1}^{i} (\overline{x}_j - \mu_0) \tag{6.53}$$

versus sample i, where $\overline{x}_j$ is the average of the jth sample. Because they combine information from several samples, cusum charts are sensitive to smaller process shifts than are Shewhart charts.

If the process remains in control at the target value μ_0, the sum defined by Eq. (6.53) is a random variable with mean zero. If the mean shifts upward to some value $\mu_1 > \mu_0$, then a positive drift will develop in the cusum chart. Conversely, if the mean shifts downward to some value $\mu_1 < \mu_0$, then a negative drift will be manifested in C_i. This effect is demonstrated in Figure 6.18, which depicts the cusum chart for the same dataset used in Figure 6.17. The upward trend after the first 20 samples is indicative of the mean shift to $\mu = 11$ described previously. This figure, however, does not represent a control chart because it lacks control limits. The methodology for establishing such limits is described in the following subsections.

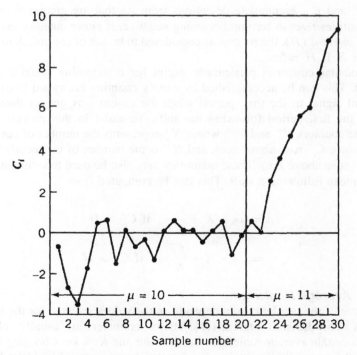

Figure 6.18. Cusum chart for before and after a mean shift from $\mu = 10$ to $\mu = 11$ [1].

6.4.5.1. Tabular Cusum Chart

The tabular form of the cusum chart may be constructed for both individual observations and for averages of rational subgroups. Let x_i be the ith observation of a normally distributed process with mean μ_0 and standard deviation σ. We can think of μ_0 as a "target" value for quality characteristic x. If the process shifts or drifts from this target value, the cusum chart should generate an alarm signal.

The tabular cusum accumulates deviations from μ_0 that are above the target with a statistic C^+ and deviations that are below the target with another statistic C^-. The quantities C^+ and C^- are called the *upper and lower cusums*, respectively. The are computed using the relations

$$C_i^+ = \max\left[0, x_i - (\mu_0 + K) + C_{i-1}^+\right] \tag{6.54}$$

$$C_i^- = \max\left[0, (\mu_0 + K) - x_i + C_{i-1}^-\right] \tag{6.55}$$

where the starting values are $C_0^+ = C_0^- = 0$. In these equations, K is called the *reference value*, and it is usually chosen to be about halfway between the target mean (μ_0) and the shifted mean that we are interested in detecting (μ_1). If the shift is expressed in terms of the standard deviation as $\mu_1 = \mu_0 + \delta\sigma$, the K is given by

$$K = \frac{\delta}{2}\sigma = \frac{|\mu_1 - \mu_0|}{2} \tag{6.56}$$

Both C^+ and C^- accumulate deviations from μ_0 that are greater than K, and both quantities reset to zero on becoming negative. If either quantity exceeds the *decision interval* (H), the process is considered to be out of control. A reasonable value for H is $H = 5\sigma$.

The tabular cusum is particularly useful for determining when a shift has occurred. This can be accomplished by simply counting backward from the out-of-control signal to the time period when the cusum was greater than zero to identify the first period following the shift. To assist in this process, we can define the counters N^+ and N^-, where N^+ represents the number of consecutive periods since C_i^+ rose above zero, and N^- is the number of consecutive periods since C_i^- rose above zero. These quantities may also be used to estimate the new process mean following a shift. This can be computed from

$$\hat{\mu} = \mu_0 + K + \frac{C_i^+}{N^+} \quad \text{if } C_i^+ > H$$

$$= \mu_0 - K - \frac{C_i^-}{N^-}, \quad \text{if } C_i^- > H \tag{6.57}$$

6.4.5.2. Average Runlength

The format of the tabular cusum depends on the values selected for the reference value (K) and decision interval (H). These parameters are usually selected to provide a certain average runlength. Let $H = h\sigma$ and $K = k\sigma$. Choosing $h = 4$–5 and $k = 0.5$ generally results in a cusum that has a reasonable ARL. Table 6.1

Table 6.1. ARL performance of tabular cusum with $k = 0.5$ and $h = 4$ or $h = 5$.

Shift in Mean (Multiple of σ)	$h = 4$	$h = 5$
0	168	465
0.25	74.2	139
0.50	26.6	38
0.75	13.3	17
1.00	8.38	10.4
1.50	4.75	5.75
2.00	3.34	4.01
2.50	2.62	3.11
3.00	2.19	2.57
4.00	1.71	2.01

provides the ARL performance of the tabular cusum for various shifts in the process mean under these conditions.

Generally, k should be chosen relative to the size of the shift to be detected. In other words, $k = 0.5\delta$, where δ is the size of the shift to be detected (in standard deviation units). This approach comes very close to minimizing the out-of-control ARL value for detecting a shift of size δ for a fixed in-control ARL. Once k is chosen, h is then selected to give the desired in-control ARL.

For a *one-sided cusum* (i.e., for either C_i^+ or C_i^-), the ARL may generally be approximated as [1]

$$\text{ARL} = \frac{\exp(-2\Delta b) + 2\Delta b - 1}{2\Delta^2} \tag{6.58}$$

for $\Delta \neq 0$, where $\Delta = \delta^* - k$, $b = h + 1.166$, and

$$\delta^* = \frac{\mu_1 - \mu_0}{\sigma} \tag{6.59}$$

where μ_0 and μ_1 are the target and shifted mean, respectively. If $\Delta = 0$, then the ARL $= b^2$. The quantity δ^* represents the shift in the mean (in standard deviation units) for which the ARL is calculated. The ARL of the two-sided cusum can be derived from the ARLs of the two one-sided statistics (ARL^+ and ARL^-) as

$$\frac{1}{\text{ARL}} = \frac{1}{\text{ARL}^+} + \frac{1}{\text{ARL}^-} \tag{6.60}$$

6.4.5.3. Cusum for Variance

It is also possible to use the cusum technique to monitor process variability. Again, let x_i be a normally distributed process measurement with mean μ_0 and standard deviation σ. Further, let y_i be the standardized value of x_i, or

$$y_i = \frac{x_i - \mu_0}{\sigma} \tag{6.61}$$

Hawkins [5] suggests creating a new standard quantity, v_i, which is sensitive to both mean and variance changes. This parameter is given by

$$v_i = \frac{\sqrt{|y_i|} - 0.822}{0.349} \tag{6.62}$$

Since the in-control distribution of v_i is approximately $N(0,1)$, the two-sided cusums can be written as

$$S_i^+ = \max[0, v_i - k + S_{i-1}^+] \tag{6.63}$$

$$S_i^- = \max[0, -k - v_i + S_{i-1}^-] \tag{6.64}$$

where $S_0^+ = S_0^- = 0$, and the values of k and h are selected in the same way as the values for controlling the process mean. The interpretation of this cusum is also the same as that of the cusum for controlling the mean. If the process standard deviation increases, the values of S_i^+ will increase and eventually exceed h, and if the standard deviation decreases, the values of S_i^- will increase and eventually exceed h.

6.4.6. Moving-Average Charts

6.4.6.1. Basic Moving-Average Chart

Suppose that $x_1, x_2, \ldots, x_n$ individual observations of a process have been collected. The moving average of span w at time i is defined as

$$M_i = \frac{x_i + x_{i-1} + \cdots + x_{i-w+1}}{w} \tag{6.65}$$

At time period i, the oldest observation is dropped and the newest one is added to the set. The variance of the moving average is

$$V(M_i) = \frac{1}{w^2} \sum_{j=i-w+1}^{i} V(x_j) = \frac{1}{w^2} \sum_{j=i-w+1}^{i} \sigma^2 = \frac{\sigma^2}{w} \tag{6.66}$$

Thus, if μ_0 is the target mean used as the centerline of the moving-average control chart, the 3σ control limits for M_i are

$$\text{UCL} = \mu_0 + \frac{3\sigma}{\sqrt{w}} \tag{6.67}$$

$$\text{LCL} = \mu_0 - \frac{3\sigma}{\sqrt{w}} \tag{6.68}$$

The control procedure then consists of calculating a new value for M_i as each new observation becomes available and plotting M_i on a control chart with limits given by Eqs. (6.67) and (6.68). Note that for samples in which $i < w$, i replaces w in these equations. This causes the control limits for the first few samples to

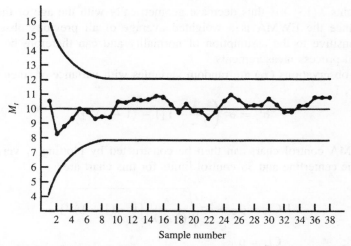

Figure 6.19. Control limits for moving-average chart for a sample dataset [1].

become variable, as is depicted in Figure 6.19. In general, the moving-average control chart is more sensitive than Shewhart charts for detecting small process shifts. However, it is not as effective in that regard as either the cusum or the EWMA (see discussion below).

6.4.6.2. Exponentially Weighted Moving-Average Chart

The *exponentially weighted moving-average* (EWMA) *control chart*, sometimes referred to as the *geometric moving average* (GMA) *chart*, is another alternative to Shewhart charts when it is desirable to detect small process shifts. The performance of the EWMA chart is comparable to that of the cusum chart. The exponentially weighted moving average is defined as

$$z_i = \lambda x_i + (1 - \lambda)z_{i-1} \tag{6.69}$$

where $0 < \lambda \le 1$ is a constant and the starting value is the process target (i.e., $z_0 = \mu_0$).

To show that the parameter z_i is a weighted average of all previous sample means, we can substitute z_{i-1} on the right side of Eq. (6.69) to obtain

$$z_i = \lambda x_i + (1 - \lambda)[\lambda x_{i-1} + (1 - \lambda)z_{i-2}]$$

$$= \lambda x_i + \lambda(1 - \lambda)x_{i-1} + (1 - \lambda)^2 z_{i-2}$$

If we continue to substitute recursively for z_{i-j} for $j = 2, 3, \ldots, t$, we obtain

$$z_i = \lambda \sum_{j=0}^{i-1} (1 - \lambda)^j x_{i-j} + (1 - \lambda)^i z_0 \tag{6.70}$$

The weights $\lambda(1 - \lambda)^j$ thus decrease geometrically with the age of the sample mean. Since the EWMA is a weighted average of all previous observations, it is insensitive to the assumption of normality and can therefore be used for individual process measurements.

If the observations (x_i) are random variables with variance σ^2, then the variance of z_i is

$$\sigma_{z_i}^2 = \sigma^2 \left(\frac{\lambda}{2 - \lambda}\right) [1 - (1 - \lambda)^{2i}] \tag{6.71}$$

The EWMA control chart can then be constructed by plotting z_i versus i (or time). The centerline and 3σ control limits for this chart are

$$UCL = \mu_0 + 3\sigma \sqrt{\frac{\lambda}{(2 - \lambda)} [1 - (1 - \lambda)^{2i}]}$$

$$CL = \mu_0 \tag{6.72}$$

$$LCL = \mu_0 - 3\sigma \sqrt{\frac{\lambda}{(2 - \lambda)} [1 - (1 - \lambda)^{2i}]}$$

Notice that the term $[1 - (1 - \lambda)^{2i}]$ in these equations approaches unity as i gets larger. The control limits therefore reach steady-state values of

$$UCL = \mu_0 + 3\sigma \sqrt{\frac{\lambda}{(2 - \lambda)}}$$

$$CL = \mu_0 \tag{6.73}$$

$$LCL = \mu_0 - 3\sigma \sqrt{\frac{\lambda}{(2 - \lambda)}}$$

This variation in control limits with i is depicted in Figure 6.20.

The EWMA method is related to the *proportional–integral–differential* (PID) approach often used in classical control problems. Note that the EWMA parameter z_i in Eq. (6.69) can be manipulated algebraically and rewritten as

$$z_i = z_{i-1} + \lambda(x_i - z_{i-1}) \tag{6.74}$$

If z_{i-1} is viewed as a forecast of the process mean in sample period i, then we can think of $x_i - z_{i-1}$ as the forecast error (e_i) for period i, or

$$z_i = z_{i-1} + \lambda e_i \tag{6.75}$$

In other words, the forecast for period i is the forecast from the previous period plus a fraction of the forecast error. The second term in Eq. (6.75) is therefore known as the *proportional* term.

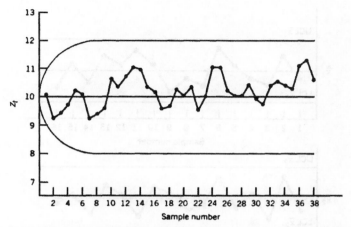

Figure 6.20. Control limits for EWMA chart with $\lambda = 0.2$ for a sample dataset [1].

We can add a second *integral* term to Eq. (6.75) to get

$$z_i = z_{i-1} + \lambda_1 e_i + \lambda_2 \sum_{j=1}^{i} e_j \qquad (6.76)$$

where λ_1 and λ_2 are coefficients that weight the error at period i and the sum of the errors accumulated up to period i, respectively. If we let $\nabla e = e_i - e_{i-1}$ represent the difference between the errors in periods i and $i - 1$, then can add a third *differential* term to Eq. (6.76) to yield

$$z_i = z_{i-1} + \lambda_1 e_i + \lambda_2 \sum_{j=1}^{i} e_j + \lambda_3 \nabla e_i \qquad (6.77)$$

In summary, the empirical control equation represented by Eq. (6.77) states that the EWMA in period i (which is a forecast of the process mean in period $i + 1$) is the sum of the current estimate of the mean (z_{i-1}), a term proportional to the forecast error, a term related to the sum of the forecast errors, and a term related to the difference between the two most recent forecast errors. The latter three terms can be thought of as *proportional, integral,* and *differential* adjustments, and the parameters λ_1, λ_2, and λ_3 are selected to provide the best forecasting performance.

6.5. MULTIVARIATE CONTROL

In many situations, it is desirable to control two or more quality characteristics simultaneously. For example, we may be interested in controlling the linewidth of a test structure on two different product wafers. Suppose that these two character-istics are represented by the random variables x_1 and x_2, which have a bivariate

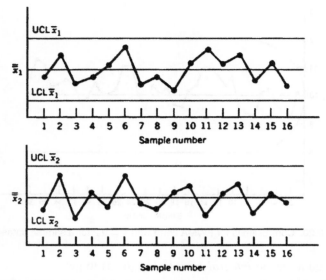

Figure 6.21. $\bar{x}$ control charts for bivariate normal process variables [1].

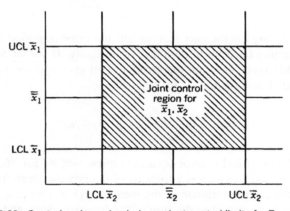

Figure 6.22. Control region using independent control limits for $\bar{x}_1$ and $\bar{x}_2$ [1].

normal distribution. Suppose also that each variable is controlled by an $\bar{x}$ chart (see Figure 6.21). Since the process is under control only if both sample means ($\bar{x}_1$ and $\bar{x}_2$) fall within their respective control limits, the joint control region for both variables is as shown in Figure 6.22.

Controlling these two process variables in this manner can be misleading. Since the probability that *either* $\bar{x}_1$ or $\bar{x}_2$ exceeds its control limits when in control is 0.0027, the joint probability that *both* variables simultaneously exceed their control limits when both are in fact in control is $(0.0027)^2 = 0.00000729$, which is significantly less than 0.0027. Moreover, the probability that both variables will plot inside the control limits when under control is $(0.9973)^2 = 0.99460729$. The use of independent $\bar{x}$ charts to control both variables simultaneously thus distorts the probability of type I error as compared to individual control charts.

Such distortion increases with the number of quality characteristics. If there are p independent quality characteristics, each with $P\{\text{type I error}\} = \alpha$, then the overall probability for type I error is

$$\alpha' = 1 - (1 - \alpha)^p \tag{6.78}$$

and the probability that all p process means will simultaneously plot inside their control limits when the process is in control is

$$P\{\text{all means plot in control}\} = (1 - \alpha)^p \tag{6.79}$$

In addition, if the p quality characteristics are not all mutually independent (which would be the case if they were related to the same product), then Eqs. (6.78) and (6.79) would not be valid, and there would be no easy way to measure the distortion in this control procedure.

6.5.1. Control of Means

Let μ_1 and μ_2 represent the mean values of x_1 and x_2, and let σ_1 and σ_2 be their respective standard deviations. The covariance between x_1 and x_2, a measure of their dependence, is denoted by σ_{12}. If $\bar{x}_1$ and $\bar{x}_2$ are the sample averages computed from a sample of size n, then the statistic

$$\chi_0^2 = \frac{n}{\sigma_1^2\sigma_2^2 - \sigma_{12}^2} \left[\sigma_2^2(\bar{x}_1 - \mu_1)^2 + \sigma_1^2(\bar{x}_2 - \mu_2)^2 - 2\sigma_{12}(\bar{x}_1 - \mu_1)(\bar{x}_2 - \mu_2) \right]$$

$$\tag{6.80}$$

has a chi-square distribution with 2 degrees of freedom. This equation can be used to develop a control chart for the process means. If the means remain under control (i.e., have not shifted), then $\chi_0^2 < \chi_{\alpha,2}^2$, where $\chi_{\alpha,2}^2$ is the upper percentage point of the chi-square distribution with 2 degrees of freedom. If, on the other hand, one of the means shifts to an out-of-control value, then $\chi_0^2 > \chi_{\alpha,2}^2$.

This control procedure can be represented graphically, as shown in Figures 6.23 and 6.24. Figure 6.23 depicts the case where x_1 and x_2 are independent ($\sigma_{12} = 0$), and the principal axes of the "control ellipse" are parallel to the $\bar{x}_1$ and $\bar{x}_2$ axes. Figure 6.24 shows the control ellipse when $\sigma_{12} \neq 0$. In both cases, sample averages yielding χ_0^2 points plotting inside the ellipse are indicative of statistical control, and points plotting outside represent out-of-control conditions.

There are two primary disadvantages of the control ellipse approach. The first is that the time sequence of the sample measurements is completely lost. The second shortcoming is the difficulty in graphically depicting the control region for more than two variables. To avoid these difficulties, it is customary to plot the values of χ_0^2 computed from Eq. (6.80) on a control chart with only an upper control limit at $\chi_{\alpha,2}^2$ (see Figure 6.25).

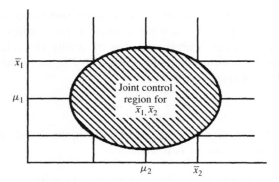

Figure 6.23. Control ellipse for two independent variables [1].

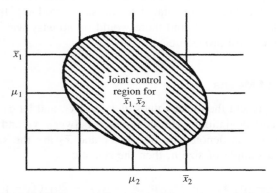

Figure 6.24. Control ellipse for two dependent variables [1].

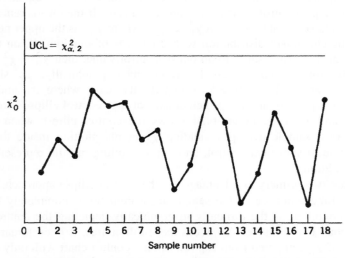

Figure 6.25. χ^2 control chart for $p = 2$ quality characteristics [1].

It is possible to extend this approach to situations where $p > 2$. Assuming again that the p random variables are multivariate normal, this procedure requires computing the sample mean of each quality characteristic from a sample of size n. This set of sample means is represented by the vector

$$\bar{\mathbf{x}} = \begin{bmatrix} \bar{x}_1 \\ \bar{x}_2 \\ \vdots \\ \bar{x}_p \end{bmatrix}$$

The test statistic to be plotted on a control chart like that of Figure 6.25 is then

$$\chi_0^2 = n(\bar{\mathbf{x}} - \boldsymbol{\mu})' \boldsymbol{\Sigma}^{-1} (\bar{\mathbf{x}} - \boldsymbol{\mu}) \tag{6.81}$$

where $\boldsymbol{\mu}' = [\mu_1, \mu_2, \ldots, \mu_p]$ and $\boldsymbol{\Sigma}$ is the covariance matrix. The upper control limit for this chart is then $\chi_{\alpha, p}^2$.

In practice, $\boldsymbol{\mu}$ and $\boldsymbol{\Sigma}$ must be estimated from samples taken when the process is under control. Suppose that m such samples are taken. The sample means and variances are

$$\bar{x}_{jk} = \frac{1}{n} \sum_{i=1}^{n} x_{ijk} \quad \begin{cases} j = 1, 2, \ldots, p \\ k = 1, 2, \ldots, m \end{cases} \tag{6.82}$$

$$S_{jk}^2 = \frac{1}{n-1} \sum_{i=1}^{n} (x_{ijk} - \bar{x}_{jk})^2 \quad \begin{cases} j = 1, 2, \ldots, p \\ k = 1, 2, \ldots, m \end{cases} \tag{6.83}$$

where x_{ijk} is the ith observation on the jth quality characteristic in the kth sample. The covariance between characteristic j and h in the kth sample is

$$S_{jhk} = \frac{1}{n-1} \sum_{i=1}^{n} (x_{ijk} - \bar{x}_{jk})(x_{ihk} - \bar{x}_{hk}) \quad \begin{cases} k = 1, 2, \ldots, m \\ j \neq h \end{cases} \tag{6.84}$$

The statistics $\bar{x}jk$, S_{jk}^2, and S_{jhk} are averaged over all m samples to obtain

$$\bar{\bar{x}}_j = \frac{1}{m} \sum_{k=1}^{m} \bar{x}_{jk} \quad j = 1, 2, \ldots, p \tag{6.85}$$

$$S_j^2 = \frac{1}{m} \sum_{k=1}^{m} S_{jk}^2 \quad j = 1, 2, \ldots, p \tag{6.86}$$

$$S_{jh} = \frac{1}{m} \sum_{k=1}^{m} S_{jhk} \quad j \neq h \tag{6.87}$$

The $\{\bar{x}_j\}$ are elements of a vector $\bar{\bar{\mathbf{x}}}$, and the $p \times p$ sample covariance matrix S is

$$\mathbf{S} = \begin{bmatrix} S_1^2 & S_{12} & S_{13} & \cdots & S_{1p} \\ & S_2^2 & S_{23} & \cdots & S_{2p} \\ & & \ddots & & \vdots \\ & & & & S_p^2 \end{bmatrix} \tag{6.88}$$

The parameters $\boldsymbol{\mu}$ and $\boldsymbol{\Sigma}$ in Eq. (6.81) are then replaced by $\bar{\bar{\mathbf{x}}}$ and $\mathbf{S}$, respectively. The test statistic is now

$$T^2 = n(\bar{\mathbf{x}} - \bar{\bar{\mathbf{x}}})'\mathbf{S}^{-1}(\bar{\mathbf{x}} - \bar{\bar{\mathbf{x}}}) \tag{6.89}$$

The T^2 statistic is know as *Hotelling's T^2*. The distribution of this T^2 statistic is related to the F distribution by the expression

$$\frac{n-p}{p(n-1)}T^2 \sim F(p, n-p) \tag{6.90}$$

The T^2 statistic can be plotted on a control chart with a UCL $= \chi_{\alpha,p}^2$. However, one difficulty that arises in the use of either the χ_0^2 or T^2 statistics in control charts is the interpretation of an out-of-control signal. Specifically, it is difficult to determine which subset of the p variables is responsible for the signal.

6.5.2. Control of Variability

Multivariate process variability is summarized by the $p \times p$ covariance matrix $\boldsymbol{\Sigma}$. One approach to controlling variability is based on the determinant of the sample covariance ($|\mathbf{S}|$), which is known as the *generalized sample variance*. Let $E(|\mathbf{S}|)$ and $V(|\mathbf{S}|)$ be the mean and variance of $|\mathbf{S}|$, respectively. It can be shown that [1]

$$E(|\mathbf{S}|) = b_1|\boldsymbol{\Sigma}| \tag{6.91}$$

$$V(|\mathbf{S}|) = b_2|\boldsymbol{\Sigma}|^2 \tag{6.92}$$

where

$$b_1 = \frac{1}{(n-1)^p} \prod_{i=1}^{p}(n-i) \tag{6.93}$$

$$b_2 = \frac{1}{(n-1)^{2p}} \prod_{i=1}^{p}(n-i)\left[\prod_{i=1}^{p}(n-j+2) - \prod_{i=1}^{p}(n-j)\right] \tag{6.94}$$

The parameters of the control chart for $|\mathbf{S}|$ are then

$$\text{UCL} = |\boldsymbol{\Sigma}|\left(b_1 + 3\sqrt{b_2}\right)$$

$$\text{CL} = b_1|\boldsymbol{\Sigma}| \tag{6.95}$$

$$\text{LCL} = |\boldsymbol{\Sigma}|\left(b_1 - 3\sqrt{b_2}\right)$$

The LCL in Eq. (6.95) is set to zero if the calculated value is less than zero. Also, in practice, Σ is usually estimated by S. In that case, $|\Sigma|$ in Eq. (6.95) is replaced by $|S|/b_1$, which is an unbiased estimator of $|\Sigma|$.

6.6. SPC WITH CORRELATED PROCESS DATA

The standard assumptions for using Shewhart control charts are that the data generated by the monitored process while it is under control are normally and independently distributed. Both the process mean (μ) and standard deviation (σ) are considered fixed and unknown. Therefore, when the process is under control, it can be represented by the model

$$x_t = \mu + \varepsilon_t \qquad t = 1, 2, \ldots \qquad (6.96)$$

where $\varepsilon_t \sim N(0, \sigma^2)$. When these assumptions hold, one may apply conventional SPC techniques to such charts.

The most critical of these assumptions is the independence of the observations. Shewhart charts do not work well if the process measurements exhibit any level of correlation over time. They will give misleading results under these conditions, usually in the form of too many false alarms. Unfortunately, the assumption of uncorrelated (or independent) observations is not satisfied for many semiconductor manufacturing processes. For example, in a CVD process, consecutive temperature measurements are often highly correlated. Automated test and inspection procedures are also examples of processes that yield measurements that are correlated in time. Alternative SPC methods must therefore be applied to these situations.

6.6.1. Time-Series Modeling

When a process measurement taken at time t depends on the value measured at time $t - 1$, the measurements are said to be *autocorrelated*. A sequence of time-oriented observations such as this is referred to as a *time series*. It is possible to measure the level of autocorrelation in a time series analytically using the *autocorrelation function*

$$\rho_k = \frac{\text{cov}(x_t, x_{t-k})}{V(x_t)} \qquad (6.97)$$

where $\text{cov}(x_t, x_{t-k})$ is the covariance of observations that are k time periods apart and $V(x_t)$ is the variance of the observations (which is assumed to be constant). Autocorrelation is usually estimated using the sample autocorrelation function

$$r_k = \frac{\sum\limits_{t=1}^{n-k} (x_t - \overline{x})(x_{t-k} - \overline{x})}{\sum\limits_{t=1}^{n} (x_t - \overline{x})^2} \qquad (6.98)$$

Autocorrelated data can be modeled using *time-series models*. For example, the quality characteristic x_t could be modeled using the expression

$$x_t = \xi + \phi x_{t-1} + \varepsilon_t \tag{6.99}$$

where ξ and $\phi(-1 < \phi < 1)$ are unknown constants, and $\varepsilon_t \sim N(0, \sigma^2)$. Equation (6.99) is called a *first-order autoregressive model*. The observations (x_t) from this model have mean $\xi(1 - \phi)$ and standard deviation $\sigma/\sqrt{1 - \phi^2}$. Observations that are k time periods apart $(x_t$ and $x_{t-k})$ have a correlation coefficient ϕ^k.

The first-order autoregressive model is clearly not the only possible model for correlated time-series data. A natural extension to Eq. (6.99) is

$$x_t = \xi + \phi_1 x_{t-1} + \phi_2 x_{t-2} + \varepsilon_t \tag{6.100}$$

which is a *second-order autoregressive model*. In autoregressive models, the variable x_t is directly dependent on previous observations. Another possibility is to model this dependence through ε_t. The simplest way to do so is

$$x_t = \mu + \varepsilon_t - \theta\varepsilon_{t-1} \tag{6.101}$$

which is called a *first-order moving-average model*. In this model, the correlation between x_t and x_{t-1} is

$$\rho_1 = -\theta/(1 + \theta^2) \tag{6.102}$$

Combinations of autoregressive and moving-average models are often useful as well. A first-order *autoregressive, moving-average* (ARMA) model is

$$x_t = \xi + \phi x_{t-1} + \varepsilon_t - \theta\varepsilon_{t-1} \tag{6.103}$$

More generally, this ARMA model may be extended to arbitrary order using

$$\hat{x}_t = \sum_{i=1}^{p} \phi_i x_{t-i} - \sum_{j=1}^{q} \theta_j e_{t-j} \tag{6.104}$$

where $\hat{x}_t$ is the model prediction of the time-series data, e_t is the residual for each timepoint (i.e., $e_t = x_t - \hat{x}_t$), ϕ_i are the autoregressive coefficients of order p, and θ_j are the moving-average coefficients of order q. The parameters in the ARMA or other time-series models may be estimated by the method of least squares (see Chapter 8). Using that technique, values of ξ, ϕ_i, or θ_j are selected that minimize the sum of squared errors (e_t^2).

A further extension of the basic ARMA model is the autoregressive *integrated* moving-average (ARIMA) model. The ARIMA model originates from the *first-order integrated moving-average model*

$$x_t = x_{t-1} + \varepsilon_t - \theta\varepsilon_{t-1} \tag{6.105}$$

While the previous models are used to describe stationary behavior (i.e., x_t wanders around a fixed mean), the model in Eq. (6.106) describes nonstationary behavior in which the process mean drifts. This situation often arises in processes in which no control actions are taken to keep the mean close to a target value. Occasionally, the original data may also show seasonal, periodic patterns. These seasonal patterns can be modeled by creating ARIMA models for seasonal means. This composite model is known as the *seasonal* ARIMA (SARIMA).

6.6.2. Model-Based SPC

One approach for dealing with autocorrelated data is to directly model the correlation with an appropriate time-series model, use that model to remove the autocorrelation from the data, and apply control charts to the residuals. This approach is known as *model-based SPC*.

Consider the first-order autoregressive model described by Eq. (6.99). Suppose that $\hat{\phi}$ is an estimator of ϕ obtained from the analysis of sample data obtained from the process. Then $\hat{x}_t$ is an estimate of x_t and the corresponding residuals $(e_t = x_t - \hat{x}_t)$ are approximately normally and independently distributed with zero mean and constant variance. Conventional Shewhart or other control charts may now be legitimately applied to the sequence of residuals. Points out of control or exhibiting unusual patterns would then be indicative of a shift in ϕ, implying that the original variable x_t was out of control.

This approach is equally valid for more complex time-series models such as the ARMA family. As an example, Figure 6.26 shows time-series data collected from a Lam Rainbow 4400 reactive-ion etching system [6]. This particular dataset

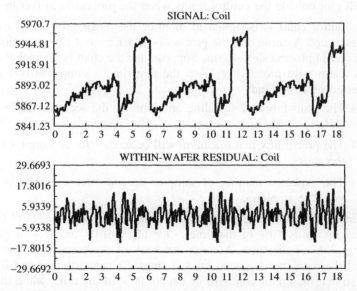

Figure 6.26. RIE RF coil position time series: (a) raw data and (b) control chart for residuals [6].

represents the signal for the coil position in the RF matching network of the RIE. Figure 6.26a shows the raw data for this signal, and Figure 6.26b shows the $\bar{x}$ control chart for the residuals after an ARMA model for the raw data has been constructed. In this case, the time-series data are under statistical control, as the residuals do not exceed the upper and lower control limits for the interval under observation.

SUMMARY

In this chapter, we have provided an overview of statistical process control, from basic control charts to advanced techniques. This overview has focused on the use of SPC to analyze quality issues and improve the performance of semiconductor manufacturing processes. As ever, the overall goal is to reduce variability and improve yield. In the next chapter, we turn our attention to statistical experimental design, an essential method for identifying the key variables influencing the quality characteristics that are monitored by SPC.

PROBLEMS

6.1. Consider Western Electric rule 3. What is the probability of four out of five consecutive points plotting beyond the 2σ warning limits?

6.2. A normally distributed quality characteristic is monitored by a control chart with 3σ limits. Derive a general expression for the probability that a point will plot outside the control limits when the process is in fact in control.

6.3. A control chart is designed to monitor the threshold voltage of NMOS transistors. Assume that the process is under control for some time before an abrupt process shift occurs. Suppose that the chart is set so that it signals an alarm with probability $1 - \beta$ the first time a sample arrives from the shifted process. Find

(a) The probability of signaling an alarm on the second sample after the shift.

(b) The probability that the alarm will be *missed* for K samples following the shift.

(c) The expected number of samples needed after the shift in order to generate an alarm.

6.4. Suppose that out of a group of 10 coins, 9 of them are "fair" (i.e., they turn up heads 50% of the time). One of them is "unfair"—it gives tails only 35% of the time. Assume that each of the coins is thrown n times and the outcome is plotted on a p chart. Calculate the control limits and n so that the unfair coin will be caught 90% of the time, while the chance of rejecting a fair coin will be at most 1%.

6.5. A particle counting device monitors wafers emerging from a plasma etcher. From previous experience, it is known that the machine generates an average of two defects per wafer. Establish a control procedure that will generate false alarms only 1% of the time (there is no lower control limit). What is the best type of control chart, and what is the necessary UCL?

6.6. Control charts for $\bar{x}$, R, and s are to be maintained for the threshold voltage of short-channel MOSFETs using sample sizes of $n = 10$. It is known that the process is normally distributed with $\mu = 0.75$ V and $\sigma = 0.10$ V. Find the centerline and control limits for each of these control charts.

6.7. Repeat the previous problem assuming that μ and σ are unknown and that we have collected 50 observations of sample size 10. These samples yielded a grand average of 0.734 V, an average s_i of 0.125 V, and an average R_i of 0.365 V.

6.8. A fabrication line used for the manufacture of analog ICs requires tight control on the relative sizes of small geometric features. In a particular case, it is required that transistors on either side of a differential pair differ less than 0.1 μm in their effective gate lengths. If that difference is normally distributed with $\mu = 0$ and $\sigma = 0.05$ μm:

(a) Calculate the process capability (C_p) and the fraction of nonconforming product when the process is in control.

(b) Suppose the mean of the process shifts and this shift doubles the fraction of nonconforming product. Calculate the sample size needed to implement a p chart that will detect this shift on the first subsequent sample with 50% certainty.

(c) Design a 3σ $\bar{x}$ and R charts that will detect the previous shift with the same 50% certainty.

6.9. The following values of saturation drain current ($I_{D,\text{sat}}$) were collected from several test wafers with a sample size of $n = 5$.

$\bar{x}$ (mA)	R (mA)
1.03	0.04
1.02	0.05
1.04	0.02
1.05	0.11
1.04	0.04
1.06	0.03
1.02	0.07
1.05	0.02
1.06	0.04
1.04	0.03

(a) Calculate the centerlines and control limits.

(b) Assuming $I_{D,\text{sat}}$ to be normally distributed, compute the standard deviation of the process.

(c) Give an estimate of the fraction nonconforming if the specification limits are 1.03 ± 0.04 mA.

(d) Suggest ways to reduce the fraction of nonconforming product.

6.10. The $\bar{x}$ and R values for 20 samples of size $n = 5$ are shown below. The specification limits of this product are 530–570.

$\bar{x}$	549	548	548	551	553	552	550	551	553	556
R	2.5	2.1	2.3	2.9	1.8	1.7	2.0	2.4	2.2	2.8
$\bar{x}$	547	545	549	552	550	548	556	546	550	551
R	2.0	3.0	3.1	2.2	2.3	2.1	1.9	1.8	2.1	2.2

(a) Construct a modified control chart with 3σ limits. Assume that if the true fraction nonconforming is as large as 1%, the process is acceptable. Is this process satisfactory?

(b) Suppose that if the true fraction nonconforming is as large as 1%, the modified control chart should detect this out-of-control condition with probability 0.9. Construct the modified chart and compare it to the chart obtained in part (a).

(c) Is this process in statistical control?

6.11. The following data represent temperature measurements from a CVD process. The target temperature is $1050°C$ and the standard deviation is $\sigma = 25°C$. Set up the cusum chart for the mean of this process. Design the cusum to quickly detect a shift of 1σ in the process mean.

Observation	T ($°C$)	Observation	T ($°C$)
1	1045	11	1139
2	1055	12	1169
3	1037	13	1151
4	1064	14	1128
5	1095	15	1238
6	1008	16	1125
7	1050	17	1163
8	1087	18	1188
9	1125	19	1146
10	1146	20	1167

6.12. Consider a process with $\mu_0 = 10$ and $\sigma = 1$. Set up 3σ EWMA control charts for $\lambda = 0.1$, 0.2, and 0.4. Discuss the effect of λ on the behavior of the control limits.

6.13. The data below come from a process with two observable quality characteristics. The data are the means of each characteristic, based on a sample
size of $n = 25$. The nominal values and covariance matrix are

$$\bar{\bar{\mathbf{x}}} = \begin{bmatrix} 55 \\ 30 \end{bmatrix} \qquad \mathbf{S} = \begin{bmatrix} 200 & 130 \\ 130 & 120 \end{bmatrix}$$

Construct the T^2 control chart using these data.

Sample	$\bar{\mathbf{x}}_1$	$\bar{\mathbf{x}}_1$
1	58	32
2	60	33
3	50	27
4	54	31
5	63	38
6	53	30
7	42	20
8	55	31
9	46	25
10	50	29
11	49	27
12	57	30
13	58	33
14	75	45
15	55	27

REFERENCES

1. D. Montgomery, *Introduction to Statistical Quality Control*, Wiley, New York, 1993.
2. A. Duncan, *Quality Control and Industrial Statistics*, Irwin, Homewood, IL, 1974.
3. Western Electric, *Statistical Quality Control Handbook*, Western Electric Corp., Indianapolis, IN, 1956.
4. S. Campbell, *The Science and Engineering of Microelectronic Fabrication*, Oxford Univ. Press, New York, 2001.
5. D. Hawkins, "Cumulative Sum Control Charting: An Underutilized SPC Tool," *Qual. Engi.* **5** (1993).
6. S. Lee, E. Boskin, H. Liu, E. Wen, and C. Spanos, "RTSPC: A Software Utility for Real-Time SPC and Tool Analysis," *IEEE Trans. Semiconduct. Manuf.* **8**(1) (Feb. 1995).

7

STATISTICAL EXPERIMENTAL DESIGN

OBJECTIVES

- Provide an overview of statistical experimental design techniques.
- Introduce the concept of analysis of variance (ANOVA).
- Define and describe various types factorial designs.
- Discuss the Taguchi method of experimental design.

INTRODUCTION

Experiments allow investigators to determine the effects of several variables on a given process or product. A *designed experiment* is a test or series of tests that involve purposeful changes to these variables in order to observe the effect of the changes on that process or product. *Statistical experimental design* is an efficient approach for systematically varying these controllable process variables and ultimately determining their impact on process or product quality. This approach is useful for comparing methods, deducing dependences, and creating models to predict effects.

Statistical process control and experimental design are closely interrelated. Both techniques can be used to reduce variability. However, SPC is a passive approach in which a process is monitored and data are collected, whereas experimental design requires active intervention in performing tests on the process under

Fundamentals of Semiconductor Manufacturing and Process Control,
By Gary S. May and Costas J. Spanos
Copyright © 2006 John Wiley & Sons, Inc.

different conditions. Experimental design can also be beneficial in implementing SPC, since designed experiments may help identify the most influential process variables, as well as their optimum settings.

Overall, experimental design is a powerful engineering tool for improving a manufacturing process. Application of experimental design techniques can lead to

- Improved yield
- Reduced variability
- Reduced development time
- Reduced cost

Ultimately, the result is enhanced manufacturability, performance, and product reliability. This chapter illustrates the use of experimental design methods in semiconductor manufacturing.

7.1. COMPARING DISTRIBUTIONS

In the method of statistical inference known as *hypothesis testing* (see Chapter 4), an investigator must evaluate a result produced by making some experimental modification of a system. The investigator must determine whether the result is explainable by mere chance or whether it is due to the effectiveness of the modification. In order to make this determination, the experimenter must identify a relevant reference set that represents a characteristic set of outcomes that could occur if the modification were completely without effect. The actual experimental outcome can then be compared with this reference set. If the experimental results are found to be exceptional, the results are considered *statistically significant*.

Consider the yield data in Table 7.1 obtained from a semiconductor manufacturing process in which two batches of 10 wafers each were fabricated using a standard method (method A) and a modified method (method B). The question to be answered from the experiment is what evidence (if any) do the data collected provide that method B is really better than method A?

To answer this question, we examine the average yields for each process. The modified method (method B) gave an average yield that was 1.30% higher than the standard method. However, because of the considerable variability in the individual test results, it might not be correct to immediately conclude that method B is superior to method A. In fact, it is conceivable that the difference observed could be due to experimental error, operator error, or even pure chance.

One approach to determining the significance of the differences between method A and method B is to use an *external reference distribution*. Suppose in this instance that additional data were available in the form of 210 past process records. These 210 past observations, plotted in Figure 7.1, were made using the standard process, method A. The key question now becomes: How often have the yield differences between the averages of successive groups of 10 wafers been at as large as 1.30%? If the answer is "frequently," we conclude that the

Table 7.1. Yield data from a hypothetical semiconductor manufacturing process [1].

Wafer	Method A Yield (%)	Method B Yield (%)
1	89.7	84.7
2	81.4	86.1
3	84.5	83.2
4	84.8	91.9
5	87.3	86.3
6	79.7	79.3
7	85.1	86.2
8	81.7	89.1
9	83.7	83.7
10	84.5	88.5
Average	84.24	85.54

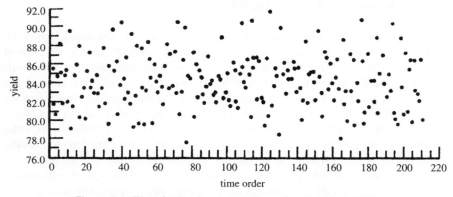

Figure 7.1. Plot of 210 prior observations of method A yield [1].

observed difference can be readily explained by the purely chance variations in the process. However, if the answer is "rarely," a better explanation is that the modification in method B has truly produced an increase in the mean yield.

Figure 7.2 shows the 191 differences between yield averages of adjacent groups of 10 observations in the database of 210 past process records. These 191 differences were obtained by comparing the averages of wafers $1-10$, $2-11$, and so on. They provide a *relevant reference set* with which the observed difference of 1.30% may be legitimately compared. Doing so, it is seen that rarely, in fact, do the differences in the reference set exceed 1.30% (specifically, in only nine cases). Using statistical terminology, we can say that in relation to this reference set, the observed difference of 1.30% is statistically significant at the $9/191 = 0.047$ level. In other words, less than 5 times in 100 would a difference as large as 1.30% be found in the reference set. Thus, it is likely that an actual difference does exist, and method B is truly better than method A.

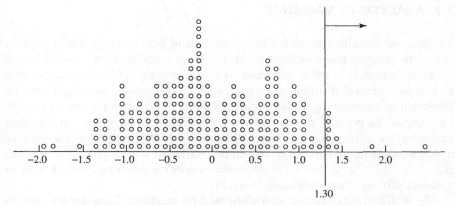

1.30

Figure 7.2. Reference distribution for historical method A yield data [1].

The external reference distribution technique can be problematic. Suppose that there is no historical database of yield obtained using method A. In this case, the proper approach to determine whether the difference between the two manufacturing processes is significant is a statistical *hypothesis test* (Section 4.5). In this case the hypothesis can be represented as

$$H_0: \mu_A = \mu_B$$

$$H_1: \mu_A \neq \mu_B \tag{7.1}$$

where μ_A and μ_B represent the mean yields for the two methods. Since the variance for this process is not known, the test statistic for this hypothesis is

$$t_0 = \frac{(\overline{y}_A - \overline{y}_B)}{s_p \sqrt{\dfrac{1}{n_A} + \dfrac{1}{n_B}}} \tag{7.2}$$

where y_A and y_B are the sample means, n_A and n_B are the number of trials in each sample (10 each in this case), and

$$s_p^2 = \frac{(n_A - 1)s_A^2 + (n_B - 1)s_B^2}{n_A + n_B - 2} \tag{7.3}$$

We use the pooled estimate of the common variance since although the variance for the process is unknown, there is no reason to suspect that the application of method A or method B will produce a different variance. The values of the sample variances [calculated using Eq. (4.2)] are $s_A = 2.90$ and $s_B = 3.65$. Using Eqs. (7.3) and (7.2) then gives values of $s_p = 3.30$ and $t_0 = 0.88$, respectively. Interpolating from Appendix D, we find that the likelihood of computing a t statistic with $n = n_A + n_B - 2 = 18$ degrees of freedom equal to 0.88 is 0.195. The value 0.195 is the *statistical significance* of the hypothesis test. This means that there is only an 19.5% chance that the observed difference between the mean yields is due to pure chance. In other words, we can be 80.5% confident that method B is really superior to method A.

7.2. ANALYSIS OF VARIANCE

The scenario described above is a useful example of how we might use hypothesis testing to compare two distributions. However, in many cases, we would like to go even further; it is often important in manufacturing applications to be able to compare several distributions simultaneously. Moreover, we might also be interested in determining which process conditions in particular have a significant impact on process quality. *Analysis of variance* (ANOVA) is an excellent technique for accomplishing these objectives. ANOVA builds on the idea of hypothesis testing and allows us to compare different sets of process conditions (i.e., "treatments"), as well as to determine whether a given treatment results in a statistically significant variation in quality.

The ANOVA procedure is best illustrated by example. Consider the data in Table 7.2, which represents hypothetical yield data measured for four different sets of process recipes (labeled A–D). Through the use of ANOVA, we will determine whether the discrepancies *between* recipes (i.e., treatments) is truly greater than the variation of the yield *within* the individual groups processed with the same recipe. We assume that the data can be treated as independent random samples from four normal populations having the same variance and differing only in their means (if at all).

Let k be the number of treatments ($k = 4$ in this case). Note that the sample size (n) for each treatment varies ($n_1 = 4$, $n_2 = n_3 = 6$, and $n_4 = 8$). The treatment means (in %) are $\bar{y}_1 = 61$, $\bar{y}_2 = 66$, $\bar{y}_3 = 68$, and $\bar{y}_4 = 61$. The total number of samples (N) is 24, and the *grand average* of all 24 samples, which is sum of all observations divided by the total number of observations, is $\bar{y} = 64\%$.

7.2.1. Sums of Squares

To perform ANOVA, several key parameters must be computed. These parameters, called *sums of squares*, serve to quantify deviations within and between different treatments. Let y_{ti} represent the ith observation for the tth treatment.

Table 7.2. Hypothetical yield (in %) for four different process recipes [1].

Recipe A	Recipe B	Recipe C	Recipe D
62	63	68	56
60	67	66	62
63	71	71	60
59	64	67	61
	65	68	63
	66	68	64
			63
			59

The sum of squares within the tth treatment is given by

$$S_t = \sum_{i=1}^{n_t} (y_{ti} - \overline{y}_t)^2 \tag{7.4}$$

where n_t is the sample size for the treatment in question and $\overline{y}_t$ is the treatment mean. The *within-treatment sum of squares* for all treatments is

$$S_R = S_1 + S_2 + \cdots + S_k = \sum_{t=1}^{k} \sum_{i=1}^{n_t} (y_{ti} - \overline{y}_t)^2 \tag{7.5}$$

In order to quantify the deviations of the treatment averages about the grand average, we use the *between-treatment sum of squares*, which is given by

$$S_T = \sum_{t=1}^{k} n_t (\overline{y}_t - \overline{y})^2 \tag{7.6}$$

Finally, the total sum of squares for all the data about the grand average is

$$S_D = \sum_{t=1}^{k} \sum_{i=1}^{n_t} (y_{ti} - \overline{y})^2 \tag{7.7}$$

Each sum of squares has an associated number of *degrees of freedom* required for its computation. The degrees of freedom for the within-treatment, between-treatment, and total sums of squares, respectively, are

$$v_R = N - k$$
$$v_T = k - 1 \tag{7.8}$$
$$v_D = N - 1$$

The final quantity needed to carry out analysis of variance is the pooled estimate of the variance quantified by each sum of squares. This quantity, known as the *mean square*, is equal to the ratio of the sum of squares to its associated number of degrees of freedom. The within-treatment, between-treatment, and total mean squares are therefore

$$s_R^2 = \frac{S_R}{v_R} = \frac{\displaystyle\sum_{t=1}^{k} \sum_{i=1}^{n_t} (y_{ti} - \overline{y}_t)^2}{N - k}$$

$$s_T^2 = \frac{S_T}{v_T} = \frac{\displaystyle\sum_{t=1}^{k} n_t (\overline{y}_t - \overline{y})^2}{k - 1} \tag{7.9}$$

$$s_D^2 = \frac{S_D}{v_D} = \frac{\displaystyle\sum_{t=1}^{k} \sum_{i=1}^{n_t} (y_{ti} - \overline{y})^2}{N - 1}$$

For a null hypothesis that there are no differences between the treatment means, the within-treatment mean square (s_R^2) and the between-treatment mean square (s_T^2) provide two estimates of the true process variance (σ^2). For the dataset in Table 7.2, using Eqs. (7.9), we obtain $s_R^2 = 5.6$ and $s_T^2 = 76.0$. The fact that the between-treatment estimate of σ^2 is much larger than the within-treatment estimate tends to discredit the null hypothesis. We are thus led to suspect that some of the between-treatment variation must be caused by real differences in the treatment means. In the following section, we show how the necessary calculations to draw this conclusion may be conveniently arranged in tabular form.

7.2.2. ANOVA Table

Once the sums of squares and mean squares have been computed, it is customary to arrange them in a tabular format called and *ANOVA table*. The general form of the ANOVA table is depicted in Table 7.3. The ANOVA table that corresponds to the via diameter data in Table 7.2 is shown in Table 7.4.

The astute reader will note that in both the "sum of squares" and "degrees of freedom" columns, the values for between and within treatments add up to give the corresponding total value. This additive property of the sum of squares arises from the algebraic identity

$$\sum_{t=1}^{k}\sum_{i=1}^{n_t}(y_{ti} - \bar{y})^2 = \sum_{t=1}^{k} n_t(\bar{y}_t - \bar{y})^2 + \sum_{t=1}^{k}\sum_{i=1}^{n_t}(y_{ti} - \bar{y}_t)^2 \qquad (7.10)$$

or equivalently, $S_D = S_T + S_R$.

The complete ANOVA table provides a mechanism for testing the hypothesis that all of the treatment means are equal. The null hypothesis in this case is thus

$$H_0: \mu_1 = \mu_2 = \mu_3 = \mu_4$$

Table 7.3. General format of the ANOVA table.

Source of Variation	Sum of Squares	Degrees of Freedom	Mean Square	F Ratio
Between treatments	S_T	$v_T = k - 1$	s_T^2	s_T^2/s_R^2
Within treatments	S_R	$v_R = N - k$	s_R^2	
Total about the grand average	S_D	$v_D = N - 1$	s_D^2	

Table 7.4. ANOVA table for yield data.

Source of Variation	Sum of Squares	Degrees of Freedom	Mean Square	F Ratio
Between treatments	$S_T = 228$	$v_T = 3$	$s_T^2 = 76.0$	$s_T^2/s_R^2 = 13.6$
Within treatments	$S_R = 112$	$v_R = 20$	$s_R^2 = 5.6$	
Total about the grand average	$S_D = 340$	$v_D = 23$	$s_D^2 = 14.8$	

If the null hypothesis were true, the ratio s_T^2/s_R^2 would follow the F distribution with n_T and n_R degrees of freedom. According to Appendix E, the significance level for the observed F ratio of 13.6 with 3 and 30 degrees of freedom is 0.000046. This means that there is only a 0.0046% chance that the means are in fact equal, and the null hypothesis is discredited. In other words, we can be 99.9954% sure that real differences exist among the four different processes used in our example.

An alternative format for the ANOVA table exists. The quantity S_D, the total sum of squares about the grand average, can also be written as

$$S_D = \sum_{t=1}^{k} \sum_{i=1}^{n_t} y_{ti}^2 - N\bar{y}^2 \tag{7.11}$$

In this expression, the latter term $(N\bar{y}^2)$ is the sum of squares due to the grand average, which is often called the *correction factor for the average*. It is denoted by S_A (i.e., $S_A = N\bar{y}^2$). The first term in the expression $\left(\sum_{t=1}^{k} \sum_{i=1}^{n_t} y_{ti}^2 \right)$ is called the *total sum of squares*, and it is denoted by S. Combining Eqs. (7.10) and (7.11), we can thus decompose the sum of squares of the original N observations into three additive terms:

$$\sum_{t=1}^{k} \sum_{i=1}^{n_t} y_{ti}^2 = N\bar{y}^2 + \sum_{t=1}^{k} n_t(\bar{y}_t - \bar{y})^2 + \sum_{t=1}^{k} \sum_{i=1}^{n_t} (y_{ti} - \bar{y}_t)^2 \tag{7.12}$$

or equivalently, $S = S_A + S_T + S_R$. The associated degrees of freedom are

$$N = 1 + (k - 1) + (N - k) \tag{7.13}$$

This representation leads to the "full" ANOVA table (Table 7.5), which specifically includes the contributions from the grand average. However, this contribution is of limited practical interest, so the ANOVA table of the form shown in Table 7.3 is usually preferred.

7.2.2.1. Geometric Interpretation

Equation (7.12) can be further explained by breaking up the yield data from Table 7.2 in the manner shown in Table 7.6. This table shows that each individual observation is composed of the following components: the grand average $(\bar{y})$; the between-treatment deviation $(\bar{y}_t - \bar{y})$; and the within-treatment deviation, or residual $(y_{ti} - \bar{y}_t)$. Each of the four entries in Table 7.6 can be considered a

Table 7.5. Full ANOVA table.

Source of Variation	Sum of Squares	Degrees of Freedom	Mean Square
Average	S_A	$\nu_A = 1$	$s_A^2 = S_A/\nu_A$
Between treatments	S_T	$\nu_T = k - 1$	$s_T^2 = S_T/\nu_T$
Within treatments	S_R	$\nu_R = N - k$	$s_R^2 = S_R/\nu_R$
Total	S	N	

Table 7.6. Arithmetic decomposition of yield data in Table 7.2 [1].

	Observations	Grand Average	Treatment Deviations	Residuals
	y_{ti}	$\bar{y}$	$\bar{y}_t - \bar{y}$	$y_{ti} - \bar{y}_t$

$$
\begin{bmatrix} 62 & 63 & 68 & 56 \\ 60 & 67 & 66 & 62 \\ 63 & 71 & 71 & 60 \\ 59 & 64 & 67 & 61 \\ & 65 & 68 & 63 \\ & 66 & 68 & 64 \\ & & & 63 \\ & & & 59 \end{bmatrix}
=
\begin{bmatrix} 64 & 64 & 64 & 64 \\ 64 & 64 & 64 & 64 \\ 64 & 64 & 64 & 64 \\ 64 & 64 & 64 & 64 \\ & 64 & 64 & 64 \\ & 64 & 64 & 64 \\ & & & 64 \\ & & & 64 \end{bmatrix}
+
\begin{bmatrix} -3 & 2 & 4 & -3 \\ -3 & 2 & 4 & -3 \\ -3 & 2 & 4 & -3 \\ -3 & 2 & 4 & -3 \\ & 2 & 4 & -3 \\ & 2 & 4 & -3 \\ & & & -3 \\ & & & -3 \end{bmatrix}
+
\begin{bmatrix} 1 & -3 & 0 & -5 \\ -1 & 1 & -2 & 1 \\ 2 & 5 & 3 & -1 \\ -2 & -2 & -1 & 0 \\ & -1 & 0 & 2 \\ & 0 & 0 & 3 \\ & & & 2 \\ & & & -2 \end{bmatrix}
$$

Vector	**Y**	=	**A**	+	**T**	+	**R**	
Sum of squares	98,644	=	93,304	+	228	+	112	
Degrees of freedom	24	=	1	+	3	+	20	

vector. Let **Y** represent the vector of observations, **A** represent the grand average, **T** represent the between-treatment deviations, and **R** represent the residuals. Using the rules of vector addition, we can write

$$\mathbf{Y} = \mathbf{A} + \mathbf{T} + \mathbf{R} \tag{7.14}$$

The sums of squares in the ANOVA table, therefore, are merely the squares of the individual vector elements summed. In other words, the sums of squares are the squared lengths of the vectors **Y**, **A**, **T**, and **R**.

The geometry of this example is illustrated graphically in Figures 7.3–7.5. In Figure 7.3, the vector **Y** is resolved into two components: **A**, which corresponds to the grand average; and **D**, whose elements are the deviations from the grand average. The vector **D** is orthogonal to **A** since $\sum_{j=1}^{N} \bar{y}(y_j - \bar{y}) = 0$. In Figure 7.4, the vector **D** is likewise resolved into two components: **T**, associated with the treatment deviations; and **R**, which corresponds to the residuals. Finally, in Figure 7.5, the observation vector **Y** is resolved into its three orthogonal components, as indicated in Eq. (7.14). The fact that these three vectors are mutually orthogonal is easily confirmed by noting that their inner products are equal to zero.

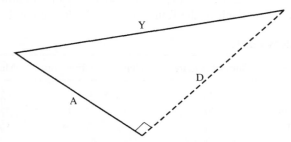

Figure 7.3. Geometric representation of the decomposition of **Y** in terms of **A** and **D** [1].

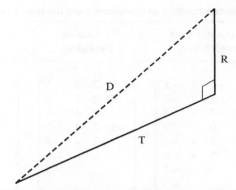

Figure 7.4. Geometric representation of the decomposition of **D** in terms of **T** and **R** [1].

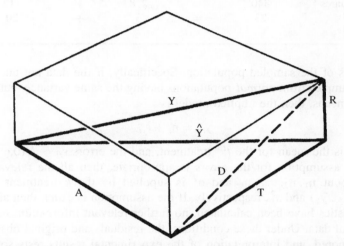

Figure 7.5. Geometric representation of ANOVA in terms of an orthogonal decomposition of **Y** in terms of **A**, **T**, and **R** [1].

The additive relationship $S = S_A + S_T + S_R$ arises from the Pythagorean theorem, which relates the square of the length of the "hypotenuse" **Y** to the sum of squares of the lengths of the three other sides: **A**, **T**, and **R**. The estimated values in the ANOVA technique are the elements of the vector $\hat{\mathbf{Y}}$, where

$$\hat{\mathbf{Y}} = \mathbf{A} + \mathbf{T} \tag{7.15}$$

As mentioned previously, the ANOVA technique is frequently applied after "elimination" of the grand average. Table 7.7 shows this approach to the analysis. The vector **D** represents the deviations from the grand average $(y_{ti} - \overline{y})$ after **A** has been subtracted from **Y**.

7.2.2.2. ANOVA Diagnostics
The ANOVA technique is appropriate for a specific implied model that links the experimental observations and the various decompositions with the underlying

Table 7.7. Arithmetic decomposition of deviations from the grand average [1].

	Deviations from Grand Average	Treatment Deviations	Residuals

$$
\begin{array}{lcccc}
 & y_{ti} - \bar{y} & \bar{y}_t - \bar{y} & y_{ti} - \bar{y}_t \\
 & \begin{bmatrix}
-2 & -1 & 4 & -8 \\
-4 & 3 & 2 & -2 \\
-1 & 7 & 7 & -4 \\
-5 & 0 & 3 & -3 \\
 & 1 & 4 & -1 \\
 & 2 & 4 & 0 \\
 & & & -1 \\
 & & & -5
\end{bmatrix}
=
\begin{bmatrix}
-3 & 2 & 4 & -3 \\
-3 & 2 & 4 & -3 \\
-3 & 2 & 4 & -3 \\
-3 & 2 & 4 & -3 \\
 & 2 & 4 & -3 \\
 & 2 & 4 & -3 \\
 & & & -3 \\
 & & & -3
\end{bmatrix}
+
\begin{bmatrix}
1 & -3 & 0 & -5 \\
-1 & 1 & -2 & 1 \\
2 & 5 & 3 & -1 \\
-2 & -2 & -1 & 0 \\
 & -1 & 0 & 2 \\
 & 0 & 0 & 3 \\
 & & & 2 \\
 & & & -2
\end{bmatrix}
\end{array}
$$

Vector	$\mathbf{D} = \mathbf{Y} - \mathbf{A}$	=	$\mathbf{T}$	+	$\mathbf{R}$
Sum of squares	340	=	228	+	112
Degrees of freedom	23	=	3	+	20

parameters of the sampled population. Specifically, if the data are uncorrelated random samples from *normal* populations having the same variance, but possibly different means, then the implied model is

$$ y_{ti} = \eta_t + \varepsilon_{ti} \tag{7.16} $$

where η_t is the mean for the tth treatment, and the errors $\varepsilon_{ti} \sim N(0,\sigma^2)$. If this normality assumption for the errors is appropriate, then all the relevant information about $\eta_1, \eta_2, \ldots, \eta_k$ and σ^2 is supplied by the k treatment averages $(\bar{y}_1, \bar{y}_2, \ldots, \bar{y}_k)$ and s_R^2, respectively. If the assumption is *exact*, then after all of these statistics have been calculated, no further relevant information remains in the original data. Under these conditions, the residuals and original observations can be ignored, and interpretation of the experimental results rests solely with the interpretation of the statistics.

However, in practice, it is unwise to proceed in this manner without further checks. The data may in fact contain information not accounted for by the model in Eq. (7.16) and therefore not revealed by the ANOVA methodology. Discrepancies of this type may be detected by studying the residuals $(y_{ti} - \hat{y}_{ti})$, which are the elements of the vector $\mathbf{R}$. These residuals are the quantities that remain after the systematic contributions from the treatment averages have been removed. When the assumptions regarding the adequacy of the model in Eq. (7.16) are true, these residuals should vary randomly. If, however, the residuals display unexplained systematic tendencies, then the model becomes suspicious.

One type of residual inspection that must be carried out is plotting an overall *dot diagram*. The dot diagram for the yield data in Table 7.2 is shown in Figure 7.6. If the normality assumption for the model errors is true, then this diagram should essentially have the appearance of a sample from a normal distribution with mean zero. (Note that considerable fluctuation in appearance will occur if the number of observations is too small.) A common discrepancy revealed by an abnormal dot diagram occurs when one or more of the residuals is much larger or smaller than the others. The plot in Figure 7.6 gives no indication of such an abnormality.

Abnormal residual behavior may also be associated with a particular treatment. To detect problems of this sort, individual dot diagrams for each treatment are prepared, as shown in Figure 7.7. Again, these plots should appear as samples from a normal distribution. The plots in Figure 7.7 do not suggest any anomalous behavior.

If the model in Eq. (7.16) is appropriate, then the residuals should also be unrelated to the levels of any known variable. In particular, they should be unrelated to the level of the response itself. This can be investigated by plotting the residuals versus the estimated response $\hat{y}_{ti}$, as shown in Figure 7.8. This plot should also appear random. If the variance increased with the value of the response, then this plot would have a "funnel-like" appearance. No such behavior is apparent in Figure 7.8.

Finally, sometimes a process may drift or the skill of the experimenter may change with time. Tendencies such as this are revealed by plotting the residuals

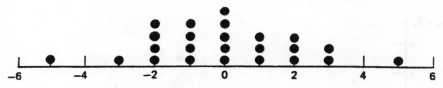

Figure 7.6. Overall dot diagram for all residuals [1].

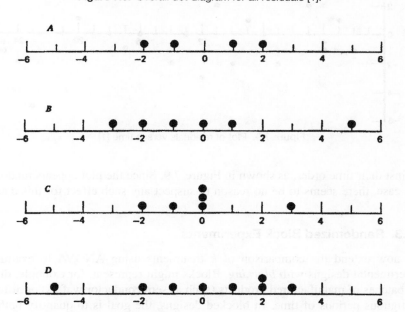

Figure 7.7. Plots of residuals for each treatment [1].

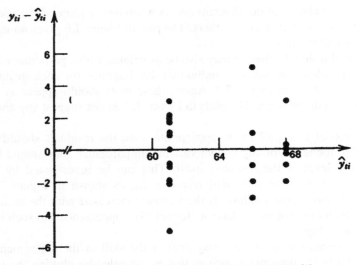

Figure 7.8. Plot of residuals versus estimated values [1].

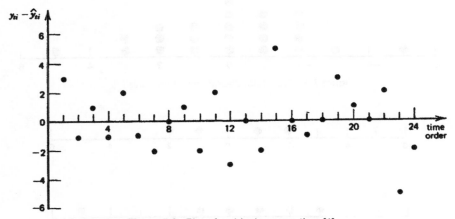

Figure 7.9. Plot of residuals versus time [1].

against their time order, as shown in Figure 7.9. Since the plot appears random in this case, there seems to be no reason to suspect any such effect for this dataset.

7.2.3. Randomized Block Experiments

We now extend the comparison of k treatments using ANOVA to examining experimental designs with *blocking*. Blocks might represent, for example, different batches of manufactured products (such as semiconductor wafers) or different contiguous periods of time. In blocked designs, the goal is to quantify both the effects of the treatments and the effect of the blocking arrangement.

As an example of a blocked experiment, consider the yield data in Table 7.8 obtained from a manufacturing process in which five batches of silicon wafers were fabricated using various methods (labeled A–D). In this case, there are $k = 4$ treatments and $n = 5$ blocks. A randomized block design of this kind serves to eliminate variations between blocks (i.e., the batches) from the comparison of treatments. It also provides a broader inductive basis than an experiment with only a single batch.

Analysis of data of this type is undertaken using the ANOVA table with the format shown in Table 7.9. In this table, $\bar{y}$ is the grand average, $\bar{y}_i$ are the block averages, and $\bar{y}_t$ are the treatment averages. The ANOVA table computed for the yield data in Table 7.8 is Table 7.10.

We are now ready to test the hypothesis that all the treatment means are equal (i.e., H_0: $\mu_A = \mu_B = \mu_C = \mu_D$). If the null hypothesis were true, the ratio s_T^2/s_R^2 would follow the F distribution with ν_T and ν_R degrees of freedom. According to Appendix E, the significance level for the observed F ratio of 1.24 with 3 and 12 degrees of freedom is 0.33. This means that there is a 33% chance that the means

Table 7.8. Yield data from a hypothetical semiconductor manufacturing process [1].

Block	method A Yield (%)	Method B Yield (%)	Method C Yield (%)	Method D Yield (%)	Block Average
Batch 1	89	88	97	94	92
Batch 2	84	77	92	79	83
Batch 3	81	87	87	85	85
Batch 4	87	92	89	84	88
Batch 5	79	81	80	88	82
Treatment Average	84	85	89	86	$\bar{y} = 86$

Table 7.9. Format for two-way ANOVA table with blocking.

Source of Variation	Sum of Squares	Degrees of Freedom	Mean Square	F Ratio
Average	$S_A = nk\bar{y}^2$	$\nu_A = 1$	$s_A^2 = S_A/\nu_A$	
Between blocks	$S_B = k\sum\limits_{i=1}^{n}(\bar{y}_i - \bar{y})^2$	$\nu_B = n-1$	$s_B^2 = S_B/\nu_B$	s_B^2/s_R^2
Between treatments	$S_T = n\sum\limits_{t=1}^{k}(\bar{y}_t - \bar{y})^2$	$\nu_T = k-1$	$s_T^2 = S_T/\nu_T$	s_T^2/s_R^2
Residuals	$S_R = \sum\limits_{t=1}^{k}\sum\limits_{i=1}^{n} \times (y_{ti} - \bar{y}_i - \bar{y}_t + \bar{y})^2$	$\nu_R = (n-1)(k-1)$	$s_R^2 = S_R/\nu_R$	
Total	$S = \sum\limits_{t=1}^{k}\sum\limits_{i=1}^{n} y_{ti}^2$	$\nu = nk$		

Table 7.10. Two-way ANOVA table for yield data.

Source of Variation	Sum of Squares	Degrees of Freedom	Mean Square	F Ratio
Average	$S_A = 147,920$	$\nu_A = 1$	$s_A^2 = 147,920$	
Between blocks	$S_B = 264$	$\nu_B = 4$	$s_B^2 = 66.0$	3.51
Between treatments	$S_T = 70$	$\nu_T = 3$	$s_T^2 = 23.3$	1.24
Residuals	$S_R = 226$	$\nu_R = 12$	$s_R^2 = 18.8$	
Total	$S = 148,480$	$\nu = 20$		

are in fact equal. In other words, we can be only 67% sure that real differences exist among the four different methods used to manufacture the wafers in this example. Thus, the four methods have not been conclusively demonstrated to give different yields.

The blocking arrangement of this experiment also allows us to test the hypothesis that the block means are equal. If this null hypothesis were true, the ratio s_B^2/s_R^2 would follow the F distribution with ν_B and ν_R degrees of freedom. According to Appendix E, the significance level for the observed F ratio of 3.51 with 4 and 12 degrees of freedom is 0.04. This means that there is only a 4% chance that the means are in fact equal. Thus, there exists a 96% chance that there are in fact differences between the batches.

7.2.3.1. Mathematical Model

The mathematical model implicit in randomized block experiments is

$$y_{ti} = \eta + \beta_i + \tau_t + \varepsilon_{ti} \tag{7.17}$$

where η is the general mean, β_i is the block effect, τ_t is the treatment effect, and ε_{ti} is the experimental error. It is assumed that $\varepsilon_{ti} \sim N(0, \sigma^2)$. Associated with this additive model is the following decomposition of the observations:

$$y_{ti} = \overline{y} + (\overline{y}_i - \overline{y}) + (\overline{y}_t - \overline{y}) + (y_{ti} - \overline{y}_i - \overline{y}_t + \overline{y}) \tag{7.18}$$

The last term, $(y_{ti} - \overline{y}_i - \overline{y}_t + \overline{y})$, is known as the *residual* because it represents what remains after the grand average, block effects, and treatment effects have all been accounted for. The model is called *additive* since, for example, if treatment τ_3 caused an increase of five units in the response and the influence of block β_4 increased the response by seven units, then the cumulative increase caused by both acting together would be $5 + 7 = 12$ units.

In vector notation, the decomposition in Eq. (7.18) can be written as follows:

$$\mathbf{Y} = \mathbf{A} + \mathbf{B} + \mathbf{T} + \mathbf{R} \tag{7.19}$$

In this equation, each of the symbols represents a vector containing $N = nk$ elements of the corresponding two-way ANOVA table. The sums of squares in the ANOVA table are once again the squares of the individual vector elements

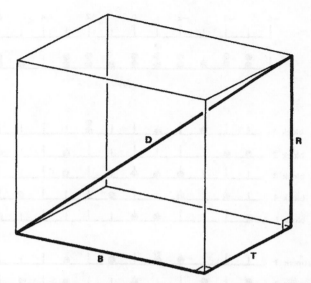

Figure 7.10. Vector decomposition for randomized block ANOVA [1].

summed. In other words, the sums of squares are the squared lengths of the vectors $\mathbf{Y}$, $\mathbf{A}$, $\mathbf{B}$, $\mathbf{T}$, and $\mathbf{R}$, or

$$S = S_A + S_B + S_T + S_R \tag{7.20}$$

The vectors $\mathbf{A}$, $\mathbf{B}$, $\mathbf{T}$, and $\mathbf{R}$, are all mutually perpendicular, as illustrated in Figure 7.10. This figure also illustrates the relationship

$$\mathbf{D} = \mathbf{B} + \mathbf{T} + \mathbf{R} \tag{7.21}$$

where $\mathbf{D} = \mathbf{Y} - \mathbf{A}$ is a vector of deviations of the data from the grand average. Since $\mathbf{B}$, $\mathbf{T}$, and $\mathbf{R}$ are mutually orthogonal, we also have

$$S_D = S_B + S_T + S_R \tag{7.22}$$

In other words, the sum of squares of the deviations from the grand average equals the sum of squares for the blocks plus the sum of squares for the treatments plus the sum of squares of the residuals.

7.2.3.2. Diagnostic Checking

Any potential inadequacies in the model proposed in Eq. (7.17) and analyzed using the randomized block ANOVA technique must be investigated by diagnostic methods similar to those discussed in Section 7.2.2.2. The residual plots for the model of the yield data in Table 7.8 are shown in Figure 7.11. The plots in (a) and (b) of this figure reveal nothing of special concern, but the plot of residuals versus predicted values in (c) shows a possible problem in its "funnel" shape, suggesting a possible relationship between the mean and variance. Such discrepancies can

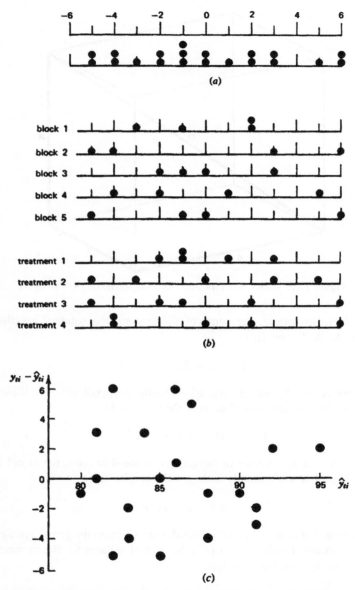

Figure 7.11. Plots of residuals, yield example: (a) overall plot; (b) plots by block and treatment; (c) $y_{ti} - \hat{y}_{ti}$ versus $\hat{y}_{ti}$ [1].

also be indicative of *nonadditivity* between the block and treatment effects. Such discrepancies can sometimes be eliminated using a suitable *transformation* of the response variable. Data transformations are discussed in greater detail in Section 7.2.4.

7.2.4. Two-Way Designs

Experimental designs with two sets of treatments (or *factors*) are referred to as *two-way designs*, and their corresponding analysis is accomplished by *two-way ANOVA*. As an example, consider the data in Table 7.11, which corresponds to the film uniformity (in %) achieved in a CVD experiment. Assume that treatments A, B, C, and D represent different gas compositions, and treatments 1, 2, and 3 represent different temperatures. This arrangement, which has been replicated four times, is also known as a 3×4 *factorial design* (see Section 7.3). There is no blocking. Both factors are of equal interest, and there is a possibility that these factors interact.

7.2.4.1. Analysis

Let y_{tij} be the nonuniformity of the jth wafer deposited at the ith temperature using the tth gas composition. The corresponding model and estimate for the nonuniformity response are given respectively by

$$y_{tij} = \eta_{ti} + \varepsilon_{ti} \tag{7.23}$$

$$= \overline{y}_{ti} + (y_{tij} - \overline{y}_{ti}) \tag{7.24}$$

If the temperature and gas composition effects do not behave additively, then

$$\eta_{ti} = \eta + \tau_t + \beta_i + \omega_{ti} \tag{7.25}$$

where τ_t is the incremental nonuniformity associated with the tth gas composition, and β_i is the increment associated with the ith temperature. Their non-additive behavior requires an additional term, ω_{ti}, which represents the *interaction effect* between the two factors. The τ_t and β_i terms are known as the *main effects*. The estimate corresponding to Eq. (7.25) is given by

$$\overline{y}_{ti} = \overline{y} + (\overline{y}_t - \overline{y}) + (\overline{y}_i - \overline{y}) + (\overline{y}_{ti} - \overline{y}_t - \overline{y}_i + \overline{y}) \tag{7.26}$$

Table 7.11. Nonuniformity (in %) of films grown by CVD [1].

Treatments	A	B	C	D
1	0.31	0.82	0.43	0.45
	0.45	1.10	0.45	0.71
	0.46	0.88	0.63	0.66
	0.43	0.72	0.76	0.62
2	0.36	0.92	0.44	0.56
	0.29	0.61	0.35	1.02
	0.40	0.49	0.31	0.71
	0.23	1.24	0.40	0.38
3	0.22	0.30	0.23	0.30
	0.21	0.37	0.25	0.36
	0.18	0.38	0.24	0.31
	0.23	0.29	0.22	0.33

The arithmetic for carrying out the data analysis closely parallels that used for the randomized block design. Group averages $(\bar{y}_{ti})$ replace the basic data, and an interaction sum of squares replaces the residual sum of squares in the randomized block analysis. In general, for n levels of some factor P ($n = 3$ temperatures in this example), k levels of another factor T ($k = 4$ gas compositions), and m replications ($m = 4$ wafers per group), the following sums of squares may be defined:

$$S_P = mk \sum_i (\bar{y}_i - \bar{y})^2 \tag{7.27}$$

$$S_T = mn \sum_t (\bar{y}_t - \bar{y})^2 \tag{7.28}$$

$$S_I = m \sum_t \sum_i (\bar{y}_{ti} - \bar{y}_t - \bar{y}_i + \bar{y})^2 \tag{7.29}$$

$$S_e = \sum_t \sum_i \sum_j (y_{tij} - \bar{y}_{ti})^2 \tag{7.30}$$

$$S = \sum_t \sum_i \sum_j (y_{tij} - \bar{y})^2 \tag{7.31}$$

Given these parameters, the ANOVA table for the nonuniformity data in Table 7.11 is shown in Table 7.12.

7.2.4.2. Data Transformation

If the model described by Eq. (7.23) is accurate and the model errors are independently and normally distributed with a constant variance [i.e., $\varepsilon_{ti} \sim N(0, \sigma^2)$], then the significance of the factors can be evaluated using the F distribution. In the CVD example, examination of Table 7.12 reveals that the effects of both the temperature and the gas composition are highly significant. For example, in the case of temperature, an F ratio of 23.27 for an F distribution with $v_p = 2$ and $v_e = 36$ degrees of freedom has less than a 0.001 significance level. This analysis of variance also indicates some suggestion of interaction between temperature and gas composition. The F ratio of 1.88 for an F distribution with $v_I = 6$ and $v_e = 36$ degrees of freedom has an approximate 0.01 significance level.

Table 7.12. ANOVA table for two-way factorial experiment.

Source of Variation	Sum of Squares ($\times 1000$)	Degrees of Freedom	Mean Square	F Ratio
Temperatures	$S_P = 1033.0$	$v_P = n - 1 = 2$	$s_P^2 = S_P/v_P = 516.5$	$s_P^2/s_e^2 = 23.27$
Gas compositions	$S_T = 922.4$	$v_T = k - 1 = 3$	$s_T^2 = S_T/v_T = 307.5$	$s_T^2/s_e^2 = 13.85$
Interaction	$S_I = 250.1$	$v_I = (n-1)(k-1)$ $= 6$	$s_I^2 = S_I/v_I = 41.7$	$s_I^2/s_e^2 = 1.88$
Error	$S_e = 800.7$	$v_e = nk(m-1) = 36$	$s_e^2 = S_e/v_e = 22.2$	
Total	$S = 3006.2$	$v = nkm - 1 = 47$		

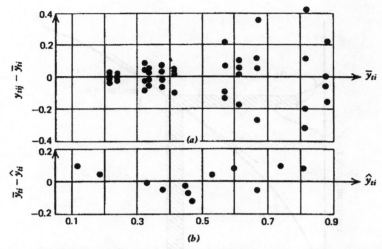

Figure 7.12. Residual diagnostics, CVD experiment: (a) plot of $y_{tij} - \bar{y}_{ti}$ versus $\bar{y}_{ti}$; (b) plot of $\bar{y}_{ti} - \hat{y}_{ti}$ versus $\hat{y}_{ti}$ [1].

Diagnostic checking of the residuals for these data, however, leads to suspicion that this model is inadequate. Figure 7.12 shows a plot of the residuals versus $\bar{y}_{ti}$. The funnel shape in Figure 7.12a suggests that the standard deviation of the data is not constant as previously assumed, but instead increases with the mean. Furthermore, ignoring the interaction term by letting $\hat{y}_{ti} = \bar{y}_t + \bar{y}_i - \bar{y}$, and plotting $\bar{y}_{ti} - \hat{y}_{ti}$ against $\hat{y}_{ti}$ reveals a curvilinear relationship, which contradicts the linearity assumption (Figure 7.12b).

In cases such as this, when σ_y is actually a function of the mean (η), it may be possible to find a convenient data transformation $Y = f(y)$ that does have a constant variance. If so, the data are said to possess a *transformable nonadditivity*. For example, suppose that σ_y is proportional to some power of η, or

$$\sigma_y \propto \eta^y \tag{7.32}$$

and the following power transformation of the data is made:

$$Y = y^\lambda \tag{7.33}$$

Then

$$\sigma_y = \theta \sigma_y \propto \theta \eta^\alpha \tag{7.34}$$

where θ is the gradient of the graph of Y versus y (see Figure 7.13). It can be shown that if Eq. (7.33) is true, then $\theta \propto \eta^{\lambda-1}$. Thus

$$\sigma_Y \propto \eta^{\lambda-1} \eta^\alpha = \eta^{\lambda+\alpha-1} \tag{7.35}$$

Therefore, Y is chosen so that σ_y does not depend on η if $\lambda = 1 - \alpha$. Some values of α with appropriate variance stabilizing transformations are presented in Table 7.13.

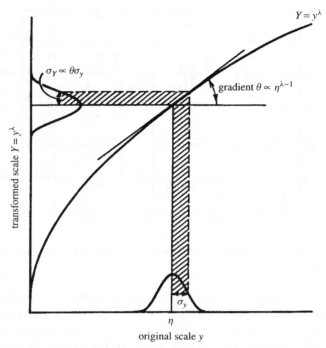

Figure 7.13. Data transformation from y to $Y = y^\lambda$ [1].

Table 7.13. Variance stabilizing data transformations when $\sigma_y \propto \eta^\alpha$ [1].

Dependence of σ_y on η	α	$\lambda = 1 - \alpha$	Variance Stabilizing Transformation
$\sigma \propto \eta^2$	2	-1	Reciprocal
$\sigma \propto \eta^{\frac{3}{2}}$	$\frac{3}{2}$	$-\frac{1}{2}$	Reciprocal square root
$\sigma \propto \eta$	1	0	Log
$\sigma \propto \eta^{\frac{1}{2}}$	$\frac{1}{2}$	$\frac{1}{2}$	Square root
$\sigma \propto$ constant	0	1	None

In order to identify an appropriate transformation for the data in the CVD experiment, α must first be determined empirically. Since $\sigma_y \propto \eta^\alpha$, it is also true that $\log \sigma_y = \text{constant} + \alpha \log \eta$. Thus, if we plot σ_y versus $\log \eta$, we obtain a straight line with a slope of α. Although σ_y and η are not known in practice, they may be estimated using s and $\bar{y}$, respectively (where s is the sample standard deviation of the data). Carrying out this procedure for the CVD data yields a slope of $\alpha \cong 2$. From Table 7.13, this implies that a reciprocal transformation is appropriate in this case. We therefore convert the entire dataset in Table 7.11 into reciprocals and repeat the analysis of variance.

Table 7.14. ANOVA for transformed and untransformed data, CVD experiment [1].

Source of Variation	Untransformed Degrees of Freedom	Untransformed Mean Square ($\times 1000$)	Transformed ($Y = y^{-1}$) Degrees of Freedom	Transformed ($Y = y^{-1}$) Mean Square ($\times 1000$)
Temperatures	2	516.5	2	1743.9
Gas compositions	3	307.5	3	680.5
Interaction	6	41.7	6	26.2
Error	36	22.2	35	24.7

A comparison of the ANOVA for the untransformed and transformed datasets is provided in Table 7.14. The fact that the data themselves have been used to choose the transformation is accounted for by reducing the number of degrees of freedom in the mean square of the error (from 36 to 35) [2]. The effects of the transformation are noteworthy. The mean squares for the transformed data are now much larger relative to the error, indicating an increase in sensitivity of the experiment. In addition, the interaction mean square, which previously gave a slight indication of statistical significance, is now closer in size to the error, contradicting that assertion. Verification of the improvement in the residual diagnostics for the transformed data is left as an exercise.

7.3. FACTORIAL DESIGNS

Experimental design is essentially an organized method of conducting experiments in order to extract the maximum amount of information from a limited number of experiments. Experimental design techniques are employed in semiconductor manufacturing applications to systematically and efficiently explore the effects of a set of input variables, or *factors* (such as processing temperature), on *responses* (such as yield). The unifying feature in statistically designed experiments is that all factors are varied simultaneously, as opposed to the more traditional "one variable at a time" technique. A properly designed experiment can minimize the number of experimental runs that would otherwise be required if this approach or random sampling was used.

Factorial experimental designs are of great practical importance for manufacturing applications. To perform a factorial experiment, an investigator selects a fixed number of *levels* for each of a number of variables (factors) and runs experiments at all possible combinations of the levels. If there are i levels for the first variable, j levels for the second, ..., and k levels for the kth, the complete set of $i \times j \times \cdots \times k$ experimental trials is called an $i \times j \times \cdots \times k$ factorial design. As mentioned previously, the data presented in Table 7.11 represent a 3×4 factorial design. In general, an $l \times m \times n$ design requires lmn runs. For example, a $2 \times 3 \times 4$ design requires 24 runs.

Two of the most important issues in factorial experimental designs are choosing the set of factors to be varied in the experiment and specifying the ranges over which variation will take place. The choice of the number of factors directly impacts the number of experimental runs (and therefore the overall cost of the experiment). The most common approach in factorial designs is the two-level factorial, which is described in Section 7.3.1.

7.3.1. Two-Level Factorials

The ranges of the process variables investigated in factorial experiments can be discretized into minimum, maximum, and "center" levels. In a *two-level factorial design*, the minimum and maximum levels of each factor (normalized to take on values -1 and $+1$, respectively) are used together in every possible combination. Thus, a full two-level factorial experiment with n factors requires 2^n experimental runs. The various factor level combinations of a three-factor experiment can be represented pictorially as the vertices of a cube, as shown in Figure 7.14.

Two-level factorial designs are important for several reasons. First, although they require relatively few trials per factor, they allow an experimenter to identify major trends and promising directions for future experimentation. These designs are also easily augmented to form more advanced designs (see Section 7.3.4). Furthermore, two-level factorials form the basis for two-level fractional factorial designs (see Section 7.3.2), which are useful for screening large numbers of factors at an early stage of experimentation. Finally, analysis of these designs facilitates the systematic analysis of the impact of *interactions* between factors. Such interactions can be obscured if the traditional "change one variable at a time" approach to experimentation, in which factors are varied individually while the remaining factors are held constant, is used. The traditional approach assumes that all of the factors act on the response additively, which is often not the case in complex processes. In addition, the factorial approach is more economical, since a n-factor traditional experiment requires a n-fold increase in the number of trials as compared to a 2^n factorial experiment.

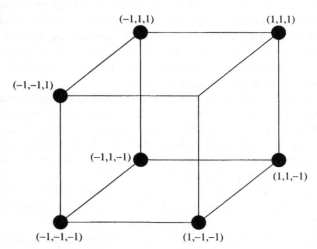

Figure 7.14. Factor combinations for a three-factor experiment represented as vertices of a cube.

7.3.1.1. Main Effects

To illustrate the use of two-level factorials, Table 7.15 shows a 2^3 factorial experiment for another CVD process. The three factors are temperature (T), pressure (P), and gas flowrate (F). The response being measured is the deposition rate (D) in angstroms per minute (Å/min). The highest and lowest levels of each factor are represented by the "+" and "−" signs, respectively. The display of levels depicted in the first three columns of this table is called a *design matrix*.

The relevant issue is what we can determine from this factorial design. For example, what do the data collected tell us about the effect of pressure on deposition rate? The effect of any single variable on the response is called a *main effect*. The method used to compute such a main effect is to find the difference between the average deposition rate when the pressure is high (i.e., runs 2, 4, 6, 8) and the average deposition rate when the pressure is low (runs 1, 3, 5, 7). Mathematically, this is expressed as

$$P = d_{p+} - d_{p-} = \tfrac{1}{4}[(d_2 + d_4 + d_6 + d_8) - (d_1 + d_3 + d_5 + d_7)] = 40.86 \tag{7.36}$$

where P is the main effect for pressure, d_{p+} is the average deposition rate when the pressure is high, and d_{p-} is the average deposition rate when the pressure is low. The manner in which we interpret this result is that the average effect of increasing pressure from its lowest to its highest level is to increase the deposition rate by 40.86 Å/min. The other main effects for temperature and flowrate are computed in a similar manner. In general, the main effect for each variable in a two-level factorial experiment is the difference between the two averages of the response (y), or

$$(\text{Main effect}) = y_+ - y_- \tag{7.37}$$

7.3.1.2. Interaction Effects

We might also be interested in quantifying how two or more factors interact. For example, suppose that the pressure effect is much greater at high temperatures than it is at low temperatures. A measure of this interaction is provided by the difference between the average pressure effect with temperature high and the average pressure effect with temperature low. By convention, *half* of

Table 7.15. 2-Level Factorial CVD Experiment.

Run	P	T	F	D (Å/min)
1	−	−	−	$d_1 = 94.8$
2	+	−	−	$d_2 = 110.96$
3	−	+	−	$d_3 = 214.12$
4	+	+	−	$d_4 = 255.82$
5	−	−	+	$d_5 = 94.14$
6	+	−	+	$d_6 = 145.92$
7	−	+	+	$d_7 = 286.71$
8	+	+	+	$d_8 = 340.52$

this difference is called the *pressure by temperature interaction,* or symbolically, the $P \times T$ interaction. This interaction may also be thought of as one-half the difference in the average temperature effects at the two levels of pressure. Mathematically, this is

$$P \times T = d_{PT+} - d_{PT-} = \tfrac{1}{4}[(d_1 + d_4 + d_5 + d_8) - (d_2 + d_3 + d_6 + d_7)] = 6.89 \tag{7.38}$$

The $P \times F$ and $T \times F$ interactions are computed in a similar fashion. Just as main effects can be viewed as a *contrast* between observations on faces of a cube like the one in Figure 7.14 (see Figure 7.15a), an interaction is a contrast between results on two diagonal planes (Figure 7.15b).

Finally, we might also be interested in the interaction of all three factors, denoted as the *pressure by temperature by flowrate* or the $P \times T \times F$ interaction. This interaction defines the average difference between any two-factor interaction at the high and low levels of the third factor. It is given by

$$P \times T \times F = d_{PTF+} - d_{PTF-} = -5.88 \tag{7.39}$$

This interaction is depicted graphically in Figure 7.15c. It is important to note that the main effect of any factor can be individually interpreted only if there is no evidence that the factor interacts with other factors.

7.3.1.3. Standard Error

When valid run replicates are made under a given set of experimental conditions, the variation between associated observations can be used to estimate the standard deviation of a single observation, and hence the standard deviation (or *standard error*) of the effects. A comparison of the size of an effect to its standard error allows one to determine the significance of the effect relative to experimental error or noise. In other words, an effect that is much larger than its standard error (in an absolute sense) is more likely to be significant, as opposed to an effect that is less than or equal to its standard error. The notion of "validity" in the context is usually accomplished by randomization of the run order. Randomization helps ensure that the variation between runs made at the same experimental conditions reflects the total variability that can be ascribed to runs made under different experimental conditions.

If there are r sets of experimental conditions replicated, and the n_i replicate runs made at the ith set provide an estimate s_i^2 of the true variance (σ^2) having $v_i = n_n - 1$ degrees of freedom, then the pooled estimate of the run variance is

$$s^2 = \frac{v_1 s_1^2 + v_2 s_2^2 + \cdots + v_r s_r^2}{v_1 + v_2 + \cdots + v_r} \tag{7.40}$$

with $v = v_1 + v_2 + \cdots + v_r$ degrees of freedom. If there are only $n_i = 2$ replicates at each of the r sets of conditions, then the formula for the ith variance reduces to $s_i^2 = d_i^2/2$ with $v_i = 1$, where d_i is the difference between the duplicate observations at the ith set of conditions. From (7.40), this implies $s^2 = \sum d_i^2/2r$.

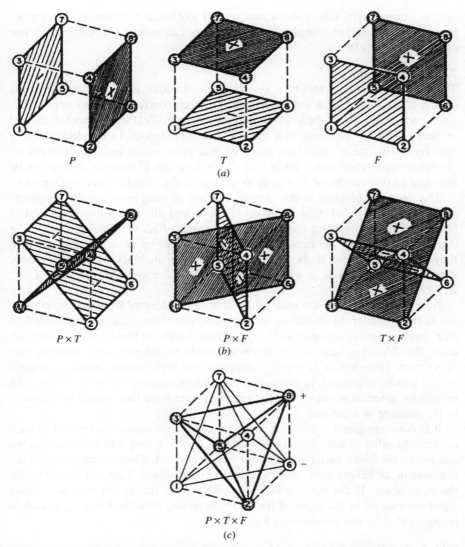

Figure 7.15. Geometric representation of contrasts corresponding to main effects (a) and two-(b) and three-factor (c) interactions [1].

Since each main effect and interaction in a two-level factorial experiment are statistics of the form $= y_+ - y_-$a, the overall variance of each effect (assuming independent errors) is given by

$$V(\text{effect}) = V(\overline{y}_+ - \overline{y}_-) = \frac{4}{N}\sigma^2 \qquad (7.41)$$

where N is the total number of runs made in conducting the factorial design or replicated factorial design and σ^2 is estimated using s^2. The standard error may be computed by taking the square root of V(effect). Equation (7.41) implies that conducting larger numbers of experiments can reduce the variance in our estimates of the effects.

7.3.1.4. Blocking

The term "blocking" refers to a systematic methodology used to eliminate the effects of parameters that the experimenter cannot control. As an example, consider once again the 2^3 factorial design used in the CVD experiment discussed in Section 7.3.1.1. Suppose that the CVD reactor needed to be cleaned every four runs. This means that each group of four runs occurs under a different set of experimental conditions. Table 7.16 shows how the 2^3 factorial design can be arranged in two blocks of four runs to neutralize the effect of reactor cleaning.

The design is blocked in this way by placing all runs in which the "product" of columns P, T, and F is minus in block 1, and all other runs are placed in block 2. This arrangement eliminates the spurious effect of cleaning since if the deposition rate of all the runs in block 2 were higher by some amount Δd than they would have been if they had been performed in block 1, then no matter what the value of Δd is, it will cancel out in the calculation of effects P, T, F, PT, PF, and TF.

Note that a tradeoff in the information that can be derived from this experiment has occurred under this blocking arrangement. The three-factor interaction effect PTF has now been *confounded* (i.e., "confused") with the block effect. Therefore, using this blocking scheme, we are now unable to independently estimate this interaction. However, it is usually assumed that higher-order interactions such as this can be neglected. In exchange, this design ensures that main effects and two-factor interactions can be more precisely measured than would be the case in the absence of blocking.

It is common practice for such a design to assign a numerical symbol to each column. In other words, $P = 1$, $T = 2$, and $F = 3$. Using this terminology, we can assign the block variable the numerical identifier **4**. Then we can think of the experiment as having four variables, the latter of which does not interact with the other three. If the new variable is produced by having its plus and minus signs correspond to the signs of the **123** interaction, then the blocking is said to be *generated* by the relationship **4 = 123**.

Table 7.16. A 2^3 factorial design in blocks of size 2.

Run	P	T	F	PT	PF	TF	PTF	Block
1	−	−	−	+	+	+	−	1
2	+	−	−	−	−	+	+	2
3	−	+	−	−	+	−	+	2
4	+	+	−	+	−	−	−	1
5	−	−	+	+	−	−	+	2
6	+	−	+	−	+	−	−	1
7	−	+	+	−	−	+	−	1
8	+	+	+	+	+	+	+	2

Table 7.17. Trial blocking scheme for 2^3 CVD experiment for blocks of size 2.

$4 = 123$	$5 = 23$	Block	45
−	+	2	−
+	+	4	+
+	−	3	−
−	−	1	+
+	−	3	−
−	−	1	+
−	+	2	−
+	+	4	+

Suppose instead in the 2^3 factorial CVD experiment that the reactor had to be cleaned every two runs. This would require four blocks of two runs, rather than two blocks of four, which means that two block generators are also required. Suppose that we initially selected the block generators $4 = 123$ and $5 = 23$. The resulting design is shown in Table 7.17. As a consequence of this arrangement, the *PTF* and *TF* effects are clearly confounded. However, there is an additional unintended consequence as well. Note in Table 7.17 that the column that represents the product of the two block generators is identical to the column for the *P* main effect! This means that this blocking scheme prevents us from identifying this main effect, which is clearly unacceptable.

Fortunately, there is a simple method to identify confounding patterns and evaluate the consequences of any proposed blocking scheme so that situations like this can be avoided. Let I be a column consisting entirely of plus signs. Thus, we can write

$$I = 11 = 22 = 33 = \cdots \text{etc.} \tag{7.42}$$

where **11**, **22**, and **33** represent the product of the elements in columns **1**, **2**, and **3**, respectively, with themselves. The effect of multiplying the elements of any column with I is to leave those elements unchanged. Now in the blocking arrangement just considered, the product of the block generators is

$$45 = 123 \times 23 = 12233 = 1II = 1 \tag{7.43}$$

which indicates that **45** is identical to column **1**, thereby clarifying the confounding inherent in this scheme.

This suggests a better blocking scheme to achieve four blocks of size two in this experiment. If we let $4 = 12$ and $5 = 13$, the *PT* and *PF* interactions are clearly confounded. Also, since $45 = 12 \times 13 = 1123 = I23 = 23$, the *TF* interaction is also confounded. However, this new blocking arrangement has the advantage of not confounding any main effect, which is much more desirable than the previous scheme.

7.3.2. Fractional Factorials

A major disadvantage of the two-level factorial design is that the number of experimental runs increases exponentially with the number of factors. To alleviate this concern, *fractional factorial* designs are often constructed by systematically eliminating some of the runs in a full factorial design. For example, a half fractional design with n factors requires only 2^{n-1} runs. Full or fractional two-level factorial designs can be used to estimate the main effects of individual factors as well as the interaction effects between factors. However, they cannot be used to estimate quadratic or higher-order effects. This is not a serious shortcoming, since higher order effects and interactions tend to be smaller than low-order effects (main effects tend to be larger than two-factor interactions, which tend to be larger than three-factor interactions, etc.). Ignoring high-order effects is conceptually similar to ignoring higher-order terms in a Taylor series expansion.

7.3.2.1. Construction of Fractional Factorials

To illustrate the use of fractional factorial designs, let $n = 5$ and consider a 2^5 factorial design. The full factorial implementation of this design would require 32 experimental runs. However, a 2^{5-1} fractional factorial design requires only 16 runs. This 2^{5-1} design is generated by first writing the design matrix for a 2^4 full factorial design in standard order. Then plus and minus signs in the four columns of the 2^4 design matrix are each "multiplied" together to form a fifth column (i.e., **5 = 1234**).

However, just as in the case of blocking, some information is lost in this fractional factorial arrangement. A 2^{5-1} design allows the estimation of 16 quantities: the mean, the 5 main effects, and the 10 two-factor interactions. The higher-order effects (the 10 three-factor interactions, 5 four-factor interactions, and single five-factor interaction) are now confounded with one of the first 16 effects. To illustrate, consider the **45** and **123** interactions. These yield the identical design sequences:

$$\mathbf{45} = - + + - + - - + - + + - + - - +$$

$$\mathbf{123} = - + + - + - - + - + + - + - - +$$

thereby indicating that these two interactions are confounded. If the 16 outputs of this 2^{5-1} fractional factorial experiment are labeled $y_1, \ldots, y_{16}$, then the symbol l_{45} denotes the linear function of these observations used estimate the **45** interaction, or

$$l_{45} = \tfrac{1}{8}(-y_1 + y_2 + y_3 - y_4 + y_5 - y_6 - y_7 + y_8 - y_9 + y_{10}$$
$$+ y_{11} - y_{12} + y_{13} - y_{14} - y_{15} + y_{16}) \tag{7.44}$$

The symbol l_{45} is called a *contrast*, since it is the difference between two averages of eight results. Since interactions **45** and **123** are confounded, the contrast l_{45} actually estimates the sum of the mean values of their effects, which is indicated by the notation $l_{45} \to 45 + 123$. However, if we accept the convention that higher-order interactions are generally less significant than lower order interactions, we would attribute the numerical value of contrast l_{45} primarily to the **45** interaction.

Recall that the 2^{5-1} design was constructed by setting $5 = 1234$. This relation is called the generator of the design. Multiplying both sides of the relation by **5**, we obtain

$$5 \times 5 = 1234 \times 5 \qquad (7.45)$$

or equivalently, $I = 12345$. The latter relation is called the *defining relation* of the fractional factorial design. The defining relation is the key to determining the confounding pattern of the design. For example, multiplying the defining relation on both sides by **1** yields **1** = **2345**, which indicates that main effect **1** is confounded with the 4-factor interaction **2345**.

Let's take another look at our CVD experiment. Suppose we only have the time and/or resources available to perform four deposition experiments, rather than the eight required for a 2^3 full factorial design. This calls for a 2^{3-1} fractional factorial alternative. This new design could be generated by writing the full 2^2 design for the pressure and temperature variables, and then multiplying those columns to obtain a third column for flowrate. This procedure is illustrated in Table 7.18. The only drawback in using this procedure is that since we have used the PT relation to define column F, we can no longer distinguish between the effects of the $P \times T$ interaction and the F main effect. These effects are therefore confounded.

7.3.2.2. Resolution

A fractional factorial design of *resolution R* is one in which no p-factor interaction is confounded with any other effect containing less than $R - p$ factors. The resolution of a design is denoted by a Roman numeral and appended as a subscript. For example, the 2^{5-1} fractional factorial discussed in the previous section is called a *resolution V* design, and is denoted as 2_V^{5-1}. In this case, main effects are confounded with four-factor interactions, and two-factor interactions are confounded with three-factor interactions. In general, the resolution of a two-level fractional factorial design is just the length of the defining relation.

7.3.3. Analyzing Factorials

Although various methods for analysis of factorial experiments that are based on simple hand calculations exist, it should be pointed out that modern analysis

Table 7.18. Illustration of 2^{3-1} fractional factorial design for CVD example.

Run	P	T	F
1	−	−	+
2	+	−	−
3	−	+	−
4	+	+	+

of statistical experiments is accomplished almost exclusively by commercially available statistical software packages. A few of the more common packages include *RS/1*, *SAS*, and *Minitab*. These packages completely alleviate the necessity of performing any tedious hand calculations. Nevertheless, a few of the more well-known hand methods are presented below.

7.3.3.1. The Yates Algorithm

It is quite tedious to calculate the effects and interactions for two-level factorial experiments using the methods described Section 7.3.1, particularly if there are more than three factors involved. Fortunately, the *Yates algorithm* provides a quicker method of computation that is also relatively easily programmed via computer. To implement this algorithm, the experimental design matrix is first arranged in what is called *standard order*. A 2^n factorial design is in standard order when the first column of the design matrix consists of alternating minus and plus signs, the second column of successive pairs of minus and plus signs, the third column of four minus signs followed by four plus signs, and so on. In general, the kth column consists of 2^{k-l} minus signs followed by 2^{k-1} plus signs.

The Yates calculations for the deposition rate data are shown in Table 7.19. Column y contains the deposition rates for each run. These are considered in successive pairs. The first four entries in column 1 are obtained by adding the pairs together, and the next four are obtained by subtracting the *top number from the bottom number* of each pair. Column 2 is obtained from column 1 in the same way, and column 3 is obtained from column 2. To obtain the experimental *effects*, one only needs to divide the column 3 entries by the *divisor*. In general, the first divisor will be 2^n, and the remaining divisors will be 2^{n-1}. The first element in the *identification* (ID) column is the grand average of all of the observations, and the remaining identifications are derived by locating the plus signs in the design matrix. The Yates algorithm provides a relatively straightforward methodology for computing experimental effects in two-level factorial designs.

7.3.3.2. Normal Probability Plots

One problem that can arise when analyzing the effects of unreplicated factorial experiments is that real and meaningful higher-order interactions do occasionally

Table 7.19. Illustration of the Yates algorithm.

P	T	F	y	(1)	(2)	(3)	Divisor	Effect	ID
−	−	−	94.8	205.76	675.70	1543.0	8	192.87	Average
+	−	−	110.96	469.94	867.29	163.45	4	40.86	P
−	+	−	214.12	240.06	57.86	651.35	4	162.84	T
+	+	−	255.82	627.23	105.59	27.57	4	6.89	PT
−	−	+	94.14	16.16	264.18	191.59	4	47.90	F
+	−	+	145.92	41.70	387.17	47.73	4	11.93	PF
−	+	+	286.71	51.78	25.54	122.99	4	30.75	TF
+	+	+	340.52	53.81	2.03	−23.51	4	−5.88	PTF

occur. In such cases, methods are needed to evaluate these effects. One way to do so is to plot the effects on *normal probability paper*.

A normal distribution is shown in Figure 7.16a. The probability of the occurrence of some value less than X is given by the shaded area P. Plotting P versus X results in the sigmoidal cumulative normal distribution curve shown in Figure 7.16b. Normal probability paper simply adjusts the vertical scale of this plot in the manner shown in Figure 7.16c, so that P versus X becomes a straight line.

Suppose that the dots in Figure 7.16 represent a random sample of 10 observations from a normal distribution. Since $n = 10$, the leftmost observation can be interpreted as representing the first 10% of the cumulative distribution. In Figure 7.16b, this observation is therefore plotted midway between zero and 10% (i.e., at 5%). Similarly, the second observation represents the next 10% of the cumulative distribution and is plotted at 15%, and so on. In general, we have

$$P_i = 100(i - 1/2)/m \tag{7.46}$$

for $i = 1, 2, \ldots, m$.

When all the sample points are plotted on normal paper, they should ideally form a straight line. However, this is only true if the effects represented by the points are *not* significant. To illustrate, consider the effects computed from a hypothetical 2^4 factorial experiment shown in Table 7.20. The $m = 15$ main effects plus interactions in this experiment represent 15 contrasts between pairs of averages containing eight observations each. If these effects are not significant,

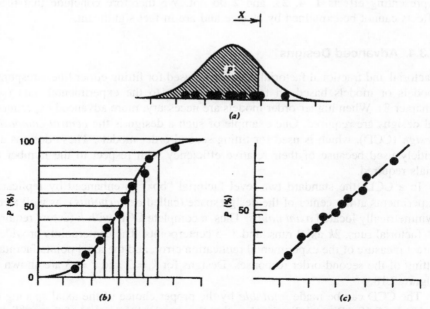

Figure 7.16. Normal probability plot concepts: (a) normal distribution; (b) ordinary graph paper; (c) normal probability paper [1].

Table 7.20. Effects and probability points for normal probability plot example [1].

i	Value of Effect	Identity of Effect	$P = 100(i - \frac{1}{2})/15$
1	−8.0	**1**	3.3
2	−5.5	**4**	10.0
3	−2.25	**3**	16.7
4	−1.25	**23**	23.3
5	−0.75	**123**	30.0
6	−0.75	**234**	36.7
7	−0.25	**34**	43.3
8	−0.25	**134**	50.0
9	−0.25	**1234**	56.7
10	0	**14**	63.3
11	0.5	**124**	70.0
12	0.75	**13**	76.7
13	1.0	**12**	83.3
14	4.5	**24**	90.0
15	24.0	**2**	96.7

they should be roughly normally distributed about zero, and they would plot on normal probability paper as a straight line. To see whether they do, we put the effects in order and plot them on normal paper, as shown in Figure 7.17. As it turns out 11 of the 15 effects fit reasonably well on a straight line, but those representing effects **1**, **4**, **23**, and **2** do not. We therefore conclude that these effects cannot be explained by chance and are in fact significant.

7.3.4. Advanced Designs

Factorial and fractional factorial designs are used for fitting either linear response models or models based on factor interactions to the experimental data (see Chapter 8). When higher-order models are necessary, more advanced experimental designs are required. One example of such a design is the *central composite design* (CCD), which is used for fitting second-order models. These designs are widely used because of their relative efficiency with respect to the number of trials required.

In a CCD, the standard two-level factorial "box" is enhanced by replicated experiments at the center of the design space (called *centerpoints*), as well as by symmetrically located *axial points*. Thus, a complete CCD with k factors requires 2^k factorial runs, $2k$ axial runs, and 3–5 centerpoints. The centerpoints provide a direct measure of the experimental replication error, and the axial points facilitate fitting of the second-order responses. Designs for $k = 2$ and $k = 3$ are shown in Figure 7.18.

The CCD can be made *rotatable* by the proper choice of the axial spacing (α in Figure 7.18). Rotatability implies that the standard deviation of the predicted response is constant at all points equidistant from the center of the design. To

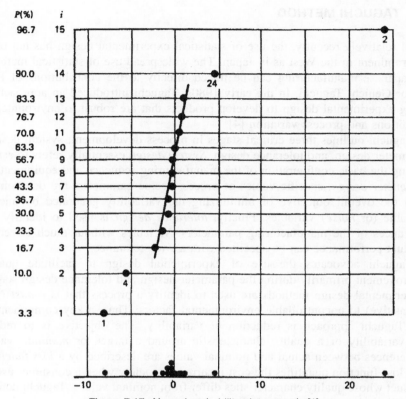

Figure 7.17. Normal probability plot example [1].

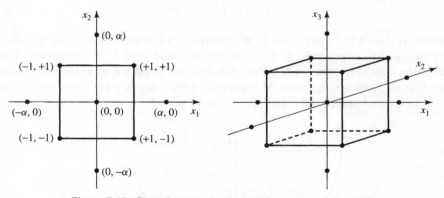

Figure 7.18. Central composite designs for $k = 2$ and $k = 3$ [3].

ensure rotatability, we select

$$\alpha = (2^k)^{1/4} \tag{7.47}$$

For the case of $k = 2$, $\alpha = 1.414$. This is the case represented on the left in Figure 7.18.

7.4. TAGUCHI METHOD

Until relatively recently, the use of statistical experimental design has not been as prominent in the West as in Japan. The widespread use of statistical methods in Japanese manufacturing can be traced directly to the contributions of Professor Genichi Taguchi. In the early 1980s, Taguchi introduced an approach to using experimental design to develop products that are robust to environmental conditions and process variation [4].

Taguchi outlines three critical stages in process development: system design, parameter design, and tolerance design. *System design* essentially refers to establishing the basic configuration of the manufacturing sequence and equipment. In *parameter design*, specific values of process recipe parameters are determined, with the overall objective of minimizing the variability generated by uncontrollable (or *noise*) variables. Finally, *tolerance design* is used to identify the tolerances of the manufacturing parameters. Variables without much effect on product performance can be specified with a wide tolerance.

Taguchi advocates the use of experimental design to facilitate quality improvement primarily during the parameter design and tolerance design stages. Experimental design methods are used to identify a process that is *robust* (i.e., insensitive) to uncontrollable environmental factors. Thus, a key component of the Taguchi approach is reduction of variability. The objective is to reduce the variability of a quality characteristic around a target, or *nominal*, value. Differences between actual and nominal values are described by a *loss function*. The loss function quantifies the cost incurred by society when a consumer uses a product whose quality characteristics differ from nominal values. Taguchi defines a quadratic loss function of the form

$$L(y) = k(y - T)^2 \tag{7.48}$$

which is shown in Figure 7.19. In this function, y represents the measured value of the quality characteristic, T is the target value, and k is a constant. This function penalizes even small excursions from the target value, as opposed to the traditional control chart-oriented approach, which attaches penalties only when y is outside of specification limits.

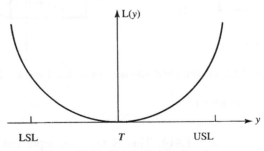

Figure 7.19. Quadratic loss function.

Taguchi's overall philosophy can be summarized by three central ideas:

1. Products and processes should be robust to variability.
2. Experimental design can be used to accomplish this.
3. Operation on target is more important to conformance to specifications.

However, a word of caution is appropriate. Although his philosophy is sound, some of the methods of statistical analysis and some of the approaches to experimental design he advocates have been shown to be unnecessarily complicated, inefficient, and even ineffective. Thus, care should be exercised in applying Taguchi's methods.

The Taguchi methodology is best illustrated by example. In the following sections, we use an example first published in the *Bell System Technical Journal* [5]. In this example, Taguchi's approach is used to optimize the photolithographic process used to form square contact windows in a CMOS microprocessor fabrication process. The purpose of the contact windows is to facilitate the interconnection between transistors. The goal is to produce windows of a size near the target dimension.

7.4.1. Categorizing Process Variables

The first step in applying the Taguchi methodology is to identify the important process variables that can be manipulated, as well as their potential working levels. These variables are categorized as either controllable or uncontrollable. The controllable factors are also referred to as either *control factors* or *signal factors*, whereas the uncontrollable factors are called *noise factors*.

For the contact window formation example, the key process steps are (1) applying the photoresist by spin coating, (2) prebaking the photoresist, (3) exposing the photoresist, (4) developing the resist, and (5) etching the windows using plasma etching. The controllable factors associated with each step are as follows:

- *Applying photoresist*—resist viscosity (B) and spin speed (C)
- *Baking*—bake temperature (D) and bake time (E)
- *Exposure*—mask dimension (A), aperture (F), and exposure time (G)
- *Development*—developing time (H)
- *Plasma etch*—etch time (I)

The operating levels of these nine factors are shown in Table 7.21. Six factors have three levels each, and three factors have only two levels. The levels of spin speed in this table are dependent on the levels of viscosity. For the 204 photoresist viscosity, the low, normal, and high levels of the spin speed are 2000, 3000, and

Table 7.21. Factors and Levels for Taguchi Example.

Label	Factor Name	Low Level	Medium Level	High Level
A	Mask dimension (μm)	—	2	2.5
B	Viscosity	—	204	206
C	Spin speed (rpm)	Low	Normal	High
D	Bake temperature ($^\circ$C)	90	105	
E	Bake time (min)	20	30	40
F	Aperture	1	2	3
G	Exposure time	20% over	normal	20% under
H	Developing time (s)	30	45	60
I	Etch time (min)	14.5	13.2	15.8

4000 rpm, respectively. For the 206 viscosity, those levels are 3000, 4000, and 5000 rpm.

Examples of potential noise factors in this study include the relative humidity or the number of particles in the cleanroom. The objective of this procedure is to determine the levels of the controllable factors that lead to windows closest to the target dimension of 3.5 μm.

7.4.2. Signal-to-Noise Ratio

Taguchi recommends analyzing the results from designed experiments using the mean response and the appropriately selected *signal-to-noise ratio* (*SN*). Signal-to-noise ratios are derived from the quadratic loss function given in Eq. (7.48). The three standard *SN*s are

$$\text{Nominal the best:} \quad SN_N = 10\log(\overline{y}/s) \tag{7.49}$$

$$\text{Larger the better:} \quad SN_L = -10\log\left(\frac{1}{n}\sum_{i=1}^{n}\frac{1}{y_i^2}\right) \tag{7.50}$$

$$\text{Smaller the better:} \quad SN_S = -10\log\left(\frac{1}{n}\sum_{i=1}^{n}y_i^2\right) \tag{7.51}$$

where s is the sample standard deviation. Each of these ratios is expressed in a decibel scale. SN_N is used if the objective is to reduce variability around a specific target, SN_L is appropriate if the system is optimized when the response is as large as possible, and SN_S is selected to optimize a system by making the response as small as possible. Factor levels that maximize the appropriate SN ratio are considered optimal. Clearly, SN_N is the right choice for the contact window formation experiment.

7.4.3. Orthogonal Arrays

A full factorial experiment to explore all possible interactions of the factors in Table 7.21 would require $3^6 \times 2^3 = 5832$ trials. Clearly, when cost of material, time, and availability of facilities are considered, the full factorial approach is prohibitively large. Taguchi recommends an alternative fractional factorial design known as the *orthogonal array*. The columns of such an array are pairwise orthogonal, meaning that for every pair of columns, all combinations of levels occur and they occur an equal number of times.

Table 7.22 shows the L_{18} orthogonal array design for the contact window formation study [5]. In this table, factors B and D are treated as a joint factor BD with levels 1, 2, and 3 representing the combinations B_1D_1, B_2D_1, and B_2D_2, respectively. This was done to accommodate the L_{18} array, which can be used to evaluate a maximum of eight factors.

The L_{18} array is a "main effects only" design that assumes that the response(s) can be approximated by a separable function. In other words, it is assumed that the response(s) can be written in terms of a sum of terms where each term is a function of a single independent variable. This type of model can yield misleading conclusions in the presence of factor interactions. However, Taguchi claims that the use of the *SN* ratio generally eliminates the need to examine interactions.

For estimating the main effects, there are 2 degrees of freedom associated with each three-level factor, one degree of freedom associated with each two-level factor, and one degree of freedom associated with the mean. Since we need at least one experiment for each degree of freedom, the minimum number of experiments required for optimizing the contact window formation process is 16. The L_{18} array has 18 trials, which provides additional precision in estimating the effects.

Table 7.22. Factor levels for the L_{18} orthogonal array.

Experiment	A	BD	C	E	F	G	H	I
1	1	1	1	1	1	1	1	1
2	1	1	2	2	2	2	2	2
3	1	1	3	3	3	3	3	3
4	1	2	1	1	2	2	3	3
5	1	2	2	2	3	3	1	1
6	1	2	3	3	1	1	2	2
7	1	3	1	2	1	3	2	3
8	1	3	2	3	2	1	3	1
9	1	3	3	1	3	2	1	2
10	2	1	1	3	3	2	2	1
11	2	1	2	1	1	3	3	2
12	2	1	3	2	2	1	1	3
13	2	2	1	2	3	1	3	2
14	2	2	2	3	2	2	1	3
15	2	2	3	1	2	3	2	1
16	2	3	1	3	2	3	1	2
17	2	3	2	1	3	1	2	3
18	2	3	3	2	1	2	3	1

7.4.4. Data Analysis

The postetch window size is the most appropriate quality measure for this experiment. Unfortunately, because of the size and proximity of the windows, the existing equipment in the paper by Phadke et al. [5] was unable to provide reproducible window size measurements. As a result, a linewidth test pattern on each chip was used to characterize the window size. The postetch line width was used as a window size metric.

Five chips were selected from each of the wafers used in the L_{18} design in Table 7.22. These five chips correspond to the top, bottom, left, right, and center of the wafers. Once again, the optimization problem posed by this experiment is to determine the optimum factor levels such that SN_N is maximum while keeping the mean on target. This problem is solved in two stages:

1. Use ANOVA techniques to determine which factors have a significant effect on SN_N. These factors are called the *control factors*. For each control factor, we choose the level with the highest SN_N as the optimum level, thereby maximizing the overall SN_N (under the separability assumption).

2. Select a factor that has the smallest effect on SN_N among the control factors. Such a factor is called a *signal factor*. Set the levels of the remaining factors (i.e., those that are neither control nor signal factors) to their nominal levels prior to the optimization experiment. Then, set the level of the signal factor so that the mean response is on target.

In cases where multiple responses exist, engineering judgment is used to resolve conflicts when different response variables suggest different levels for any single factor.

For each trial in Table 7.21, the mean, standard deviation, and SN_N for the postetch line width were computed. The following linear model was used to analyze these data

$$y_i = \mu + x_i + e_i \tag{7.52}$$

where y_i is the SN_N for experiment i, μ is the overall mean, x_i is the sum of the main effects of all eight factors in experiment i, and e_i is the random error in experiment i. The ANOVA table for SN_N and the mean postetch line width are shown in Tables 7.23 and 7.24, respectively. In Table 7.24, a *pooled ANOVA* was derived by pooling the sum of squares for those factors whose sums of squares were smaller than the error sum of squares (D, E, F, and I) with the error sum of squares. The "percent contribution" in the last column of Table 7.24 is a Taguchi metric that is equal to the total sum of squares explained by a factor after an appropriate estimate of the error sum of squares has been removed from it. A larger percent contribution implies that more can be expected to be achieved by changing the level of that factor.

Table 7.23 indicates that none of the nine process factors has a significant effect on SN_N for postetch linewidth. Thus, none of these may be considered

Table 7.23. Postetch linewidth ANOVA for SN_N.

Factor	Degrees of Freedom	Sum of Squares	Mean Square	F Ratio
A	1	0.005	0.005	0.02
B	1	0.134	0.134	0.60
C	1	0.003	0.003	0.01
D	2	0.053	0.027	0.12
E	2	0.057	0.028	0.13
F	2	0.085	0.043	0.19
G	2	0.312	0.156	0.70
H	2	0.156	0.078	0.35
I	2	0.008	0.004	0.02
Error	2	0.444	0.222	
Total	17	1.257		

Table 7.24. Pooled ANOVA for mean postetch linewidth.

Factor	Degrees of Freedom	Sum of Squares	Mean Square	F Ratio	% Contribution
A	1	0.677	0.677	16.92	8.5
B	1	2.512	2.512	63.51	32.9
C	2	1.424	0.712	17.80	17.9
G	2	1.558	0.779	19.48	19.6
H	2	0.997	0.499	12.48	12.2
Error	9	0.356	0.040		8.9
Total	17	7.524			100.0

control factors in this experiment. However, all the factors in Table 7.24 (viscosity, exposure, spin speed, mask dimension, and developing time) were significant at a 95% confidence level for the mean value of this response. The mean linewidth for each factor is shown in Figure 7.20.

To keep the process on target, a signal factor must be selected that has a significant effect on the mean, but little effect on SN_N. Changing the signal factor then affects only the mean. In this experiment, exposure time (G) was selected as the signal factor. This factor was adjusted to obtain the optimum linewidth and therefore, window size. This adjustment resulted in a factor of 2 decrease in window size variation and a factor of three decrease in the number of windows not printed. Thus the Taguchi methodology was proven to be effective in this example.

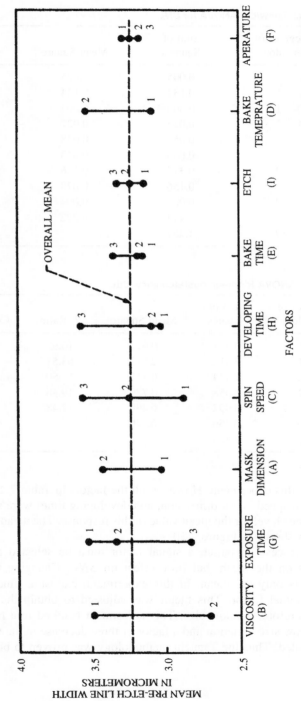

Figure 7.20. Mean postetch linewidth; The mean for each factor is indicated by a dot, and the number next to the dot indicates the factor level [5].

SUMMARY

In this chapter, we have provided an overview of statistical experimental design by introducing the concept of analysis of variance and describing various types designs, including two-level factorial designs and the Taguchi methodology. Important topics also included the analysis of such experiments and the use of various analytical and graphical methods to interpret experimental results. In the next chapter, we will examine how data generated from design experiments may be used to construct models that predict process behavior.

PROBLEMS

7.1. To compare two photolithography processes (A and B), 4 of 8 wafers were randomly assigned to each. The electrically measured linewidth of several NMOS transistors gave the following averages (in μm):

A:	1.176	1.230	1.146	1.672
B:	1.279	1.000	1.146	1.176

Assuming that the processes have the same standard deviation, calculate the significance for the comparison of means.

7.2. Suppose that there are now four photolithography processes to compare (A, B, C, and D). Using 15 wafers, the measurements are as follows (in μm):

	I	II	III	IV
A	1.176	1.230	1.146	1.672
B	1.279	1.000	1.146	1.176
C	0.954	1.079	1.204	—
D	0.699	1.114	1.114	—

Calculate the full ANOVA table and find the level of significance for rejecting the hypothesis of equality. Explain any assumptions and perform the necessary diagnostics on the residuals.

7.3. The following data are for the throughput, as measured by the number of wafer lots produced per day by different operators (A, B, C, and D) on different machines (each operator used each machine on two different days):

Machine	A	B	C	D
1	18(9), 17(76)	16(11), 18(77)	17(22), 20(72)	27(3), 27(73)
2	17(1), 13(71)	18(3), 18(73)	20(57), 16(70)	28(2), 23(78)
3	16(3), 17(77)	17(7), 19(70)	20(25), 16(73)	31(33), 30(72)
4	15(2), 17(72)	21(4), 22(74)	16(5), 16(71)	31(6), 24(75)
5	17(17), 18(84)	16(10), 18(72)	14(39), 13(74)	28(7), 22(82)

Eighty-four working days were needed to collect the data. The numbers in parentheses refer to the days on which the results were obtained. For example, on the first day, operator A produced 17 lots using machine 2, and on the 84th day, operator A produced 18 lots using machine 5. On some days (such as the third day), more than one item of data was collected, and on other days (such as day 40), no data was collected. Analyze the data, stating all assumptions and conclusions.

7.4. Consider the data in Table 7.11. Carry out analysis of variance using the data transformation $Y = y^{-1}$. Consider whether in the new response metric there is evidence of model inadequacy. Compare the treatment averages for the two different representations of the response.

7.5. The following single-replicate 2^3 factorial design was used to develop a nitride etch process. State any assumptions you make, and analyze this experiment.

Temperature (°F)	Concentration (%)	Catalyst	Yield (%)
160	20	1	60
180	20	1	77
160	40	1	59
180	40	1	68
160	20	2	57
180	20	2	83
160	40	2	45
180	40	2	85

7.6. **(a)** Why do we "block" experimental designs?

(b) Write a 2^3 factorial design.

(c) Write a 2^3 factorial design in four blocks of two runs each such that the main effects are not confounded with the blocks.

7.7. Consider a 2^{8-4} fractional factorial design.

(a) How many variables does the design have?

(b) How many runs are involved in the design?

(c) How many levels are used for each variable?

(d) How many independent block generators are there?

(e) How many words are there in the defining relations (counting I)?

7.8. Construct a 2^{7-1} fractional factorial design. Show how the design may be divided into eight blocks of eight runs each so that no main effect or two-factor interaction is confounded with any block effect.

7.9. A 2^3 factorial design on a CVD system was replicated 20 times with the following results:

P	T	F	$\overline{y}$
$-$	$-$	$-$	7.76 ± 0.53
$+$	$-$	$-$	10.13 ± 0.74
$-$	$+$	$-$	5.86 ± 0.47
$+$	$+$	$-$	8.76 ± 1.24
$-$	$-$	$+$	9.03 ± 1.12
$+$	$-$	$+$	14.59 ± 3.22
$-$	$+$	$+$	9.18 ± 1.80
$+$	$+$	$+$	13.04 ± 2.58

In this tabulation, $\overline{y}$ is the deposition rate in μm/min and the number following the $\pm$ sign is the standard deviation of $\overline{y}$. The variables P, T, and F represent pressure, temperature, and flowrate. Analyze the results. Should a data transformation be made?

REFERENCES

1. G. Box, W. Hunter, and J. Hunter, *Statistics for Experimenters*, Wiley, New York, 1978.
2. G. Box and D. Cox, "An Analysis of Transformations," *J. Roy. Stat. Soc. B.* **26**, 211 (1964).
3. D. Montgomery, *Introduction to Statistical Quality Control*, Wiley, New York, 1993.
4. G. Taguchi and Y. Wu, Control Japan Quality Control Organization, Nagoya, Japan, 1980.
5. M. Phadke, R. Kackar, D. Speeney, and M. Grieco, "Off-Line Quality Control in Integrated Circuit Fabrication Using Experimental Design," *Bell Syst. Tech. J.*, (May–June 1983).

8

PROCESS MODELING

OBJECTIVES

- Provide an overview of statistical modeling techniques such as regression and response surface methods.
- Introduce the concept of principal-component analysis (PCA).
- Discuss new modeling methods based on artificial intelligence techniques.
- Describe methods of model-based process optimization.

INTRODUCTION

As discussed in Chapter 7, a designed experiment is an extremely useful tool for discovering key variables that influence quality characteristics. Statistical experimental design is a powerful approach for systematically varying process conditions and determining their impact on output parameters that measure quality. Data derived from such experiments can then be used to construct *process models* of various types that enable the analysis and prediction of manufacturing process behavior.

The models so derived may be used to visualize process behavior in the form of a *response surface*. The proper fit is obtained using statistical regression techniques such as the method of *least squares* (also known as *linear regression analysis*). The goal of regression analysis is to develop a quantitative model

Fundamentals of Semiconductor Manufacturing and Process Control,
By Gary S. May and Costas J. Spanos
Copyright © 2006 John Wiley & Sons, Inc.

(usually in the form of a polynomial) that predicts a relationship between input factors and a given response. An accurate model should minimize the difference between the observed values of the response and its own predictions.

Over time, several novel methods have been developed to augment regression modeling. For example, *principal component analysis* (PCA) is a useful statistical technique for streamlining a multidimensional dataset to facilitate subsequent modeling. Dimensionality reduction through PCA is achieved by transforming a data to a new set of variables (i.e., the *principal components*), which are uncorrelated and ordered such that the first few retain most of the variation present in the original dataset.

Approaches that utilize *artificial intelligence* (AI) methods such as *neural networks* or *fuzzy logic* are capable of performing highly complex mappings on noisy and/or nonlinear experimental data, thereby inferring very subtle relationships between diverse sets of input and output parameters. Moreover, these techniques can also generalize well enough to learn overall trends in functional relationships from limited training data.

Process modeling permits an engineer to manipulate and optimize the process efficiency with a minimum amount of experimentation. A well-developed process model can in turn be used to generate a recipe of the process deposition conditions to obtain particular desired responses. In effect, this required that the neural process model be used "in reverse" to predict the necessary operating conditions to achieve the desired film characteristic. This chapter explores various process modeling methodologies, from traditional regression analysis to more contemporary AI-based approaches for deriving predictive models in semiconductor manufacturing applications. We then explore various optimization (or recipe synthesis) procedures.

8.1. REGRESSION MODELING

Raw experimental data have limited meaning in and of themselves; they are most useful in relation to some conceptual model of the process being studied. Once such data have been obtained from a designed experiment (see Chapter 7), the results may be summarized in the form of a *response surface*. The proper fit for a response surface is obtained using statistical regression techniques. When the formulation of the response surface is such that the outcome is a linear function of the unknown parameters, these parameters can be estimated by the method of *least squares* (also known as *linear regression analysis*). Linear regression analysis is a statistical technique for modeling and investigating the relationship between two or more variables. The goal of regression analysis is to develop a quantitative model that predicts this relationship between controllable input factors and a given response.

In general, suppose that there is a single *dependent variable* or *response* y that is related to k *independent variables*, say, $x_1, x_2, \ldots, x_k$. Assume that the dependent variable y is a random variable, and the independent variables $x_1, x_2, \ldots, x_k$

are exactly known or can be measured with negligible error. The independent variables are controllable by the experimenter. The relationship between these variables is characterized by a mathematical model called a *regression equation*. More precisely, we speak of the regression of y on $x_1, x_2, \ldots, x_k$. This regression model is fitted to a set of data. In some instances, the experimenter will know the exact form of the true functional relationship between y and $x_1, x_2, \ldots, x_k$, say, $y = f(x_1, x_2, \ldots, x_k)$. However, in most cases, the true functional relationship is unknown, and the experimenter must derive an appropriate function to approximate the function f. A polynomial model is often employed as the approximating function. An accurate model should minimize the difference between the observed values of the response and its own predictions. In addition to predicting the response, such a model can also be used for process optimization or process control purposes.

8.1.1. Single-Parameter Model

The simplest polynomial response surface is merely a straight line. Models fit to a straight line are derived using *linear regression*.[1] Consider fitting experimental data to a straight line that passes through the origin. Although rather elementary, this example illustrates the basic principles of least squares.

Suppose that we are studying the etch rate of a wet etchant, and we collect $n = 9$ observations of the data shown in Table 8.1, where x is the time in minutes and y is the thickness of film etched away. Physical considerations indicate that a simple proportional relationship between x and y is reasonable; that is, the relationship between x and y should be described by a straight line through the origin, or

$$y_u = \beta x_u + \varepsilon_u \qquad u = 1, 2, \ldots, n \tag{8.1}$$

Table 8.1. Hypothetical etching data.

Observation (u)	Time [x_u (min)]	Thickness [y_u (μm)]
1	8	6.16
2	22	9.88
3	35	14.35
4	40	24.06
5	57	30.34
6	73	32.17
7	78	42.18
8	87	43.23
9	98	48.76

[1]The term "linear" refers to the fact that the regression equation is linear to the unknown parameters. In this sense, as we will see, linear regression is also capable of deriving models that are non-linear to the regressors.

where β is a constant of proportionality (i.e., the slope of the "best fit" line) and the ε_u are random, independent experimental errors with zero mean and constant variance [i.e., $\varepsilon_u \sim N(0, \sigma^2)$]. The response or output variable y is the dependent variable, and the input variable x is the independent variable or the *regressor*. The objective of regression analysis is to find an estimate of b that minimizes the difference between the measured values of y and the predictions of Eq. (8.1).

According to the method of least squares, the best-fit model is the one that minimizes the quantity

$$S(\beta) = \sum_{u=1}^{n}(y_u - \beta x_u)^2 = \sum(y - \hat{y})^2 \tag{8.2}$$

where $\hat{y} = \beta x$ is an estimate of y and the subscripts have been dropped to simplify the notation. The curve represented by this equation is a parabola, so the goal is to find the value of β at the minimum of the parabola. Let b be the value of β at the minimum point. Using the rules of calculus, we can find b by simply taking the derivative of S with respect to β and setting the derivative equal to zero, or

$$\frac{dS}{d\beta} = 2\sum(y - \hat{y})x = 2\sum(y - bx)x = 0 \tag{8.3}$$

since $\hat{y} = bx$ at the minimum point. Solving (8.3) for b yields

$$b = \frac{\sum xy}{\sum x^2} \tag{8.4}$$

Using the etching data in Table 8.1, we compute $b = 0.501$ μm/min. This value has been substituted into $\hat{y} = bx$ and plotted in Figure 8.1.

8.1.1.1. Residuals
Once the least-squares estimate (b) of the unknown coefficient (β) has been obtained, the estimated response $\hat{y}_u = bx_u$ can be computed for each x_u. These estimated responses can be compared with the observed values (y_u). The differences between the estimated and observed values ($y_u - \hat{y}_u$) are known as

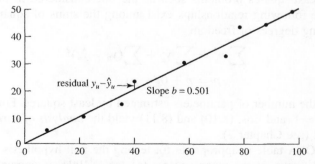

Figure 8.1. Plot of data fitted to least-squares line, etching example [1].

residuals. The *sum of squares* of the residuals is given by

$$S_R = S(b) = \sum_{u=1}^{n} (y_u - \hat{y}_u)^2 \tag{8.5}$$

In this example, $S_R = 64.67 \ \mu m^2$. As discussed in Chapter 7, it is important to examine the residuals individually and collectively for inadequacies in the model.

8.1.1.2. Standard Error

If the one-parameter linear model is adequate, then an estimate (s^2) of the experimental error variance (σ^2) can be obtained by dividing the residual sum of squares by its number of degrees of freedom. The number of degrees of freedom will generally equal the number observations less the number of parameters estimated. Since only a single parameter is estimated in this model, there are $n - 1 = 8$ degrees of freedom in this example. An estimate of σ^2 is therefore

$$s^2 = \frac{S_R}{n-1} = 8.08 \tag{8.6}$$

The corresponding estimated variance for b is then [1]

$$V(b) = \frac{s^2}{\sum_{u=1}^{n} x_u^2} = 0.00023 \tag{8.7}$$

Thus, the *standard error* of b is $SE(b) = \sqrt{V(b)} = 0.015 \ \mu m/min$. This metric can be used to perform a hypothesis test (see Chapter 4) to determine whether the true value of β is equal to some specific value β^* using the test statistic

$$t_0 = \frac{b - \beta^*}{SE(b)} \tag{8.8}$$

which is distributed according to the t distribution with $n - 1 = 8$ degrees of freedom. The $1 - \alpha$ confidence interval for β is bounded by

$$b \pm [t_{\alpha/2} \times SE(b)] \tag{8.9}$$

8.1.1.3. Analysis of Variance

For linear least-squares problems such as the one considered in the preceding sections, the following relationships exist among the sums of squares and their corresponding degrees of freedom

$$\sum y_u^2 = \sum \hat{y}_u^2 + \sum (y_u - \hat{y}_u)^2 \tag{8.10}$$

$$n = p + (n - p) \tag{8.11}$$

where p is the number of parameters estimated by least squares. For the etching problem, $p = 1$, and Eqs. (8.10) and (8.11) yield the *analysis of variance* shown in Table 8.2 (see Chapter 7).

This ANOVA table is appropriate for testing the null hypothesis that $\beta^* = 0$. To do so, the ratio of the mean squares ($s_M^2/s_R^2 = 1094$) is compared with the

Table 8.2. ANOVA table for etch data.

Source of Variation	Sum of Squares	Degrees of Freedom	Mean Square	F Ratio
Model	$S_M = 8836.64$	1	$s_M^2 = 8836.64$	$s_M^2/s_R^2 = 1094$
Residual	$S_R = 64.67$	8	$s_R^2 = 8.08$	
Total	$S_T = 8901.31$	9		

value of the F distribution with 1 and 8 degrees of freedom. The ratio for this example is overwhelmingly significant, indicating that there is little probability that β^* is in fact zero. This test is exactly equivalent to applying the t test implied by Eq. (8.8).

8.1.2. Two-Parameter Model

Many modeling situations require more than a single parameter. Consider the data in Table 8.3 representing the level of impurities in a polymer dielectric layer as a function of the concentration of a certain monomer and a certain dimer. Here, the appropriate model is

$$y = \beta_1 x_1 + \beta_2 x_2 + \varepsilon \tag{8.12}$$

where y is the percent impurity concentration, x_1 is the percent concentration of the monomer, x_2 is the percent concentration of the dimer, and $\varepsilon \sim N(0, \sigma^2)$.

The best-fit model in this case is the one that minimizes the quantity

$$S(\beta) = \sum (y - \beta_1 x_1 - \beta_2 x_2)^2 \tag{8.13}$$

Since there are two parameters, this equation now represents a plane rather than a line. We could find the values of β_1 and β_2 that minimize $S(\beta)$ (i.e., b_1 and b_2, respectively) using the same calculus-based approach as we used for the single-parameter model. Alternatively, we can also use what are called the *normal equations* to compute these values. If we let $\hat{y} = b_1 x_1 + b_2 x_2$, this approach utilizes the fact that the vector of *residuals* (i.e., the vector composed of the

Table 8.3. Hypothetical polymer impurity data.

Observation	Monomer Concentration [x_1 (%)]	Dimer Concentration [x_2 (%)]	Impurity Concentration [y (%)]
1	0.34	0.73	5.75
2	0.34	0.73	4.79
3	0.58	0.69	5.44
4	1.26	0.97	9.09
5	1.26	0.97	8.59
6	1.82	0.46	5.09

values of $y - \hat{y}$ for each of the n observations) has the property of being normal (at right angles) to each vector of x values when the least squares estimate is used.

In this model, there are two regressors, x_1 and x_2. The normal equations in this case are

$$\sum(y - \hat{y})x_1 = 0 \qquad \sum(y - \hat{y})x_2 = 0 \qquad (8.14)$$

or

$$\sum(y - b_1x_1 - b_2x_2)x_1 = 0 \qquad \sum(y - b_1x_1 - b_2x_2)x_2 = 0 \quad (8.15)$$

Simplifying further gives

$$\sum yx_1 - b_1 \sum x_1^2 - b_2 \sum x_1x_2 = 0 \qquad \sum yx_1 - b_1 \sum x_1x_2 - b_2$$

$$\sum x_2^2 = 0 \qquad (8.16)$$

Solving these two equations simultaneously using the data in Table 8.3 yields $b_1 = 1.21$ and $b_2 = 7.12$. The fitted surface appears in Figure 8.2. In general,

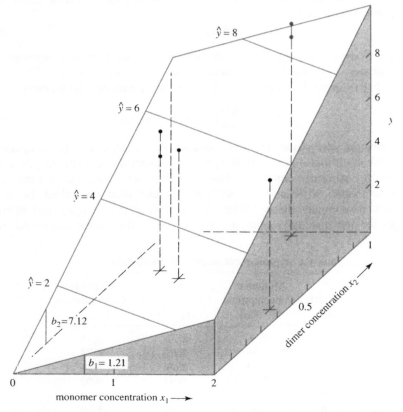

Figure 8.2. Fitted plane $\hat{y} = 1.21x_1 + 7.12x_2$ for polymer impurity example [1].

this method can be applied to a linear equation of the form given by Eq. (8.12) with an arbitrary number of regressor variables.

8.1.2.1. Analysis of Variance

The first step in evaluating the adequacy of the model presented above is inspection of the residuals (i.e., $y - \hat{y}$). However, with such a small dataset, only gross discrepancies would be revealed by such an analysis. In this case, no such discrepancies are evident.

Another type of analysis is also appropriate when some of the experimental runs have been replicated. In this case, runs 1 and 2 are replicates, as are runs 4 and 5. The sum of squares associated with these replicate runs is

$$S_E = \frac{(y_1 - y_2)^2}{2} + \frac{(y_4 - y_5)^2}{2} \tag{8.17}$$

This sum of squares, which has 2 degrees of freedom, is part of the overall residual sum of squares (S_R) and is a measure of the "pure" experimental error. The remaining part of the residual sum of squares is given by

$$S_L = S_R - S_E \tag{8.18}$$

This quantity measures the experimental error plus any contribution from possible lack of fit of the model. A comparison of the mean squares derived from S_E and S_L can therefore be used to check the lack of fit. These concepts are summarized in the ANOVA given in Table 8.4.

In this example, the close agreement between the two mean squares (as indicated by the F ratio near unity) gives no reason to suspect a significant lack of fit. An examination of Appendix E reveals that a mean-square ratio greater than 1.2 can be expected about 45% of the time with this small number of degrees of freedom. It can therefore be concluded that the fit for this model is adequate.

8.1.2.2. Precision of Estimates

According to the assumption that the model is adequate, an estimate of the error variance of the model is

$$s^2 = \frac{S_R}{n - p} = 0.33 \tag{8.19}$$

Table 8.4. ANOVA table for polymer impurity data.

Source of Variation	Sum of Squares	Degrees of Freedom	Mean Square	F Ratio
Model	$S_M = 266.59$	2		
Residual	$S_R = 1.33$			
Lack of fit	$S_L = 0.74$	2	$s_L^2 = 0.37$	$s_L^2/s_E^2 = 1.2$
Pure error	$S_E = 0.59$	2	$s_E^2 = 0.30$	
Total	$S_T = 267.92$	6		

To estimate the variances of b_1 and b_2, the correlation (ρ) between these parameters must first be computed using

$$\rho = \frac{-\sum x_1 x_2}{\sqrt{\sum x_1^2 x_2^2}} = -0.825 \tag{8.20}$$

The variances are then given by

$$V(b_1) = \frac{1}{(1-\rho)^2} \frac{s^2}{\sum x_1^2} = 0.147 \tag{8.21}$$

$$V(b_2) = \frac{1}{(1-\rho)^2} \frac{s^2}{\sum x_2^2} = 0.285$$

Since the standard error for each parameter is just the square root of its variance, $\text{SE}(b_1) = 0.383$, and $\text{SE}(b_2) = 0.534$. Given these values for standard error, it is possible to define $(1 - \alpha)$ confidence limits for each parameter using Eq. (8.9).

8.1.2.3. Linear Model with Nonzero Intercept

Consider the problem of fitting data to a linear model that does not pass through the origin. The equation of such a line is given by

$$y = \beta_0 + \beta x + \varepsilon \tag{8.22}$$

where the intercept $\beta_0 \neq 0$. This model is just a special case of the model given in Eq. (8.12), with the following substitutions:

$$\beta_1 = \beta_0, \quad x_1 = 1, \quad \beta_2 = \beta, \quad \text{and} \quad x_2 = x$$

The "variable" $x_1 = 1$ is referred to as an *indicator variable*. For this model, the normal equations [Eqs. (8.14)–(8.16)] simplify to

$$b_0 n + b \sum x = \sum y \tag{8.23}$$

$$b_0 \sum x + b \sum x^2 = \sum xy$$

and the solutions are

$$b = \frac{\sum (x - \bar{x})(y - \bar{y})}{\sum (x - \bar{x})^2} \tag{8.24}$$

$$b_0 = \bar{y} - b\bar{x}$$

where n is the number of points to be fitted, $\bar{x} = \frac{1}{n}\sum_{i=1}^{n} x_i$, and $\bar{y} = \frac{1}{n}\sum_{i=1}^{n} y_i$.

To illustrate this situation, consider the hypothetical data in Table 8.5, which represents particle counts in a class 100 cleanroom as a function of equipment utilization. Applying Eq. (8.24) to these data yields $b = 11.66$ and $b_0 = 65.34$. Therefore, the appropriate linear model in this case is $\hat{y} = 65.34 + 11.66x$. This

Table 8.5. Hypothetical particle data.

Observation	Equipment Utilization [x (Arbitrary Units)]	Particle Count [y (ft^{-3})]
1	2.00	89
2	2.50	97
3	2.50	91
4	2.75	98
5	3.00	100
6	3.00	104
7	3.00	97

line is plotted in Figure 8.3. Given the expression for b_0 in Eq. (8.24), this line can also be written as

$$\hat{y} = b_0 + bx = \overline{y} - b\overline{x} + bx = \overline{y} + b(x - \overline{x}) = a + b(x - \overline{x}) \qquad (8.25)$$

where $a = \overline{y}$. For the current example, $\hat{y} = 96.57 + 11.66(x - 2.28)$.

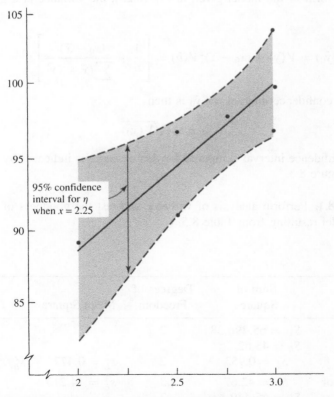

Figure 8.3. Fitted line $\hat{y} = 65.34 + 11.66x$ for cleanroom particle example [1].

The precision of the estimates for this model can be evaluated using an approach similar to that outlined in Section 8.1.2.2. Assuming that the model is adequate, an estimate of the error variance of the model is

$$s^2 = \frac{S_R}{n - p} = 8.72 \tag{8.26}$$

where $S_R = \sum(y - \hat{y})^2 = 43.62$, $n = 7$, and $p = 2$. The standard errors for the coefficients are

$$SE(b_0) = s \left[\frac{1}{n} + \frac{\bar{x}^2}{\sum(x - \bar{x})^2} \right]^{1/2} = 8.64$$

$$SE(b) = \frac{s}{\sqrt{\sum(x - \bar{x})^2}} = 3.22 \tag{8.27}$$

$$SE(a) = \frac{s}{\sqrt{n}} = 1.12$$

Using the form of the model given in Eq. (8.25), the variance at a given point (x_0, y_0) is

$$V(\hat{y}_0) = V(\bar{y}) + (x_0 - \bar{x})^2 V(b) = \left[\frac{1}{n} + \frac{(x_0 - \bar{x})^2}{\sum(x - \bar{x})^2} \right] s^2 \tag{8.28}$$

A $(1 - \alpha)$ confidence interval for $\hat{y}_0$ is then

$$\hat{y}_0 \pm t_{\alpha/2} \sqrt{V(\hat{y}_0)} \tag{8.29}$$

A 95% confidence interval computed for this example is indicated by the dotted lines in Figure 8.3.

Example 8.1. Perform analysis of variance and test the goodness of fit for the linear model resulting from Table 8.5.

Solution:

Source of Variation	Sum of Squares	Degrees of Freedom	Mean Square	F Ratio
Model	$S_M = 65,396.38$	2		
Residual	$S_R = 43.62$			
Lack of fit	$S_L = 0.953$	2	$s_L^2 = 0.477$	$s_L^2/s_E^2 = 0.034$
Pure error	$S_E = 42.67$	3	$s_E^2 = 14.213$	
Total	$S_T = 65,440$	7		

The ANOVA table is shown above. Note that since observations 2 and 3, as well as 5, 6, and 7 are replicates:

$$S_E = \frac{(y_2 - y_3)^2}{2} + \frac{(y_5 - y_6)^2}{3} + \frac{(y_6 - y_7)^2}{3} + \frac{(y_5 - y_7)^2}{3} = 42.67$$

The ratio of lack of fit to pure error mean squares is only 0.034. A true lack of fit would be indicated by a much larger value of this ratio (see Appendix E). We can be more than 99% confident that this model fits the data.

8.1.3. Multivariate Models

The method of least squares described above can be used in general for modeling any process in which the estimated parameters of the model (β_1, β_2, etc.) are *linear*. A model is linear in its parameters if it can be written in the form

$$\hat{y} = \beta_0 + \beta_1 x_1 + \cdots + \beta_p x_p \tag{8.30}$$

where the x terms are the quantities known for each experimental run and are not functions of the β terms. The models discussed in Sections 8.1.1 and 8.1.2 are clearly of the linear type. Another example of a model that is linear in its parameters is the polynomial model:

$$\hat{y} = \beta_0 + \beta_1 x + \beta_2 x^2 + \cdots + \beta_p x^p \tag{8.31}$$

A polynomial model with $p \geq 2$ would be used when the process has been observed to exhibit higher order effects that are inadequately captured by a straight-line model. Another example is the sinusoidal model

$$\hat{y} = \beta_0 + \beta_1 \sin \theta + \beta_2 \cos \theta \tag{8.32}$$

where θ is varied in the different experimental runs. This type of model might be appropriate for a process known to exhibit periodic or cyclical behavior. In general, one could develop any functional relationship between the independent and dependent variables in a set of experimental data by simply substituting an arbitrary function for the x values in Eq. (8.30). For example, if we let $x_1 = \log \xi_1$ and $x_2 = e^{\xi_2} \xi_3$, then we obtain the model

$$\hat{y} = \beta_0 + \beta_1 \log \xi_1 + \beta_2 \frac{e^{\xi_2}}{\xi_3} \tag{8.33}$$

where ξ_1, ξ_2, and ξ_3 are known for each experimental trial.

When limited to models that are linear in their estimated coefficients, standard matrix algebra provides a convenient approach to solving least-squares regression problems. For example, in matrix notation, Eq. (8.30) or (8.31) can be rewritten as

$$\hat{\mathbf{y}} = \mathbf{X}\mathbf{b} \tag{8.34}$$

where $\hat{\mathbf{y}}$ is the $n \times 1$ vector of predicted values for the response, $\mathbf{X}$ is the $n \times p$ matrix of independent variables, and $\mathbf{b}$ is the $p \times 1$ vector of parameters to be estimated. Under these circumstances, the normal equations can be written as

$$\mathbf{X}^T(\mathbf{y} - \hat{\mathbf{y}}) = 0 \tag{8.35}$$

where T represents the transpose operation. Substituting Eq. (8.34) yields

$$\mathbf{X}^T(\mathbf{y} - \mathbf{X}\mathbf{b}) = 0 \tag{8.36}$$

If we assume that $\mathbf{X}^T\mathbf{X}$ has an inverse, solving Eq. (8.36) for $\mathbf{b}$ yields

$$b = [\mathbf{X}^T\mathbf{X}]^{-1}\mathbf{X}^T\mathbf{y} \tag{8.37}$$

In general, the variance–covariance matrix for the estimates is

$$V(\mathbf{b}) = [\mathbf{X}^T\mathbf{X}]^{-1}\sigma^2 \tag{8.38}$$

if the experimental variance σ^2 is known. Otherwise, assuming the form of the model is appropriate, σ^2 can be estimated using $s^2 = S_R/(n - p)$, where S_R is the residual sum of squares.

Although the β values in these models can be found by calculus-based methods or using the normal equations, computer programs are now widely available for this purpose. Such programs have become virtually indispensable for model building, as well as for use in model validation and verification. This is especially true when the model form is not known, and several functional forms must be analyzed and compared in terms of prediction and lack-of-fit characteristics.

Example 8.2. Assume that the yield (y) of a given process varies according to process condition x according to the relationship in Table 8.6.
Use Eq. (8.37) to fit these yield data to the quadratic model

$$\hat{y} = \beta_0 + \beta_1 x + \beta_2 x^2$$

Table 8.6. Hypothetical yield data.

Observation	Process Condition [x (Arbitrary Units)]	Yield [y (%)]
1	10	73
2	10	78
3	15	85
4	20	90
5	20	91
6	25	87
7	25	86
8	25	91
9	30	75
10	35	65

Solution: The matrices needed are

$$
\mathbf{X} = \begin{array}{ccc} x_0 & x & x^2 \end{array}
\begin{bmatrix}
1 & 10 & 100 \\
1 & 10 & 100 \\
1 & 15 & 225 \\
1 & 20 & 400 \\
1 & 20 & 400 \\
1 & 25 & 625 \\
1 & 25 & 625 \\
1 & 25 & 625 \\
1 & 30 & 900 \\
1 & 35 & 1225
\end{bmatrix}
\qquad
\mathbf{y} = \begin{bmatrix}
73 \\ 78 \\ 85 \\ 90 \\ 91 \\ 87 \\ 86 \\ 91 \\ 75 \\ 65
\end{bmatrix}
\qquad
\mathbf{b} = \begin{bmatrix} b_0 \\ b_1 \\ b_2 \end{bmatrix}
$$

$$
\mathbf{X}^T\mathbf{X} = \begin{bmatrix}
10 & 215 & 5225 \\
215 & 5225 & 138{,}125 \\
5225 & 138{,}125 & 3{,}873{,}125
\end{bmatrix}
\qquad
\mathbf{X}^T\mathbf{y} = \begin{bmatrix}
821 \\
17{,}530 \\
418{,}750
\end{bmatrix}
$$

Solving for **b** yields

$$
\mathbf{b} = \begin{bmatrix} 35.66 \\ 5.26 \\ -0.128 \end{bmatrix}
$$

so the appropriate quadratic equation is $\hat{y} = 35.66 + 5.26x - 0.128x^2$. This curve is plotted in Figure 8.4.

8.1.4. Nonlinear Regression

While a vast array of regression problems can be approximated by linear regression models (i.e., models that are linear in the parameters to be estimated), there are also models that must be nonlinear to their estimated parameters. Consider, for example, the exponential model

$$
\hat{y} = \beta_1(1 - e^{-\beta_2 x}) \tag{8.39}
$$

where x is known for each experimental trial. This model clearly cannot be written in the form of Eq. (8.30). It is, therefore, an example of a model that is nonlinear in its parameters. Fortunately, however, the general concept of least squares can still be applied to fit such models. However, while in linear regression we have an exact, closed-form solution, for most nonlinear regression problems we have an approximate, iterative solution. Further, some of the statistical

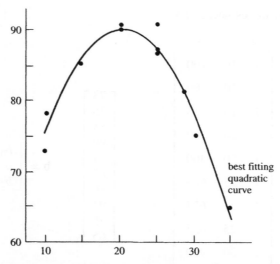

Figure 8.4. Fitted curve $\hat{y} = 35.66 + 5.26x - 0.128x^2$ for yield example [1].

assumptions that allowed us to use ANOVA in selecting model forms and in validating the results may be weak and subject to some speculation.

As an example, suppose that the number of particles generated by a particular process over time is given according to Table 8.7. Furthermore, assume that physical considerations suggest that the exponential model given by Eq. (8.39) should describe the phenomenon. The sum of squares in this case is given by

$$S = \sum_{u=1}^{n}[y_u - \beta_1(1 - e^{-\beta_2 x_u})]^2 \tag{8.40}$$

The estimated values of β_1 and β_2 that minimize S are $b_1 = 213.8$ and $b_2 = 0.5473$. Substituting these values into Eq. (8.39) gives the fitted least-squares curve shown in Figure 8.5.

Today, such curve fitting is not typically done by hand. On the contrary, modern computer software packages (such as *RS/Explore* [2]) exist that are capable

Table 8.7. Hypothetical particle data.

Observation	Particle Count [y (ft^{-3})]	Day (x)
1	109	1
2	149	2
3	149	3
4	191	5
5	213	7
6	224	10

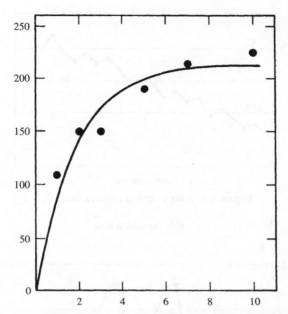

Figure 8.5. Fitted curve for particle count example [1].

of carrying out the iterative computations necessary to locate the values of the coefficients that minimize the sum of squares for nonlinear models. In such programs, the user need only supply the experimental data and the functional form of the model to be fitted. The program is able to find the coefficients, as well as their standard errors and confidence regions. The theoretical underpinning of that analysis is usually based on assuming that, at least locally, the impact of the estimated parameters on the model outcome is approximately linear.

8.1.5. Regression Chart

The concept of linear regression can be a useful tool for statistical process control (see Chapter 6). Some processes monitored using SPC exhibit specific time varying behavior. Such a trend is illustrated in Figure 8.6. Behavior like this may be a result of gradual tool wear, settling or separation of components in a chemical process, human operator fatigue, or seasonal influences (such as temperature changes). When trends are present in data, traditional control charts are inadequate. However, a device that is useful for monitoring such processes is the *regression chart*.

Consider a polysilicon CVD process in which the process chamber is regularly cleaned. After each cleaning, the polysilicon deposition rate in the chamber is effectively reset, resulting in periodic behavior in the data. On a traditional control chart, the behavior of the deposition rate would look approximately like that shown in Figure 8.7.

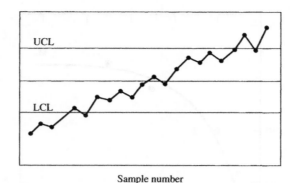

Sample number

Figure 8.6. A time varying process trend [3].

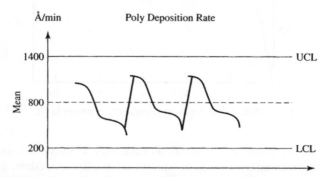

Figure 8.7. Polysilicon deposition rate plotted on standard Shewhart control chart.

This is an ideal situation for the application of a regression chart. In this case, the deposition rate data can be transformed in such a way that they are plotted as a function of the number of samples since the chamber was last cleaned, rather than as an absolute function of time or sample number. In other words, the data can be fit to a regression line of the form

$$\hat{y} = a + b(x - \bar{x}) \tag{8.41}$$

where x is the number of samples processed (or "runs") since the last chamber cleaning. Note that this model is in exactly the same form as that of Eq. (8.25); thus, the same analysis applies. The coefficients are

$$b = \frac{\sum (x - \bar{x})(y - \bar{y})}{\sum (x - \bar{x})^2} \tag{8.42}$$

$$a = \bar{y}$$

An estimate of the error variance of the model is

$$s^2 = \frac{S_R}{n - p} \tag{8.43}$$

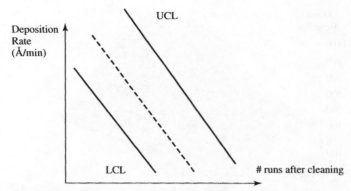

Figure 8.8. Regression chart for polysilicon deposition rate data.

where $S_R = \sum(y - \hat{y})^2$, n is the number of samples used to build the model, and $p = 2$. The standard errors for the coefficients are

$$\text{SE}(b) = \frac{s}{\sqrt{\sum(x - \overline{x})^2}} \tag{8.44}$$

$$\text{SE}(a) = \frac{s}{\sqrt{n}}$$

The data are now in a form in which they can be monitored using a regression chart whose control limits can be set at $\pm 3\sigma$ from the model prediction line. Such a chart is illustrated in Figure 8.8.

8.2. RESPONSE SURFACE METHODS

Response surface methodology (RSM) is a general technique used in the empirical study of relationships between measured responses and independent input variables. A response surface is usually a polynomial whose coefficients are extracted by means of a least-squares fit to experimental data. The concept of the response surface and an example analytical representation is shown in Figure 8.9 for a function of two variables, x_1 and x_2. The response surface method is quite powerful since, in addition to modeling, RSM also focuses on using the models developed to find the optimum operating conditions for the process under investigation.

8.2.1. Hypothetical Yield Example

To illustrate a typical application of the RSM procedure, consider an experiment whose goal is to select the settings of time (t) and temperature (T) that produces the maximum yield for a given hypothetical process. The conditions used for this process prior to the experiment were $t = 75$ min and $T = 130°C$. Assume that time can be varied from 70 to 80 min and temperature from 127.5 to 132.5°C.

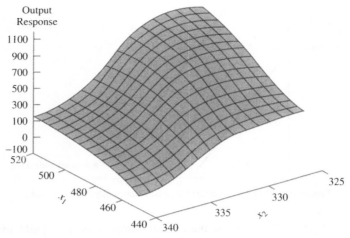

$$Response = \hat{y} = \beta_0 + \beta_1 x_1 + \beta_2 x_2 + \beta_{11} x_x^2 + \beta_{22} x_2^2 + \beta_{12} x_1 x_2 + \cdots$$

Figure 8.9. Example of response surface and analytical representation.

Table 8.8. Results from factorial design.

Run	Time (min)	Temperature (°C)	x_1	x_2	Yield (%)
1	70	127.5	−1	−1	54.3
2	80	127.5	+1	−1	60.3
3	70	132.5	−1	+1	64.6
4	80	132.5	+1	+1	68.0
5	75	130.0	0	0	60.3
6	75	130.0	0	0	64.3
7	75	130.0	0	0	62.3

The first step is to perform a 2^2 factorial experiment with three replications at the center of the design space. This design is shown in Table 8.8 and also indicated by the crosses at the lower left corner of Figure 8.10. This first-order design allows efficient fitting of the polynomial model

$$\hat{y} = \beta_0 + \beta_1 x_1 + \beta_2 x_2 \tag{8.45}$$

where x_1 represents time, x_2 is temperature, and y is the yield. The levels of the variables in normalized units (shown in columns 3 and 4 of the table) are

$$x_1 = \frac{t - 75}{5} \qquad x_1 = \frac{T - 130}{2.5} \tag{8.46}$$

This design is selected because at this first stage of the investigation, we might be some distance away from the maximum yield. In this case, it is likely that the local characteristics of the yield surface can be roughly represented by this planar

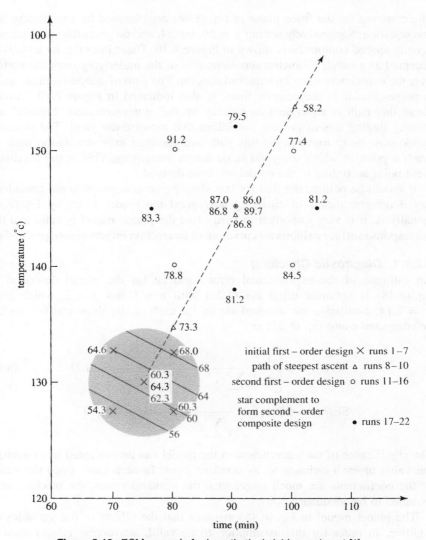

Figure 8.10. RSM example for hypothetical yield experiment [1].

model. If this is correct, estimating β_1 and β_2 allows us to follow a direction of increasing yield "up the hillside" formed by the planar response surface.

The least-squares estimate of β_1 is

$$b_1 = \tfrac{1}{4}(-54.3 + 60.3 - 64.6 + 68.0) = 2.35 \qquad (8.47)$$

Similarly, we compute $b_2 = 4.50$. The least-squares estimate of β_0 is the average of all seven observations, or 62.01. We thus obtained the fitted equation

$$\hat{y} = 62.01 + 2.35x_1 + 4.50x_2 \qquad (8.48)$$

The contours for the fitted plane in Eq. (8.48) are obtained by substituting into this equation. Successively setting $\hat{y} = 56, 60, 64$, and 68 gives the set of parallel equally spaced contour lines shown in Figure 8.10. These lines can be tentatively accepted as a rough geometric representation of the underlying response surface over the experimental region explored thus far. The path of steepest ascent, which is perpendicular to the contour lines, is also indicated in Figure 8.10. Moving along this path is equivalent to moving up the aforementioned "hillside" and thereby finding sets of process conditions that increase the yield. The objective would now be to move along this path and continue experimentation until we reach a point at which the yield is no longer increasing. This is the maximum yield point according to the model we have derived.

It should be pointed out that for this simple example, we have not considered any diagnostic means of checking how good the model given by Eq. (8.48) actually is. It is very important to verify that the planar model is valid and that the response surface exhibits no curvature or interaction effects before proceeding.

8.2.1.1. Diagnostic Checking

An estimate of the experimental error variance for the model described by Eq. (8.48) is obtained using Eq. (8.26) with $n = 7$ and $p = 3$, which gives $s^2 = 2.14$. Similarly, the standard errors for each of the three coefficients can be computed using Eq. (8.27) as

$$\mathrm{SE}(b_0) = s\left[\frac{1}{n} + \frac{\overline{x}^2}{\sum(x-\overline{x})^2}\right]^{1/2} = \frac{s}{\sqrt{n}} = 0.21 \tag{8.49}$$

$$\mathrm{SE}(b_1) = \mathrm{SE}(b_2) = \frac{s}{\sqrt{\sum(x-\overline{x})^2}} = \frac{s}{2} = 0.73$$

The significance of each coefficient in the model can be evaluated by comparing the value of each estimate to its standard error. In each case, since the values of the coefficients are much larger than the standard errors, the model can be assumed to be adequate.

The planar model in Eq. (8.48) assumes that the effects of the variables are additive. In order for this assumption to be valid, *interaction* effects must be checked. Interaction effects can be accounted for by evaluating an additional model coefficient, β_{12}, for the cross-product term $x_1 x_2$ in the model. Based on the data in Table 8.8, an estimate of this term is given by

$$b_{12} = \tfrac{1}{4}(54.3 - 60.3 - 64.6 + 68.0) = -0.65 \tag{8.50}$$

The standard error of this estimate is the same as that for b_1 and b_2, or 0.73. Since the magnitude of the interaction coefficient is less than its standard error, the interaction is deemed insignificant.

Yet another check on the local planarity of the model is accomplished by comparing the average response for the four points of the 2^2 factorial $(\overline{y}_f)$ with the average at the center of the design $(\overline{y}_c)$. If we envision the design as resting on a saucerlike surface, then the difference between these two parameters is a measure of the overall *curvature* of that surface. It can be shown that if β_{11} and β_{22} are the coefficients of the terms x_1^2 and x_2^2, this curvature measure will be an estimate of $\beta_{11} + \beta_{22}$. The estimate of the overall curvature is

$$b_{11} + b_{22} = \tfrac{1}{4}(54.3 + 60.3 + 64.6 + 68.0) - \tfrac{1}{3}(60.3 + 62.3 + 64.3) = -0.50 \tag{8.51}$$

The standard error for this estimate is $s/\sqrt{2} = 1.03$. Therefore, there is no reason to question the adequacy of the planar model based on the curvature metric.

8.2.1.2. Augmented Model

The path of steepest ascent, which is perpendicular to the contour lines, is indicated by the dotted arrow in Figure 8.10. This path is determined by starting at the center of the experimental space and moving $b_2 = +4.50$ units along x_2 for every $b_1 = +2.35$ units along x_1. Experiments 8, 9, and 10, represented by the triangles in Figure 8.10, are trials conducted along this path, and the associated yields derived for each trial are presented next to the triangles. The yield increases at trial 8 ($y = 73.3\%$), but then decreases for the large jump made to trial 9 ($y = 58.2\%$) The best results are achieved at experiment 10 ($y = 86.8\%$), suggesting that subsequent experiments should be performed in the vicinity of this point.

As the experimental region of interest ascends the response surface, the possibility increases that first-order effects will become smaller and a second-order model will be needed to more accurately represent the response. Expanding the original experimental design makes sense in this case, since a second-degree approximation should provide a better approximation over a larger region than a first-degree approximation. Therefore, a new 2^2 factorial design with two centerpoints at trial 10 is performed with the following new coded variables:

$$x_1 = \frac{t - 90}{10} \qquad x_1 = \frac{T - 145}{5} \tag{8.52}$$

This new design is indicated by the open circles in Figure 8.10. The data obtained from this new set of experiments are shown in rows 11–16 of Table 8.9.

The first order model obtained from the second 2^2 factorial experiment is

$$\hat{y} = 84.73 - 2.025x_1 + 1.325x_2 \tag{8.53}$$

An estimate of the experimental error variance for this model is once again obtained using Eq. (8.26), with $n = 6$ and $p = 3$, which gives $s^2 = 45.45$. The interaction and curvature terms, respectively, are

$$b_{12}' = -4.88 \pm 3.37, \qquad b_{11}' + b_{22}' = -5.28 \pm 4.78 \tag{8.54}$$

where the values following the $\pm$ symbol represent the standard errors for each estimate. In this case, the magnitudes of the interaction and curvature coefficients are comparable to their respective standard errors. Thus, the first-order model given by Eq. (8.53) is not adequate to represent the local response function.

Table 8.9. Results from augmented factorial design.

Run	Time (min)	Temperature (°C)	x_1	x_2	Yield (%)
11	80	140	-1	-1	78.8
12	100	140	$+1$	-1	84.5
13	80	150	-1	$+1$	91.2
14	100	150	$+1$	$+1$	77.4
15	90	145	0	0	89.7
16	90	145	0	0	86.8
17	76	145	$-\sqrt{2}$	0	83.3
18	104	145	$+\sqrt{2}$	0	81.2
19	90	138	0	$-\sqrt{2}$	81.2
20	90	152	0	$+\sqrt{2}$	79.5
21	90	145	0	0	87.0
22	90	145	0	0	86.0

Since the first-degree polynomial is inadequate, in the new experimental region, the second degree polynomial approximation

$$\hat{y} = \beta_0 + \beta_1 x_1 + \beta_2 x_2 + \beta_{11} x_x^2 + \beta_{22} x_2^2 + \beta_{12} x_1 x_2 \qquad (8.55)$$

should now be considered. To estimate the six coefficients in this model, the second 2^2 factorial experiment (trials 11–16) is augmented with a *central composite circumscribed* (CCC) design consisting of four axial (or "star") points and two additional center points (trials 17–22 in Table 8.9). These additional trials are shown as dark circles in Figure 8.10. The second-order equation fit by least-squares methods to the model in Eq. (8.55) is

$$\hat{y} = 87.36 - 1.39 x_1 + 0.37 x_2 - 2.15 x_x^2 - 3.12 x_2^2 - 4.88 x_1 x_2 \qquad (8.56)$$

The contours for this fitted equation are shown in Figure 8.11. In many semiconductor manufacturing applications, second-order models are the highest-order models required to describe the responses of interest.

8.2.2. Plasma Etching Example

We now consider an actual case study in the use of response surface methodology in a semiconductor manufacturing application [4]. Plasma etch modeling from a fundamental physical standpoint has had limited success. Physically based models attempt to derive self-consistent solutions to first-principle equations involving continuity, momentum balance, and energy balance inside a high-frequency, high-intensity electric field. This is accomplished by means of computationally expensive numerical simulation methods that typically produce outputs such as profiles of the distribution of electrons and ions within the plasma sheath. However, although detailed simulation is useful for equipment design and optimization, it is subject to many simplifying assumptions. Because of the extremely

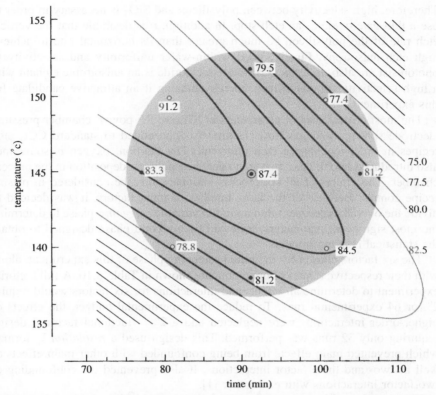

Figure 8.11. Contours of fitted second-order equation and data from second-order design for hypothetical yield experiment [1].

complex nature of particle dynamics within a plasma, the connection between these microscopic models and macroscopic parameters such as etch rate is difficult to distinguish.

In this example, the response characteristics of a CCl_4-based plasma process used to etch doped polysilicon were examined via a 2^{6-1} fractional factorial experiment, which was followed by a supplemental central composite design. The effects of variation in RF power, pressure, electrode spacing, CCl_4 flow, He flow, and O_2 flow on several output variables, including etch rate, selectivity, and process uniformity, were investigated. The factorial experiment was used for variable *screening* to isolate the most significant input factors. The supplemental phase of the experiment enabled the development of polynomial models of etch behavior using response surface methods.

8.2.2.1. Experimental Design
An example of a fabrication step in which reactive-ion etching is essential is in the definition of polysilicon features for MOS circuits. This step often requires that a relatively thick polysilicon gate be etched down to a thin silicon dioxide layer.

Therefore, high selectivity between polysilicon and SiO_2 is necessary in order to use a thin gate oxide as an etch stop. In addition, it is desirable that the vertical etch rate of the polysilicon be much greater than its horizontal rate to achieve high etch anisotropy. Finally, good within-wafer uniformity and selectivity to photoresist are also desirable. Carbon tetrachloride is an anisotropic etchant with a high selectivity for polysilicon, thereby making it an attractive candidate for this experiment.

The most critical control parameters in RIE are RF power, chamber pressure, electrode spacing, and gas flow. Helium is often added to standard CCl_4 etch recipes in order to enhance etch uniformity. In addition, oxygen is sometimes also introduced into the gas mixture to decrease polymer deposition in the process chamber. The effects of all six process variables must be considered in plasma recipe control. Because of the large number of input factors, it was decided to divide the overall experiment into an initial variable *screening* phase to determine the most significant parameters, followed by a second phase designed to obtain the statistical response models.

The six factors chosen for the initial screening phase of this experiment, along with their respective ranges of operation, are shown in Table 8.10. A full factorial experiment to determine all effects and interactions for six factors would require 2^6, or 64 experimental runs. To reduce the experimental budget, the effects of higher-order interactions were neglected and a 2^{6-1} fractional factorial design requiring only 32 runs was performed. This design used a *resolution V* format, which prevented main effects from being confounded with other main effects as well as two- and three-factor interactions. It also prevented the confounding of two-factor interactions with each other [1].

The experimental runs were performed in two blocks of 16 trials each in such a way that no main effects or first-order interactions were confounded with any hidden time effects (such as unscheduled equipment maintenance during the experiment). Three centerpoints were added to the design to provide a check for model nonlinearity. The experimental sequence was randomized in order to avoid biases due to equipment aging during the experiment.

Analysis of the first stage of the experiment revealed significant nonlinearity in nearly all responses, which indicated the necessity of quadratic models. Also, none of the input factors were found to have a statistically insignificant effect

Table 8.10. Range of input factors.

Parameter	Range	Units
RF power (R_f)	300–400	watts
Pressure (P)	200–300	mTorr
Electrode spacing (G)	1.2–1.8	cm
CCl_4 flow (CCl_4)	100–150	sccm[a]
He flow (He)	50–200	sccm
O_2 flow (O_2)	10–20	sccm

[a] Standard cubic centimeters.

on all of the responses of interest. Thus, none were omitted from the response surface models derived in the subsequent phase. To obtain these models, the data gathered were augmented with a second experiment that employed a CCC design. In this design, the two-level factorial "box" was enhanced by further replicated experiments at the center (to provide a direct measure of the equipment and measurement replication error) as well as symmetrically located axial points. A complete CCC design for six factors requires a total of 91 runs. In order to reduce the size of the experiment and combine it with the results from the screening phase, a half-replicate design was again employed. The entire second phase required a total of 18 additional runs.

8.2.2.2. Experimental Technique

Etching was performed on a test structure designed to facilitate the simultaneous measurement of the vertical etch rates of polysilicon, SiO_2, and photoresist, as well as the lateral etch rate of polysilicon on the same wafer. The patterns were fabricated on 4-in-diameter silicon wafers. Approximately 1.2 μm of phosphorus-doped polysilicon was deposited over 0.5 μm of SiO_2 by low-pressure chemical vapor deposition (LPCVD). The oxide was grown in a steam ambient at 1000°C. One micrometer of photoresist was spun on and baked for 60 s at 120°C. Polysilicon lines for scanning electron microscopy (SEM) photos were patterned with a low-temperature oxide (LTO) mask deposited at 450°C by LPCVD. The etching equipment consisted of a Lam Research Corporation Autoetch 490 single-wafer parallel-plate system operating at 13.56 MHz.

Film thickness measurements were performed on five points per wafer (as in Figure 8.12) before and after etching. Vertical etch rates were calculated by dividing the difference between the pre- and postetch thicknesses by the etch time. The lateral etch rate for polysilicon was determined using SEM photos by measuring the difference between the pre- and postetch linewidths under the assumption that the preetch width was that at the base of the polysilicon line (see Figure 8.13).

Expressions for the selectivity of the polysilicon with respect to oxide (S_{ox}) and with respect to resist (S_{ph}), along with percent anisotropy (A) and percent

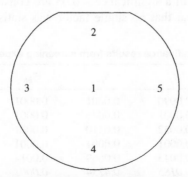

Figure 8.12. Wafer measurement sites [4].

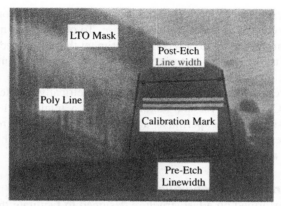

Figure 8.13. SEM photos of typical polysilicon lines used to determine the sidewall slope and lateral etch rate for the anisotropy calculation [4].

nonuniformity (U), respectively, are

$$S_{ox} = R_p/R_{ox} \tag{8.57}$$

$$S_{ph} = R_p/R_{ph} \tag{8.58}$$

$$A = (1 - L_p/R_p) \times 100 \tag{8.59}$$

$$U = \frac{|R_{pc} - R_{pe}|}{R_{pc}} \times 100 \tag{8.60}$$

where R_p is the mean vertical polysilicon etch rate over the five points, R_{ox} is the mean oxide etch rate, R_{ph} is the mean resist etch rate, L_p is the lateral polysilicon etch rate, R_{pc} is the poly etch rate at the center of the wafer, and R_{pe} is the mean polysilicon etch rate of the four points located about one inch from the edge.

8.2.2.3. Analysis
The experimental data were analyzed using the *R/S Discover* commercial software package [5]. Table 8.11 shows the significance level for each of the main effects. Only factors with a significance <0.05 are considered significant. From these results, it was clear that no single factor was statistically insignificant for

Table 8.11. Statistical significance results from screening experiment [4].

Factor	R_p	S_{ox}	S_{ph}	U	A
Pressure	0.0090	0.0001	0.0001	0.0677	*0.3008*
RF power	0.0001	0.0046	0.0001	0.0493	*0.5119*
CCl₄ flow	0.0032	0.0410	0.0001	0.0672	*0.5244*
He flow	0.0001	0.0001	0.0001	0.0002	*0.0157*
O₂ flow	0.0043	0.0669	0.0014	0.9581	*0.6418*
Electrode spacing	*0.0185*	*0.4134*	*0.0001*	*0.0107*	*0.4634*

Table 8.12. ANOVA table for etch rate model [4].

Source	DF	Sum of Squares	Mean Square	F Ratio	Significance
Total	52	24,717,141	475,329.63		
Regression	13	20,983,554	1,614,120.00	16.86	0.000
Residual	39	3,733,587	95,732.99		
Lack of Fit	31	3,402,778	109,767.03	2.66	0.075
Error	8	330,809	41,351.11		

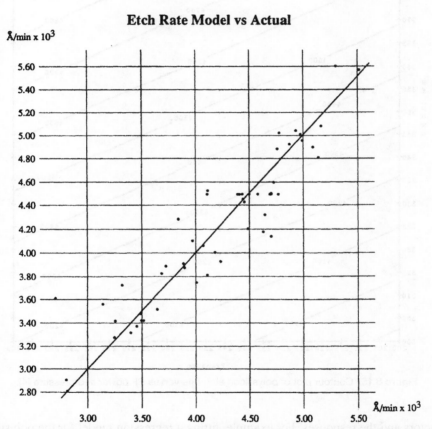

Figure 8.14. Scatterplot of etch rates predicted by the RSM model versus actual experimental values [4].

all five responses of interest. For example, although the electrode spacing had little effect on the etch selectivity with respect to the silicon dioxide mask, it had a dramatic impact on etch rate and uniformity.

The additional 18 runs that constituted the second phase of the experiment yielded quadratic models that describe the precise interaction between input

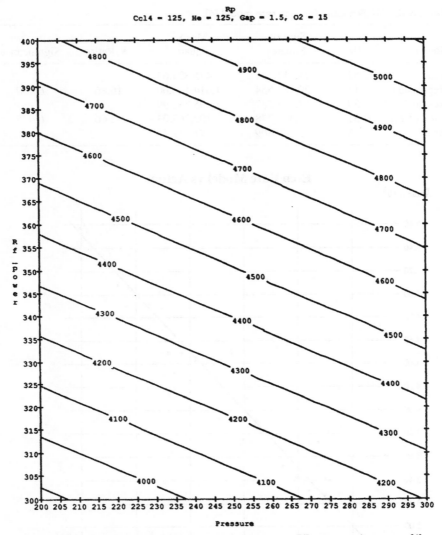

Figure 8.15. Contour plot of polysilicon etch rate versus RF power and pressure [4].

factors and the responses. For example, fitting a regression model for the polysilicon etch rate resulted in the following expression:

$$R_p = -245 - 4.24P + 11.0Rf + 0.742CCl_4 + 11.2He + 523G$$
$$+ 35.9O_2 - 0.034P * He$$
$$+ 7.82P * G + 0.085Rf * CCl_4 - 8.36Rf * G - 0.132(CCl_4)^2$$
$$+ 0.059CCl_4 * He - 0.059He^2 \tag{8.61}$$

where R_p is in Å/min and the units of every other parameter are given in Table 8.10. The ANOVA table for the etch rate model is shown in Table 8.12.

The F test indicates that this model is highly significant, since the regression mean square, which is the amount of variation explained by the proposed model, is significant. This fact is confirmed by the F-ratio statistic. If the regression mean square was not significant, then this ratio would be distributed according to the F distribution with 15 and 37 degrees of freedom. The value 16.86, however, is highly unlikely to occur in the $F_{15,37}$ distribution. The lack-of-fit F test reveals little evidence that the inclusion of additional terms would improve this model, since a lack of fit as large as 2.66 occurs 7.5% of the time in the $F_{29,8}$ distribution. Therefore, most of the residual of the model originates from experimental error.

A scatterplot of the predicted etch rate values versus the corresponding experimental values is shown in Figure 8.14. The straight line in this plot represents the region of perfect agreement between model and experiment. Although this etch rate model is fairly complex, a few interesting relationships emerge. In Figure 8.15, for example, R_p surfaces are plotted against RF power and chamber pressure with all other parameters set at their nominal values. For high process throughput, etch rate should preferably be as high as possible. This occurs at high power and high pressure.

Similar polynomial models and observations were derived for the other etch responses. These models were shown to describe the operation of the characterized equipment very precisely. Unlike computationally expensive physically based simulators, which are often impractical because of their slowness and lack of precision, the empirical models derived in this case study can be used for a variety of manufacturing purposes, including process optimization, control, and diagnosis (refer to Chapters 9 and 10).

8.3. EVOLUTIONARY OPERATION

As alluded to in the previous section, response surface methodology can also provide a useful construct for process monitoring and optimization in addition to modeling. In many situations, a strong relationship between one or more controllable process variables and an observed response variable can be exploited for this purpose. Suppose that a process engineer wishes to maximize the yield of a given process, and that the yield is a function of two controllable process variables, x_1 and x_2, or

$$y = f(x_1, x_2) + \varepsilon \tag{8.62}$$

where ε is random error. Even after a set of optimal levels for x_1 and x_2 that maximizes yield and provides acceptable values for all other quality characteristics has been identified, if the process operates continuously at these levels, it may gradually drift away from the optimum because of variations in the incoming raw materials, environmental changes, personnel, and so on.

Evolutionary operation (EVOP) was proposed by Box in 1957 as a method for continuous operation and monitoring of a process with the goal of moving

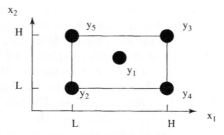

Figure 8.16. Factorial design for EVOP [4].

the operating conditions toward the optimum or following drift [6]. EVOP is, in effect, an online application of statistical experimental design. The EVOP procedure consists of systematically introducing small changes in the levels of the operating variables. EVOP requires that each independent process variable be assigned a "high" (H) and "low" (L) level. For x_1 and x_2, the four possible combinations of high and low levels are shown in Figure 8.16. This arrangement is just a 2^2 factorial design with a centerpoint. Typically, the design would be centered at the best current estimate of the optimum operating conditions.

Let y_i (where $i = 1, \ldots, 5$) be the observed values of the response variable corresponding to the various combinations of x_1 and x_2. After one observation at each point in the design, a cycle is defined as completed. Recall from Chapter 7 that the main effect of a factor is defined as the average change in response produced by a change from the low level to the high level of the factor. Thus, the main effects for x_1 and x_2 are given by

$$x_1(\text{effect}) = \tfrac{1}{2}[(y_3 + y_4) - (y_2 + y_5)] \tag{8.63}$$

$$x_2(\text{effect}) = \tfrac{1}{2}[(y_3 + y_5) - (y_2 + y_4)] \tag{8.64}$$

If the change from the low to the high level of x_1 produces an effect that is different at the two levels of x_2, then there is interaction between x_1 and x_2. The interaction effect is given by

$$x_1 \times x_2(\text{effect}) = \tfrac{1}{2}[(y_2 + y_3) - (y_4 + y_5)] \tag{8.65}$$

After n EVOP cycles, there will be n observations at each of the five design points. The effects and interaction are then computed by replacing the individual observations in Eqs. (8.63)–(8.65) by the averages ($\overline{y}_i$) of the n observations at each point. After several cycles have been completed, one or more process variables (or their interactions) may seem to have a significant effect on the response. When this occurs, a decision to change the operating conditions to improve the response may be appropriate. When improved conditions are detected, a *phase* is completed.

In testing the significance of process variables and interactions, an estimate of experimental error is also required. This estimate of the experimental error usually comes from experimental replications as the process is monitored. By comparing the response at the centerpoint with the corner points in the factorial design, the presence of curvature in the response can be evaluated. If the process has really been optimized, then the response at the center should be significantly better than the responses at the corner points. In theory, EVOP can be applied to an arbitrary number of process variables, but in practice, only two or three variables are usually considered.

The EVOP process is best illustrated by means of an example. Montgomery [3] provides an excellent example for a process whose yield (in %) is a function of temperature (x_1) and pressure (x_2). Suppose that the current operating conditions are $x_1 = 250°C$ and $x_2 = 145$ mTorr. The EVOP procedure uses the design shown in Figure 8.16. A cycle is completed by running each design point in numerical order. The yields for the first cycle $(n = 1)$ are shown in line (3) of Table 8.13, which is the EVOP calculation sheet. At the end of the first cycle, no estimate of the standard deviation can be made. The calculation of the main effects and interaction are shown in Table 8.14. In this table, the "change in mean" (CIM) effect is given by

$$\text{CIM} = \tfrac{1}{5}[(\bar{y}_2 + \bar{y}_3 + \bar{y}_4 + \bar{y}_5 - 4\bar{y}_1)] \tag{8.66}$$

Table 8.13. EVOP calculation sheet for first cycle ($n = 1$) [3].

Line		1	2	3	4	5	
1	Previous cycle sum						Previous sum $S =$
2	Previous cycle average						Previous average $S =$
3	New observations	84.5	84.2	84.9	84.5	84.3	New $S =$ Range $\times f_{5,n} =$
4	Differences (2 − 3)						Range of line 4 $=$
5	New sums $(1 + 3)$	84.5	84.2	84.9	84.5	84.3	New sum $S =$
6	New averages $[\bar{y}_i = (v)/n]$	84.5	84.2	84.9	84.5	84.3	New average $S = \dfrac{\text{New sum } S}{n - 1}$

Table 8.14. EVOP calculation of effects and error limits for first cycle ($n = 1$) [3].

Calculation of Effects	Calculation of Error Limits
Temperature effect $= \tfrac{1}{2}[(\bar{y}_3 + \bar{y}_4) - (\bar{y}_2 + \bar{y}_5)] = 0.45$	For new average: $\dfrac{2}{\sqrt{n}}S =$
Pressure effect $= \tfrac{1}{2}[(\bar{y}_3 + \bar{y}_5) - (\bar{y}_2 + \bar{y}_4)] = 0.25$	For new effects: $\dfrac{2}{\sqrt{n}}S =$
Interaction effect $= \tfrac{1}{2}[(\bar{y}_2 + \bar{y}_3) - (\bar{y}_4 + \bar{y}_5)] = 0.15$	
CIM $= \tfrac{1}{5}[(\bar{y}_2 + \bar{y}_3 + \bar{y}_4 + \bar{y}_5 - 4\bar{y}_1)] = 0.02$	For new effects: $\dfrac{1.78}{\sqrt{n}}S =$

The quantities in the EVOP calculation sheet follow directly from analysis of the 2^2 factorial design. For example, the variance of the first main effect is simply

$$V[\tfrac{1}{2}(\bar{y}_3 + \bar{y}_4 - \bar{y}_2 - \bar{y}_5)] = \tfrac{1}{4}(\sigma_{\bar{y}_3}^2 + \sigma_{\bar{y}_4}^2 + \sigma_{\bar{y}_2}^2 + \sigma_{\bar{y}_5}^2) = \tfrac{1}{4}(4\sigma_{\bar{y}}^2) = \sigma^2/n \quad (8.67)$$

where σ^2 is the variance of the observations (y). The variance for the other main effect, as well as the interaction effect, is calculated in the same way. Thus, 2σ error limits on any effect (corresponding to 95% of the variation) would be $\pm 2\sigma/\sqrt{n}$. Similarly, the variance of the change in mean is

$$V(\text{CIM}) = V[\tfrac{1}{5}(\bar{y}_2 + \bar{y}_3 + \bar{y}_4 + \bar{y}_5 - 4\bar{y}_1)] = \tfrac{1}{25}(4\sigma_{\bar{y}}^2 + 16\sigma_{\bar{y}}^2) = \tfrac{20}{25}\sigma^2/n$$
$$(8.68)$$

Therefore, 2σ error limits on the CIM are $\pm 2\sigma/\sqrt{0.8n} = \pm 1.78\sigma/\sqrt{n}$.

The standard deviation is estimated using the range method. If $y_i(n)$ is the ith observation in cycle n, then $\bar{y}_i(n)$ is the corresponding average of $y_i(n)$ after n cycles. The quantities in row 4 of the EVOP sheet in Table 8.13 are the differences $y_i(n) - \bar{y}_i(n-1)$. The variance of these differences is $\sigma^2[n/(n-1)]$. The range of the differences (R_D) is related to the estimate of the distribution of the differences by $\hat{\sigma}_D = R_D/d_2$. Since $R_D/d_2 = \hat{\sigma}\sqrt{n/(n-1)}$, we have

$$\hat{\sigma} = \sqrt{\frac{(n-1)}{n}} \frac{R_D}{d_2} = (f_{k,n})R_D \equiv S \quad (8.69)$$

where $k = 5$ is the number of points used in the experimental design (i.e., a 2^2 factorial plus one centerpoint) and the values of $f_{k,n}$ are given in Table 8.15. The quantity S can be used to estimate the standard deviation of the EVOP operations.

Using this computation framework to proceed, the data corresponding to the second cycle of EVOP for this example are shown in Tables 8.16 and 8.17. Since none of the effects in Table 8.17 exceeds their error limits, the true effect is likely close to zero, and no changes in operating conditions are recommended.

The results of a third EVOP cycle are shown in Tables 8.18 and 8.19. In this cycle, the effect of pressure now exceeds its error limit, and the temperature effect is equal to the error limit. Thus, a change in operating conditions is probably justified. In light of the results, it seems reasonable to begin a new EVOP phase centered at point (column) 3. Thus, $x_1 = 225°C$ and $x_2 = 150$ mTorr become the center of the 2^2 design in the next phase.

Table 8.15. Values of $f_{k,n}$ [3].

$n =$		2	3	4	5	6	7	8	9	10
$k =$	5	0.30	0.35	0.37	0.38	0.39	0.40	0.40	0.40	0.41
	9	0.24	0.27	0.29	0.30	0.31	0.31	0.31	0.32	0.32
	10	0.23	0.26	0.28	0.29	0.30	0.30	0.30	0.31	0.31

Table 8.16. EVOP calculation sheet for second cycle ($n = 2$) [3].

	1	2	3	4	5	
1 Previous cycle sum	84.5	84.2	84.9	84.5	84.3	Previous sum $S =$
2 Previous cycle average	84.5	84.2	84.9	84.5	84.3	Previous average $S =$
3 New observations	84.9	84.6	85.9	83.5	84.0	New $S =$ range $\times f_{5,n} =$ 0.60
4 Differences (2 − 3)	−0.4	−0.4	−1.0	1.0	0.3	Range of line 4 = 2.0
5 New sums (1 + 3)	169.4	168.8	170.8	168.0	168.3	New sum $S = 0.60$
6 New averages $[\overline{y}_i = (\mathbf{v})/n]$	84.70	84.40	85.40	84.00	84.15	New average $S = \dfrac{\text{newsum } S}{n - 1} = 0.60$

Table 8.17. EVOP calculation of effects and error limits for second cycle ($n = 2$) [3].

Calculation of Effects	Calculation of Error Limits
Temperature effect $= \frac{1}{2}[(\overline{y}_3 + \overline{y}_4) - (\overline{y}_2 + \overline{y}_5)] = 0.43$	For new average: $\dfrac{2}{\sqrt{n}} S = 0.85$
Pressure effect $= \frac{1}{2}[(\overline{y}_3 + \overline{y}_5) - (\overline{y}_2 + \overline{y}_4)] = 0.58$	For new effects: $\dfrac{2}{\sqrt{n}} S = 0.85$
Interaction effect $= \frac{1}{2}[(\overline{y}_2 + \overline{y}_3) - (\overline{y}_4 + \overline{y}_5)] = 0.83$	
CIM $= \frac{1}{5}[(\overline{y}_2 + \overline{y}_3 + \overline{y}_4 + \overline{y}_5 - 4\overline{y}_1)] = -0.17$	For new effects: $\dfrac{1.78}{\sqrt{n}} S = 0.76$

Table 8.18. EVOP calculation sheet for third cycle ($n = 3$) [3].

	1	2	3	4	5	
1 Previous cycle sum	169.4	168.8	170.8	168.0	168.3	Previous sum $S = 0.60$
2 Previous cycle average	84.70	84.40	84.50	84.00	84.15	Previous average $S = 0.60$
3 New observations	85.0	84.0	86.6	84.9	85.2	New $S =$ range $\times f_{5,n} = 0.56$
4 Differences (2 − 3)	−0.3	0.4	−1.2	−0.9	−1.05	Range of line 4 = 1.60
5 New sums (1 + 3)	254.4	252.8	257.4	252.9	253.5	New sum $S = 1.16$
6 New averages $[\overline{y}_i = (\mathbf{v})/n]$	84.80	84.27	85.80	84.30	84.50	New average $S = \dfrac{\text{new sum } S}{n - 1} = 0.58$

Table 8.19. EVOP calculation of effects and error limits for third cycle ($n = 3$) [3].

Calculation of Effects	Calculation of Error Limits
Temperature effect $= \frac{1}{2}[(\bar{y}_3 + \bar{y}_4) - (\bar{y}_2 + \bar{y}_5)] = 0.67$	For new average: $\dfrac{2}{\sqrt{n}} S = 0.67$
Pressure effect $= \frac{1}{2}[(\bar{y}_3 + \bar{y}_5) - (\bar{y}_2 + \bar{y}_4)] = 0.87$	For new effects: $\dfrac{2}{\sqrt{n}} S = 0.67$
Interaction effect $= \frac{1}{2}[(\bar{y}_2 + \bar{y}_3) - (\bar{y}_4 + \bar{y}_5)] = 0.64$	
CIM $= \frac{1}{5}[(\bar{y}_2 + \bar{y}_3 + \bar{y}_4 + \bar{y}_5 - 4\bar{y}_1)] = -0.07$	For new effects: $\dfrac{1.78}{\sqrt{n}} S = 0.60$

8.4. PRINCIPAL-COMPONENT ANALYSIS

Principal-component analysis (PCA) is a modeling technique designed to estimate variability and reduce the dimensionality of a dataset that contains a large number of interrelated variables [7]. The objective of PCA is to perform this dimensionality reduction while retaining as much of the variation present in the original dataset as possible. Reduction is accomplished by transforming the original dataset into a new set of variables (i.e., the principal components), which are uncorrelated.

Consider a vector $\mathbf{x}$ that consists of p random variables. Let Σ be the covariance matrix of $\mathbf{x}$, which can be estimated using Eq. (6.88). Then, for $k = 1, 2, \ldots, p$, the kth principal component (PC) is given by

$$z_k = \alpha_k^T \mathbf{x} \tag{8.70}$$

where α_k is an eigenvector of Σ corresponding to its kth largest eigenvalue, and T represents the transpose operation. Furthermore, if α_k is chosen to have unit length (i.e., $\alpha_k^T \alpha_k = 1$), the variance of $z_k = \lambda_k$.

Dimensionality reduction through PCA is achieved by transforming the raw data to a new set of coordinates (*i.e.*, selected eigenvectors), which are uncorrelated and ordered such that the first few retain most of the variation present in the original dataset. Generally, if the eigenvalues are ordered from largest to smallest, then the first few PCs will account for most of the variation in the original vector $\mathbf{x}$. A simplified example of PCA with two measurement variables, x_1 and x_2, is presented in Figure 8.17.

Principal components are identified by first finding a linear function $\alpha_1^T \mathbf{x}$ of the elements of x that has maximum variance, where

$$\alpha_1^T \mathbf{x} = \alpha_{11} x_1 + \alpha_{12} x_2 + \cdots + \alpha_{1p} x_p = \sum_{j=1}^{p} \alpha_{1j} x_j \tag{8.71}$$

The next step is to find another linear function $\alpha_2^T \mathbf{x}$ that is uncorrelated with (or orthogonal to) $\alpha_1^T \mathbf{x}$ and also has maximum variance. This process is repeated k times, and $\alpha_k^T \mathbf{x}$ is referred to as the kth principal component.

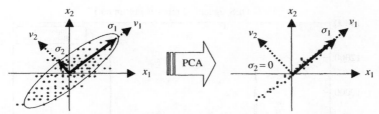

Figure 8.17. Illustration of principal-component analysis for two variables: x_1 and x_2. In this illustration, v_1 and v_2 are eigenvectors, and σ_1 and σ_2 are the corresponding standard deviations.

The variance of $\alpha_1^T \mathbf{x}$ is $\alpha_1^T \Sigma \alpha_1$. The eigenvector α_1 that maximizes this variance, subject to the normalization constraint $\alpha_1^T \alpha_1 = 1$, is found using the method of *Lagrange multipliers*. Using this technique requires maximizing the quantity $\alpha_1^T \Sigma \alpha_1 - \lambda(\alpha_1^T \alpha_1 - 1)$, where λ is known as the Lagrange multiplier. Differentiating this quantity with respect to α_1 and setting the result equal to zero gives

$$\Sigma \alpha_1 - \lambda \alpha_1 = 0 \tag{8.72}$$

or

$$(\Sigma - \lambda \mathbf{I}_p)\alpha_1 = 0 \tag{8.73}$$

where $\mathbf{I}_p$ is the $(p \times p)$ identity matrix. Thus, λ is an eigenvalue of Σ, and α_1 is the corresponding eigenvector. To determine which of the p eigenvectors is maximum, we simply select the one corresponding to the largest λ. This procedure is repeated for each PC.

As previously mentioned, although as many as p principal components can be identified in this manner, most of the variation in $\mathbf{x}$ is usually accounted for by the first one to three PCs. The cumulative percentage of total variation (β) is typically employed as a criterion for choosing the number of PCs. The definition of the cumulative percentage of variation is

$$\beta_k = \left(\sum_{l=1}^{k} \lambda_l \Big/ \sum_{l=1}^{c} \lambda_l \right) \cdot 100(\%) \tag{8.74}$$

where c is the total number of eigenvalues and λ_l is the lth diagonal element of the eigenvalue matrix (k denotes the number of subset of principal components).

To illustrate the use of PCA to solve a practical semiconductor manufacturing problem, consider the work of Hong et al. [8], who used PCA to estimate the variation in optical emission spectroscopy data generated during reactive-ion etching. Although OES is an excellent tool for monitoring plasma emission intensity during etching, a primary issue with its use is the large dimensionality of the spectroscopic data. To alleviate this concern, Hong implemented PCA as a mechanism for feature extraction to reduce the dimensionality of OES data. OES data were generated from a central composite experiment designed to characterize RIE process variation during the etching of benzocyclobutene (BCB) in a SF_6/O_2 plasma, with controllable input factors consisting of the two gas flows,

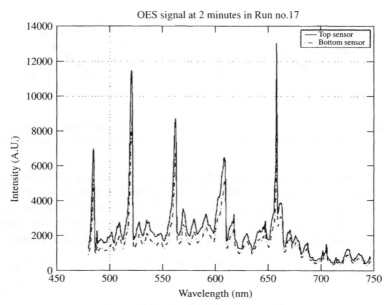

Figure 8.18. Typical OE spectrum [8].

RF power, and chamber pressure. The OES data, consisting of 226 wavelengths sampled every 20 s, were compressed into five principal components using PCA. A sample OE spectrum from this experiment is shown in Figure 8.18.

In the Hong et al. study [8], the raw OES data were three-dimensional, where the three dimensions were trial, wavelength, and time. By unfolding the 3D dataset into 2D matrix, a *multiway principal component analysis* (MPCA) of the 3D OES data was accomplished. A total of 27 experimental trials (i.e., 27 processed wafers) were conducted. Emission intensity was recorded and stored in a local computer. The OES system collected 2048 data points over the wavelength range from 186.58 to 746.85 nm. The following equation represents the selected dataset that was expanded into an $r \times c$ matrix $\mathbf{X}$:

$$\mathbf{X} = \begin{bmatrix} (x_1, \ldots, x_w)_{o=1}^{r=1}, & \ldots, & (x_1, \ldots, x_w)_{o=10}^{r=1} \\ \vdots & \ddots & \vdots \\ (x_1, \ldots, x_w)_{o=1}^{r=27}, & \ldots, & (x_1, \cdots, x_w)_{o=10}^{r=27} \end{bmatrix} \qquad (8.75)$$

where the index $r = 1, 2, \ldots, 27$ represents the process run, $w = 226$ is the wavelength, $o = 1, 2, \ldots, 10$ is one of 10 consecutive observations collected every 20 s, and $c = 2260$ is the total number of columns. The data matrix $\mathbf{X}$ was then *mean-centered* with respect to each column using the equation:

$$m_{ij} = x_{ij} - \bar{x}_j \quad \text{for } 1 \leq i \leq r \quad \text{and} \quad 1 \leq j \leq c \qquad (8.76)$$

where m_{ij} and x_{ij} are the components of the mean-centered matrix $\mathbf{M}$ and the raw sample matrix $\mathbf{X}$, respectively, $\bar{x}_j$ is the mean of column j over the 27 rows.

The covariance of matrix $\mathbf{\Sigma}$ was then computed using

$$\mathbf{\Sigma} = \left(\frac{1}{c-1}\right)\mathbf{M}^T\mathbf{M} \tag{8.77}$$

Using *singular value decomposition*, the covariance matrix $\mathbf{\Sigma}$ was decomposed into a linear combination of the eigenvectors and eigenvalues using

$$\mathbf{\Sigma} = \mathbf{U}\mathbf{\Lambda}\mathbf{U}^T \tag{8.78}$$

where $\mathbf{U} = [\mathbf{u}_1, \mathbf{u}_2, \ldots, \mathbf{u}_c]$ is a $c \times c$ orthogonal (unitary) matrix containing c eigenvectors and $\mathbf{\Lambda}$ is the diagonal matrix of eigenvalues such that $\lambda_1 \geq \lambda_2 \geq \cdots \geq \lambda_c \geq 0$. Using singular value decomposition, the eigenvectors were arranged in descending order according to the magnitude of the eigenvalues. In this case, as expected, the first few eigenvectors capture most of the variation in the original data. Using this approach, the mean-centered OES dataset was compressed into $k = 5$ principal-component vectors by transposing $\mathbf{M}$ onto the selected new set of coordinates, or

$$\hat{\mathbf{A}} = \mathbf{M}\hat{\mathbf{U}} \tag{8.79}$$

where $\hat{\mathbf{U}}$ is a $c \times k$ orthogonal matrix. A plot of the magnitude of the eigenvector corresponding to the first PC (which accounted for 99.27% of the variation in the original OES dataset) versus wavelength is shown in Figure 8.19. The 27 five-element principal-component vectors derived in this manner can be used to identify the most significant wavelengths in the spectrum, or as a compressed representation of the raw OES data for modeling this RIE process.

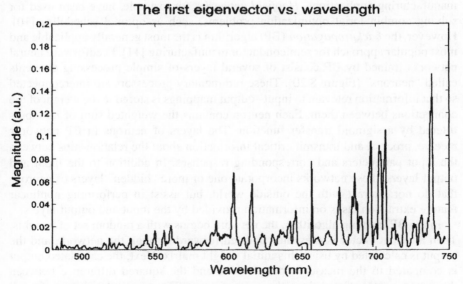

Figure 8.19. Plot of eigenvector corresponding to the first PC versus wavelength [8].

8.5. INTELLIGENT MODELING TECHNIQUES

More recently, the use of *computational intelligence* in various manufacturing applications has enhanced manufacturing process control, throughput, and yield. The semiconductor manufacturing arena is no exception to this trend. Artificial neural networks, fuzzy logic, and other techniques have emerged as powerful tools for assisting IC-CIM systems in performing various process monitoring, modeling, control, and diagnostic functions [9]. This section provides a brief introduction to two key computational intelligence tools—neural networks and fuzzy logic—and discusses the manner in which these tools have been used in modeling semiconductor manufacturing processes.

8.5.1. Neural Networks

Because of their inherent learning capability, adaptability, and robustness, artificial neural networks are used to solve problems that have heretofore resisted solutions by other more traditional methods. Although the name "neural network" stems from the fact that these systems crudely mimic the behavior of biological neurons, the neural networks used in microelectronics manufacturing applications actually have little to do with biology. However, they share some of the advantages that biological organisms have over standard computational systems. Neural networks are capable of performing highly complex mappings on noisy and/or nonlinear data, thereby inferring very subtle relationships between diverse sets of input and output parameters. Moreover, these networks can also generalize well enough to learn overall trends in functional relationships from limited training data.

Several neural network architectures and training algorithms are eligible for manufacturing applications. *Hopfield networks*, for example, have been used for solving combinatorial optimization problems, such as optimal scheduling [10]. However, the *backpropagation* (BP) algorithm is the most generally applicable and most popular approach for semiconductor manufacturing [11]. Feedforward neural networks trained by BP consist of several layers of simple processing elements called "neurons" (Figure 8.20). These rudimentary processors are interconnected so that information relevant to input–output mappings is stored in the weight of the connections between them. Each neuron contains the weighted sum of its inputs filtered by a sigmoid transfer function. The layers of neurons in BP networks receive, process, and transmit critical information about the relationships between the input parameters and corresponding responses. In addition to the input and output layers, these networks incorporate one or more "hidden" layers of neurons that do not interact with the outside world, but assist in performing nonlinear feature extraction tasks on information provided by the input and output layers.

In the BP learning algorithm, the network begins with a random set of weights. Then an input vector is presented and fed forward through the network, and the output is calculated by using this initial weight matrix. Next, the calculated output is compared to the measured output data, and the squared difference between these two vectors determines the system error. The accumulated error for all

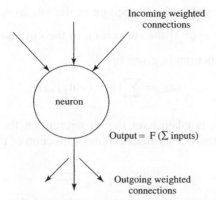

Figure 8.20. Schematic of a single neuron. The output of the neuron is a function of the weighted sum of its inputs, where *F* is a sigmoid function. Feedforward neural networks consist of several layers of interconnected neurons [9].

the input–output pairs is defined as the Euclidean distance in the weight space that the network attempts to minimize. Minimization is accomplished via the *gradient descent* approach, in which the network weights are adjusted in the direction of decreasing error. It has been demonstrated that if a sufficient number of hidden neurons are present, a three-layer BP network can encode any arbitrary input–output relationship [12].

The structure of a typical BP network appears in Figure 8.21. Referring to this figure, let

$$w_{ijk} = \text{weight between the } j\text{th neuron in layer } (k-1)$$
$$\text{and the } i\text{th neuron in layer } k$$

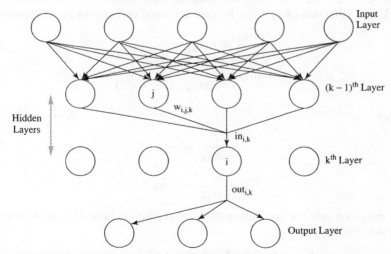

Figure 8.21. BP neural network showing input, output, and hidden layers, as well as interconnection strengths (weights), inputs, and outputs of neurons in different layers [9].

$$in_{ik} = \text{input to the } i\text{th neuron in the } k\text{th layer}$$

$$out_{ik} = \text{output of the } i\text{th neuron in the } k\text{th layer.}$$

The input to a given neuron is given by

$$in_{ik} = \sum_{j} (w_{ijk} \cdot out_{j,k-1}) \qquad (8.80)$$

where the summation is taken over the all neurons in the previous layer. The output of a given neuron is a sigmoidal transfer function of the input expressed as

$$out_{ik} = \frac{1}{1 + e^{-in_{ik}}} \qquad (8.81)$$

Error is calculated for each input–output pair as follows. Input neurons are assigned a value, and computation occurs by a forward pass through each layer of the network. Then the computed value at the output is compared to its desired value, and the square of the difference between these two vectors provides a measure of the error (E) using

$$E = 0.5 \sum_{j=1}^{q} (d_j - out_{jn})^2 \qquad (8.82)$$

where n is the number of layers in the network, q is the number of output neurons, d_j is the desired output of the jth neuron in the output layer, and out_{jn} is the calculated output of that same neuron.

After a forward pass through the network, error is propagated backward from the output layer. Learning occurs by minimizing error through modification of the weights one layer at a time. The weights are modified by calculating the derivative of E and following the gradient that results in a minimum value. From Eqs. (8.80) and (8.81), the following partial derivatives are computed as

$$\frac{\partial (in)_{ik}}{\partial w_{ijk}} = out_{j,k-1} \qquad (8.83)$$

$$\frac{\partial (out)_{ik}}{\partial w_{ijk}} = out_{j,k-l}(1 - out_{ik})$$

Now let

$$\frac{\partial E}{\partial (in)_{ik}} = -\delta_{ik} \qquad (8.84)$$

$$\frac{\partial E}{\partial (out)_{ik}} = -\phi_{ik}$$

Using the chain rule for computing derivatives, the gradient of error with respect to the weights is given by

$$\frac{\partial E}{\partial w_{ijk}} = \left(\frac{\partial E}{\partial (in)_{ik}} \right) \left(\frac{\partial (in)_{ik}}{\partial w_{ijk}} \right) = -\delta_{ik} \cdot out_{j,k-1} \qquad (8.85)$$

In the previous expression, $\text{out}_{j,k-1}$ is available from the forward pass. The quantity δ_{ik} is calculated by propagating the error backward through the network. Consider that for the output layer

$$-\delta_{in} = \frac{\partial E}{\partial(\text{in})_{in}} = \left(\frac{\partial E}{\partial(\text{out})_{in}}\right)\left(\frac{\partial(\text{out})_{in}}{\partial(\text{in})_{in}}\right) \tag{8.86}$$

$$= (d_j - \text{out}_{jn})(\text{out}_{in})(1 - \text{out}_{in})$$

where the expressions in Eqs. (8.82) and (8.83) have been substituted. Likewise, the quantity ϕ_{in} is given by

$$-\phi_{in} = (d_j - \text{out}_{jn}) \tag{8.87}$$

Consequently, for the inner layers of the network

$$-\phi_{ik} = \frac{\partial E}{\partial(\text{out})_{ik}} = \sum_j \left(\frac{\partial E}{\partial(\text{in})_{j,k+1}}\right)\left(\frac{\partial(\text{in})_{j,k+1}}{\partial(\text{out})_{ik}}\right) \tag{8.88}$$

where the summation is taken over all neurons in the $(k+1)$th layer. This expression can be simplified using Eqs. (8.80) and (8.84) to yield

$$\phi_{ik} = \sum_j (\delta_{j,k+1} \cdot w_{ij,k+1}) \tag{8.89}$$

Then δ_{ik} is determined from Eq. (8.86) as

$$\delta_{ik} = \phi_{ik}(\text{out}_{ik})(1 - \text{out}_{ik}) \tag{8.90}$$

$$= \text{out}_{ik}(1 - \text{out}_{ik})\sum_j (\delta_{j,k+1} \cdot w_{ij,k+1})$$

Note that ϕ_{ik} depends only on the δ in the $(k+1)$th layer. Thus, ϕ for all neurons in a given layer can be computed in parallel. The gradient of the error with respect to the weights is calculated for one pair of input–output patterns at a time. After each computation, a step is taken in the opposite direction of the error gradient. This procedure is repeated until convergence is achieved.

The ability of neural networks to learn input–output relationships from limited data is quite beneficial in semiconductor manufacturing, where many highly nonlinear fabrication processes exist, and experimental data for process modeling are expensive to obtain. Several researchers have reported noteworthy successes in using neural networks to model the behavior of a few key fabrication processes [9]. In so doing, the usual strategy is to perform a series of statistically designed characterization experiments, and then to train BP neural nets to model the experimental data. The experiments conducted to characterize the process typically consist of a factorial or fractional factorial exploration of the input parameter space, which may be subsequently augmented by a more advanced experimental design. Each set of input conditions in the design corresponds to a particular set of measured process responses. This input–output mapping is precisely what the neural network learns.

As an example of the neural network process modeling procedure, Himmel and may used BP neural networks to model the same plasma etching process and dataset described in Section 8.2.2 [13]. In modeling applications, the input layer of neurons receives the external information for the network to process. This corresponds to the six input parameters (power, pressure, electrode spacing, and the three gas flows). The output layer transmits processed information to the outside world, and thus corresponds to the etch responses (etch rate, uniformity, etc.). The hidden layer of neurons can be regarded as representing the fundamental physical and chemical properties of the plasma system. Such properties include electron density, electron temperature, and reactive species concentrations. The system inputs have a direct influence on these fundamental quantities, which in turn affect the output responses. Since they represent the physical properties of the plasma itself, the proper number of hidden-layer neurons is not known in advance. Thus, the number of hidden neurons is varied to achieve maximum performance.

Himmel compared the previously derived response surface models to BP neural network models and found that the neural network models exhibited 40–70% better accuracy (as measured by RMS error) than RSM models and required fewer training experiments. Furthermore, the results of this study also indicated that the generalizing capabilities of neural network models were superior to their conventional statistical counterparts. This fact was verified by using both the RSM and "neural" process models to predict previously unobserved experimental data (or test data). Neural networks showed the ability to generalize with an RMS error 40% lower than the statistical models even when built with less training data.

8.5.2. Fuzzy Logic

Fuzzy logic techniques represent yet another alternative for developing empirical models of semiconductor manufacturing processes. Fuzzy logic depends conceptually on *fuzzy sets*, which were first introduced by Lotfi Zadeh to manipulate information that possesses uncertainty [14]. Fuzzy sets are a generalization of conventional set theory that provide a systematic way to represent vagueness, as well as data structures that are an intuitively plausible way to formulate and solve various problems in pattern recognition.

The basic ideas behind fuzzy sets are relatively simple. Suppose that a car is approaching a red light and a driving instructor must advise a student when to apply the brakes. Rather than saying "Begin braking 74 ft from the crosswalk," the instructor would more likely say, "Apply the brakes pretty soon." The former instruction is too precise to be implemented. This example illustrates the utility of vagueness in natural language used in everyday life. This type of imprecision or uncertainty is known as "fuzziness."

Conventional (or "crisp") sets contain objects that satisfy precise properties required for membership in the set. For example, the set of numbers H from 6 to 8 is crisp and well defined by its *membership function* $m_H(x)$ as

$$m_H(x) = \begin{cases} 1 & 6 \le x \le 8 \\ 0 & \text{otherwise} \end{cases} \tag{8.91}$$

Whereas for crisp sets the membership status is captured by a binary variable, for fuzzy sets we refer to the degree of membership (or "grade") of an input variable in a given set. The membership grade has a value between zero and one, where "0" indicates no membership of the variable in the set, and "1" reflects full membership. The crisp set H and the graph of m_H are shown on the left side of Figure 8.22.

Now consider the fuzzy set F of real numbers that are close to seven. Because the property "close to seven" is imprecise, there is not unique membership function m_F for F. However, a number of intuitive properties are plausible candidates for defining such a function. These include, for example

1. Normality: $m_F(7) = 1$.
2. Monotonicity: the closer x is to 7, the closer m_F is to 1.
3. Symmetry: numbers equally distant from 7 should have equal memberships.

Either of the functions on the right side of Figure 8.22 have these properties and would thus be usefully representative of F. Because of its continuity, the triangular membership function m_{F2} in the lower right is the type most often used.

So, the membership function maps the degree to which an object belongs to a given set onto the range [0, 1] (see Figure 8.23), and its values measure how well the object satisfies imprecisely defined properties. It is therefore the fundamental idea in fuzzy set theory. Zadeh defined several classical operations that allow fuzzy sets to be manipulated. For two fuzzy sets A and B, and corresponding membership functions m_A and m_B, these operations are

$$\text{Equality:} \quad A = B \Leftrightarrow m_A(x) = m_B(x) \tag{8.92}$$

$$\text{Containment:} \quad A \subset B \Leftrightarrow m_A(x) \le m_B(x) \tag{8.93}$$

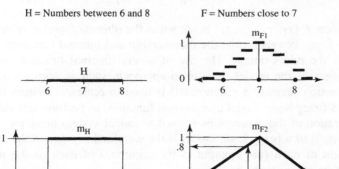

Figure 8.22. Membership functions for crisp and fuzzy sets [15].

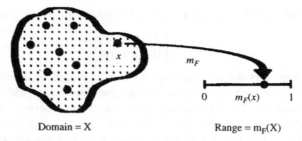

Figure 8.23. Mapping operation of membership functions [15].

$$\text{Complement:} \qquad m_{\overline{A}}(x) = 1 - m_A(x) \tag{8.94}$$

$$\text{Intersection:} \qquad m_{A \cap B}(x) = \min[m_A(x), m_B(x)] \tag{8.95}$$

$$\text{Union:} \qquad m_{A \cup B}(x) = \max[m_A(x), m_B(x)] \tag{8.96}$$

Other authors have subsequently proposed other useful functions for these basic operations [15], but these five are the most important and nearly ubiquitous in real fuzzy logic applications.

The use of fuzzy logic for process modeling was introduced by Takagi and Sugeno [16]. In general, a fuzzy logic model consists of four major elements: membership functions, internal functions, rules, and outputs. The input variable is first normalized (usually onto the interval [0, 1]). A group of membership functions, one from each input variable, constitutes a *fuzzy cell*. For the ith cell, the fuzzy membership function $A_l^i(x_l)$ computes the membership grade (u) for each of k input variables x_l (where $l = 1, \ldots, k$). The total number of cells is $n = r_1 \times r_2 \times \cdots \times r_k$, where r_l is the number of membership functions for x_l.

With respect to each fuzzy cell, fuzzy rules are used to map input variables to output conditions. If F_j^i is a fuzzy set, then the rule of the ith cell is of the form

If x_1 is F_1^i and x_2 is F_2^i and $\cdots$ and x_k is F_k^i, then

$$f^i(x_1, x_2, \ldots, x_k) = p_0^i + p_1^i x_1 + p_2^i x_2 + \cdots + p_k^i x_k$$

The function $f^i(x_1, x_2, \ldots, x_k)$ is known as the *internal function* with parameters $p_0^i, p_1^i, \ldots, p_k^i$. For each rule, the membership and internal functions are used to determine the rule's output. The use of several internal functions accounts for the fuzziness of the model. In a crisp approach, such as regression analysis, a single function (usually a polynomial) is used to represent system behavior. In contrast, a fuzzy logic model uses several functions to perform this mapping, and the integration of those functions is used to model system behavior.

The output of a fuzzy logic model is the weighted average of the rule outputs. The weight of each rule's output is the minimum of each of the membership grades of its antecedents, or

$$w^i = \min[A_1^i(x_1), A_2^i(x_2), \ldots, A_k^i(x_k)] \tag{8.97}$$

where i is the index of the n fuzzy rules. The output corresponding to the jth pair of input variables is

$$\hat{y}_j(x_{1j}, x_{2j}, \ldots, x_{kj}) = \frac{\sum_{i=1}^{n} \min[A_1^i(x_{1j}), A_2^i(x_{2j}), \ldots, A_k^i(x_{kj})] \times (p_0^i + p_1^i x_{1j} + p_2^i x_{2j} + \cdots + p_k^i x_{kj})}{\sum_{i=1}^{n} \min[A_1^i(x_{1j}), A_2^i(x_{2j}), \ldots, A_k^i(x_{kj})]} \quad (8.98)$$

The model output can be compared with experimental data to calculate the coefficients of the internal function. When the model is established (i.e., when the coefficients are known), it can be used to predict process behavior. The total number of coefficients of the internal functions to be derived is $r_1 \times r_2 \times \cdots \times r_k \times (k + 1)$. These coefficients are determined by minimizing the sum of squares of the errors between the experimental data and the outputs of the fuzzy logic model.

Xie et al. successfully used this approach to model the epitaxial growth of silicon by CVD [17]. For a horizontal reactor employing a gas mixture of silane and hydrogen, three input variables were considered: mean gas velocity (V_0),

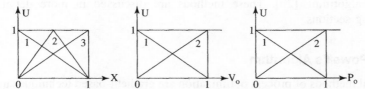

Figure 8.24. Membership function used to model the CVD process in the paper by Mie et al. [17].

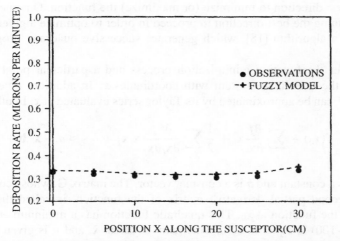

Figure 8.25. Comparison between observations and predictions for fuzzy logic model of CVD process [17].

partial silane pressure (P_0), and axial position on the graphite susceptor (x). A total of 63 data points were used to train the fuzzy model, and the membership functions used for the input variables are shown in Figure 8.24. The internal functions were linear functions of these variables. The CVD deposition rates predicted by the fuzzy model developed were within 5% of those predicted by a previously established empirical model (Figure 8.25).

8.6. PROCESS OPTIMIZATION

A formal, systematic methodology that facilitates the design of specific sets of process conditions (or "recipes") to achieve desired process objectives is necessary to optimize a given unit process. These process objectives are generally specific locations on the multidimensional response surfaces that geometrically depict the variation of process output characteristics with respect to input variables. Recipe generation can be achieved by employing previously described process models in conjunction with response surface exploration schemes. These schemes include traditional approaches such as Powell's algorithm [18] or Nelder and Mead's simplex algorithm [19], as well as more advanced techniques like genetic algorithms [20]. These methods are discussed in more detail in the following sections.

8.6.1. Powell's Algorithm

Classical methods of process optimization are gradient-based techniques using the "hill-climbing" approach. In this case, the "hills" are multidimensional response surfaces such as that depicted in Figure 8.9. For function minimization (or maximization) in n-dimensional space using these methods, it is important to find the best direction to minimize (or maximize) the function. One method used in determining the best direction to proceed in order to optimize a given function is Powell's algorithm [18], which generates successive quadratic programming problems.

Consider the function minimization process and a particular point x_0 at the origin of the coordinate system with coordinates x. In addition, note that any function f can be approximated by its Taylor series evaluated at x. In other words

$$f(\mathbf{x}) = f(\mathbf{x}_0) + \sum_i \frac{\partial f}{\partial \mathbf{x}_i} \mathbf{x}_i + \frac{1}{2} \sum_{i,j} \frac{\partial^2 f}{\partial \mathbf{x}_i \partial \mathbf{x}_j} \mathbf{x}_i \mathbf{x}_j + \cdots \approx a + \mathbf{x}^T \mathbf{b} + \frac{1}{2} \mathbf{x}^T \mathbf{G} \mathbf{x}$$

(8.99)

where a is a constant and b is a constant vector. The matrix $\mathbf{G}$, whose components are the second partial derivative matrix of the function, is called the *Hessian matrix* of the function at x_0. This quadratic function has a minimum at the point where (8–100) is equal to zero. This point is called $\hat{\mathbf{x}}$, and it is given by

$$\hat{\mathbf{x}} = -\mathbf{G}^{-1}\mathbf{b}$$

(8.100)

Subject to certain continuity conditions, a function can be approximated in the region of the point x_0 by

$$\phi(\mathbf{x}) = f(\mathbf{x}_0) + (\mathbf{x} - \mathbf{x}_0)^T \nabla f(\mathbf{x}_0) + \tfrac{1}{2}(\mathbf{x} - \mathbf{x}_0)^T \mathbf{G}(\mathbf{x}_0)(\mathbf{x} - \mathbf{x}_0) \qquad (8.101)$$

where $\mathbf{G}(\mathbf{x}_0)$ is the Hessian matrix at x_0.

A reasonable approximation to the minimum of $f(\mathbf{x})$ is the minimum of $\phi(\mathbf{x})$. If the latter is at x_m, we have

$$\nabla f(\mathbf{x}_0) + \mathbf{G}(\mathbf{x}_0)(\mathbf{x}_m - \mathbf{x}_0) = 0 \qquad (8.102)$$
$$\therefore \ \mathbf{x}_m = \mathbf{x}_0 - \mathbf{G}^{-1}(\mathbf{x}_0)\nabla f(\mathbf{x}_0) = \mathbf{x}_0 - \mathbf{G}^{-1}(\mathbf{x}_0)\mathbf{g}(\mathbf{x}_0)$$

Thus, the iterative equation from point x_i to the next approximation is

$$\mathbf{x}_{i+1} = \mathbf{x}_i - \mathbf{G}^{-1}(\mathbf{x}_i)\mathbf{g}(\mathbf{x}_i) \qquad (8.103)$$

Both $|\mathbf{g}(\mathbf{x}_{i+1})|$ and $|\mathbf{x}_{i+1} - \mathbf{x}_i|$ should be checked as termination criteria. Note that the search direction is not $-\mathbf{g}(\mathbf{x}_i)$, but $-\mathbf{G}^{-1}(\mathbf{x}_i)\mathbf{g}(\mathbf{x}_i)$ if second derivatives are taken into account.

The direction of search at each stage is thus a crucial factor in the efficiency of iterative search methods because the evaluation and inversion of the Hessian matrix require significant computation, especially for "implicit" optimization functions whose derivatives must be estimated by means of perturbations. For a quadratic function of n variables such as Eq. (8.99), the best direction for optimization is a direction that is *conjugate* to the previous search direction. Two directions $\mathbf{p}$ and $\mathbf{q}$ are said to be conjugate with respect to the symmetric positive-definite matrix $\mathbf{G}$ if

$$\mathbf{p}^T \mathbf{G} \mathbf{q} = 0 \qquad (8.104)$$

If line minimizations of a function are performed in one direction and successively redone along a conjugate set of directions, then any previously explored direction need not be repeated.

The goal of this method is to come up with a set of n linearly independent, mutually conjugate directions. Then, one pass of n line minimizations will put the algorithm exactly at the minimum of a quadratic form such as Eq. (8.99). For functions that are not exactly quadratic, this approach will not result in identifying the exact minimum, but repeated cycles of n line minimizations will, in due course, converge to the minimum.

The evaluation and inversion of the Hessian matrix in Eq. (8.103) at each step involves significant computation. Powell first discovered a direction set method that does produce n mutually conjugate directions without calculation of Hessian matrix. The procedure is as follows:

1. First, initialize the set of directions x_i to the basis vectors.
2. Repeat the following sequence of steps until convergence (typically defined as achieving suitable small error metrics):

- Save the starting position as x_0.
- For $i = 1, \ldots, n$, move x_{i-1} to the minimum along direction u_i, and call this point x_i.
- For $i = 1, \ldots, n-1$, set $u_i \leftarrow u_{i+1}$.
- Set $u_n \leftarrow x_n - x_0$.
- Move $\mathbf{x}_n$ to the minimum along direction $\mathbf{u}_n$ and call this point $\mathbf{x}_0$.

Powell showed that, for a quadratic form such as Eq. (8.99), k iterations of the basic procedure listed above produces a set of directions u_i whose last k members are mutually conjugate. Therefore, n iterations of the basic procedure, which amounts to $n(n+1)$ line minimizations in all, will exactly minimize a quadratic form.

8.6.2. Simplex Method

A regular *simplex* is a set of $n+1$ mutually equidistant points in n-dimensional space. In two dimensions, the simplex is an equilateral triangle, and in three dimensions, it is a regular tetrahedron. The idea of the simplex method of optimization is to compare the values of the function at the vertices of the simplex and move the simplex toward the optimal point during the iterative process. The original simplex method maintained a regular simplex at each stage. Nelder and Mead proposed several modifications to the method that allow the simplices to become nonregular [19]. The result is a very robust direct search method that is extremely powerful for up to five variables.

The movement of the simplex in this method is achieved by the application of three basic operations: *reflection, expansion* and *contraction*. Nelder and Mead's minimization procedure is as follows:

1. Start with points $x_1, x_2, \ldots, x_{n+1}$ and find

$$f_1 = f(\mathbf{x}_1), \ f_2 = f(\mathbf{x}_2), \ \cdots \ f_{n+1} = f(\mathbf{x}_{n+1}) \tag{8.105}$$

2. Next, find the highest function value f_h, the next highest function value f_g and the lowest function value f_l and corresponding points x_h, x_g, and x_l.
3. Find the centroid of all the points except $\mathbf{x}_h$. Call the centroid x_o, where

$$\mathbf{x}_0 = \frac{1}{n} \sum_{i \neq h} \mathbf{x}_i \tag{8.106}$$

Evaluate $f(\mathbf{x}_0) = f_0$.

4. It would seem reasonable to move away from x_h. Therefore, we reflect x_h in x_o to find x_r and find $f(\mathbf{x}_r) = f_r$. If α is the *reflection* factor, find $\mathbf{x}_r$ such that

$$\mathbf{x}_r - \mathbf{x}_0 = \alpha(\mathbf{x}_0 - \mathbf{x}_h) \tag{8.107}$$

$$\therefore \ \mathbf{x}_r = (1 + \alpha)\mathbf{x}_0 - \alpha\mathbf{x}_h$$

5. Now compare f_r and f_l.

   a. If $f_r < f_l$, the lowest function value has not yet been obtained. The direction from x_o to x_r appears to be a good one to move along. We therefore expand in this direction to find x_e and evaluate $f(\mathbf{x}_e) = f_e$. With an *expansion* factor $\gamma(0 < \gamma < 1)$, we have

   $$\mathbf{x}_e - \mathbf{x}_0 = \gamma(\mathbf{x}_r - \mathbf{x}_0) \qquad (8.108)$$

   $$\therefore \ \mathbf{x}_e = \gamma\mathbf{x}_r + (1 - \gamma)\mathbf{x}_0$$

      i. If $f_e < f_l$, replace x_h by x_e and test the $(n + 1)$ points of the simplex for convergence to the minimum. If convergence has been achieved, stop; if not, return to step 2.

      ii. If $f_e \geq f_l$, abandon $\mathbf{x}_e$. We have evidently moved too far in the direction x_o to x_r, which we know gave improvement (step 5a). Test for convergence, and if it is not achieved, return to step 2.

   b. If $f_r > f_l$ but $f_r \leq f_g$, x_r is an improvement on the two worst points of the simplex, and we replace x_h by x_r. Test for convergence, and if it is not achieved, return to step 2.

   c. If $f_r > f_l$ and $f_r > f_g$, proceed to step 6.

6. Compare f_r and f_h.

   a. If $f_r > f_h$, proceed directly to the *contraction* step 6b. If $f_r < f_h$, replace x_h by x_r and f_h by f_r. Remember $f_r > f_g$ from step 5c. Proceed to step 6b.

   b. In this case $f_r > f_h$, so it would appear that we have moved too far in the direction x_h to x_o. We rectify this by finding x_c (and f_c) by a *contraction* step. (Figure 8.26 illustrates reflection, expansion, and contraction). If $f_r > f_h$, proceed directly to contraction and find x_c from

   $$\mathbf{x}_c - \mathbf{x}_0 = \beta(\mathbf{x}_h - \mathbf{x}_0) \qquad (8.109)$$

   $$\therefore \ \mathbf{x}_c = \beta\mathbf{x}_h + (1 - \beta)\mathbf{x}_0$$

   where $\beta(0 < \beta < 1)$ is the contraction coefficient. If, however, $f_r < f_h$, replace x_h by x_r and contract. Thus we find $\mathbf{x}_c$ from (see Figure 8.27)

   $$\mathbf{x}_c - \mathbf{x}_0 = \beta(\mathbf{x}_r - \mathbf{x}_0) \qquad (8.110)$$

   $$\therefore \ \mathbf{x}_c = \beta\mathbf{x}_r + (1 - \beta)\mathbf{x}_0$$

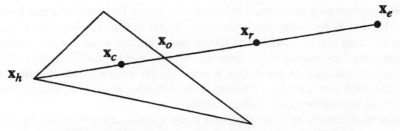

Figure 8.26. Illustration of reflection, expansion, and contraction.

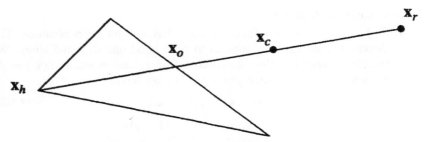

Figure 8.27. Illustration of contraction if $f_r < f_h$.

7. Compare f_c and f_h.

   a. If $f_c < f_h$, replace x_h by x_c, check for convergence, and if convergence is not achieved, return to step 2.

   b. If $f_c > f_h$, all efforts to find a value $< f_h$ have failed, proceed to step 8.

8. Reduce the size of the simplex by halving the distance of each point of the simplex from x_l to the point generating the lowest function value. Thus, x_i is replaced by

$$x_l + \tfrac{1}{2}(x_i - x_l) = \tfrac{1}{2}(x_i + x_l) \tag{8.111}$$

   Then calculate f_i for $i = 1, 2, \ldots, (n + 1)$, test for convergence, and if convergence has not been met, return to step 2.

9. The test of convergence is based on the standard deviation (σ) of the $(n + 1)$ function values being less than some predetermined small value (ε). Thus, we calculate

$$\sigma^2 = \frac{1}{n + 1} \sum_{i=1}^{n+1} (f_i - \hat{f})^2 \tag{8.112}$$

   where $\hat{f} = \sum f_i / (n + 1)$. If $\sigma < \varepsilon$, all function values and points are very close together near the minimum x_l.

There remain some important details to clarify. The first concerns the values of α, β, and γ. Nelder and Mead recommend $\alpha = 1$, $\beta = 0.5$, $\gamma = 2$ [19]. This recommendation appears to allow the method to work efficiently in many different situations.

If one of the x_i must be nonnegative in a minimization problem, then the simplex method may be adapted in one of two ways. The scale of x can be transformed (such as by using the logarithm) so that negative values are excluded, or the function can be modified to take a large positive value for all negative x. Alternatively, any points trespassing the simplex border will be followed automatically by contraction moves that will eventually keep it inside. In either case, an actual minimum with $x = 0$ would be inaccessible in general, though one can achieve an arbitrarily close approximation.

Constraints involving more than one x can accounted for by using the second technique, provided that an initial simplex can be found inside the permitted region from which to start the process. Linear constraints that reduce the dimensionality of the field of search can be included by choosing the initial simplex to satisfy the constraints and to reduce the dimensions accordingly. Thus, to minimize $y = (x_1, x_2, x_3)$ subject to $x_1 + x_2 + x_3 = \mathbf{K}$, we could choose an initial simplex with vertices $(\mathbf{K}, 0, 0)$, $(0, \mathbf{K}, 0)$, and $(0, 0, \mathbf{K})$, and treating the search as two-dimensional. In particular, any x_i may be held constant by setting its value to that constant for all vertices of the initial simplex.

8.6.3. Genetic Algorithms

Genetic algorithms (GAs) were first proposed for solving optimization problems by John Holland at the University of Michigan in 1975 [20]. GAs are guided stochastic search techniques inspired by the mechanics of genetics. They use three genetic operations found in natural genetics to guide their trek through the search space: *reproduction, crossover*, and *mutation*. Using these operations, GAs are able to search through large, irregularly shaped spaces quickly, requiring only objective function value information (detailing the quality of possible solutions) to guide the search. This is a desirable characteristic, considering that the majority of traditional search techniques require derivative information, continuity of the search space, or complete knowledge of the objective function to guide the search. Furthermore, GAs take a more global view of the search space than many methods currently encountered in engineering optimization.

In computing terms, a genetic algorithm maps a problem on to a set of binary strings, each string representing a potential solution. The GA then manipulates the most promising strings in searching for improved solutions. A GA operates typically through a simple cycle of four stages:

1. Creation of a "population" of strings
2. Evaluation of each string
3. Selection of "best" strings
4. Genetic manipulation, to create the new population of strings

In each computational cycle, a new generation of possible solutions for a given problem is produced. At the first stage, an initial population of potential solutions is created as a starting point for the search process. Each element of the population is encoded into a string (the "chromosome"), to be manipulated by the genetic operators. In the next stage, the performance (or *fitness*) of each individual of the population is evaluated with respect to the constraints imposed by the problem. According to the fitness of each individual string, a selection mechanism chooses "mates" for the genetic manipulation process. The selection policy is responsible for assuring survival of the most "fit" individuals.

Binary strings are typically used in coding genetic search, although alphanumeric strings can be used as well. One successfully used method of coding multivariate optimization problems is concatenated, multiparameter fixed-point coding. If $x \in [0, 2^l]$ is the parameter of interest (where l is the length of the string), the decoded unsigned integer x can be mapped linearly from $[0, 2^l]$ to a specified interval $[U_{min}, U_{max}]$. In this way, both the range and precision of the decision variables can be controlled. The precision (p) of this mapped coding is

$$p = \frac{U_{max} - U_{min}}{2^l - 1} \tag{8.113}$$

To generate a multiparameter coding, the necessary number of single parameters as can be concatenated. Each coding may have its own sublength (i.e., its own U_{min} and U_{max}). Figure 8.28 shows an example of two-parameter coding with four bits in each parameter. The ranges of the first and second parameter are 2–5 and 0–15, respectively.

The string manipulation process employs genetic operators to produce a new population of individuals ("offspring") by manipulating the genetic "code" possessed by members ("parents") of the current population. It consists of three operations: reproduction, crossover, and mutation. Reproduction is the process by which strings with high fitness values (i.e., good solutions to the optimization problem under consideration) receive appropriately large numbers of copies in the new population. A popular method of reproduction is elitist roulette wheel selection. In this method, those strings with large fitness values F_i are assigned a proportionately higher probability of survival into the next generation. This probability is defined by

$$P_i = \frac{F_i}{\sum F} \tag{8.114}$$

Thus, an individual string whose fitness is n times better than another will produce n times the number of offspring in the subsequent generation. Once the strings have reproduced, they are stored in a "mating pool" awaiting the actions of the crossover and mutation operators.

The crossover operator takes two chromosomes and interchanges part of their genetic information to produce two new chromosomes (see Figure 8.29). After the crossover point has been randomly chosen, portions of the parent strings (P1 and P2) are swapped to produce the new offspring (O1 and O2) on the basis of a specified crossover probability. Mutation is motivated by the possibility that the initially defined population might not contain all the information necessary to solve the problem. This operation is implemented by randomly changing a fixed number of bits every generation according to a specified mutation probability (see Figure 8.30).

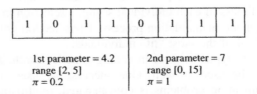

Figure 8.28. Illustration of multiparameter coding.

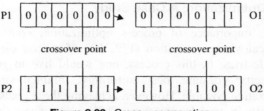

Figure 8.29. Crossover operation.

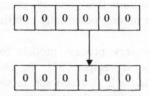

Figure 8.30. Mutation operation.

Typical values for the probabilities of crossover and bit mutation range from 0.6 to 0.95 and 0.001 to 0.01, respectively. Higher mutation and crossover rates disrupt good "building blocks" (*schemata*) more often, and for smaller populations, sampling errors tend to wash out the predictions. For this reason, the greater the mutation and crossover rates and the smaller the population size, the less frequently predicted solutions are confirmed.

8.6.4. Hybrid Methods

Recipe generation is essentially a procedure for searching a multidimensional response space in order to locate an optimum. Each algorithm described above represents an approach to performing such a search. In both Powell's algorithm and the simplex method, however, the initial starting searching point has a profound effect on overall performance. With an improper initial starting point, both algorithms are more likely to be trapped in local optima, and this is why they are both considered "local" optimization methods. However, if the proper initial point is given, the search is very fast. On the other hand, genetic algorithms can explore the overall domain area very fast, and this is why they are known as "global" optimizers. Unfortunately, although they are quick to reach the vicinity of the global optimum, converging to the global optimum point is very slow.

It has been suggested and demonstrated that hybrid combinations of genetic (global) algorithms with one of the local algorithms can sometimes offer improved results in terms of both speed and accuracy [21]. Hybrid algorithms start with genetic algorithms to initially sample the response surface and find the general vicinity of the global optimum. After some number of generations, the best point found using the GA is handed over to a local optimization algorithm as a starting point. With this initial point, both Powell's algorithm and simplex method can quickly locate the optimum.

8.6.5. PECVD Optimization: A Case Study

To illustrate the importance of process optimization, consider the plasma-enhanced chemical vapor deposition (PECVD) of silicon dioxide films used as interlayer dielectrics. In this process, one would like to grow a film with the lowest dielectric constant, best uniformity, minimal stress, and lowest impurity concentration possible. However, achieving these goals usually requires a series of tradeoffs in growth conditions. Optimized process models can help a process engineer navigate complex response surfaces and provide the necessary combination of process conditions (temperature, pressure, gas composition, etc.) or find the best compromise among potentially conflicting objectives to produce the desired results.

Han has used neural network process models for the PECVD process to synthesize other novel process recipes [21]. To characterize the PECVD of SiO_2 films, he first performed a 2^{5-1} fractional factorial experiment with three centerpoint replications. Data from these experiments were used to develop neural process models for SiO_2 deposition rate, refractive index, permittivity, film stress, wet etch rate, uniformity, silanol (SiOH) concentration, and water concentration. A recipe synthesis procedure was then performed to generate the necessary deposition conditions to obtain specific film qualities, including zero stress, 100% uniformity, low permittivity, and minimal impurity concentration.

Han compared five optimization methods to generate PECVD recipes: (1) genetic algorithms, (2) Powell's method, (3) Nelder and Mead's simplex algorithm, (4) a hybrid combination of genetic algorithms and Powell's method, and (5) a hybrid combination of genetic algorithms and the simplex algorithm. The desired output characteristics of the PECVD SiO_2 film to be produced are reflected by the following fitness function

$$F = \frac{1}{1 + \sum_r |K_r(y_d - y)|} \tag{8.115}$$

where r is the number of process responses, y_d are the desired process responses, y are the process outputs dictated by the current choice of input parameters, and K_r is a constant which represents the relative importance of the rth process response. The optimization procedures were stopped after a fixed number or iterations or when F was within a predefined tolerance.

For genetic algorithms, the probabilities of crossover and mutation were set to 0.6 and 0.01, respectively. A population size of 100 was used in each generation. Each of the five process input parameters was coded as a 40-bit string, resulting in a total chromosome length of 200 bits. Maximization of F continued until a final solution was selected after 500 generations. Search in the simplex method is achieved by applying three basic operations: reflection, expansion, and contraction. Nelder and Mead recommend values of 1, 2, and 0.5 for α, γ, and β, respectively [19]. These were the values used in the simplex and hybrid simplex–genetic method.

Table 8.20. PECVD recipe synthesis results [21].

Method	% Nonuniformity	Permittivity	Stress (MPa)	H_2O (Wt%)	SiOH (Wt%)
GA	3.13	4.28	−209.3	1.99	5.08
Simplex	0.66	4.37	−173.4	1.28	3.17
Powell	0.26	4.26	−233.6	1.78	2.66
GA/simplex	1.64	4.38	−216.3	1.25	4.51
GA/Powell	5.05	4.11	−264.3	1.19	4.01
Weight	1	100	1	50	50
Goal	0%	Minimum	0	0	0

It was expected that hybrid methods would readily improve accuracy compared with genetic search alone if initiated immediately after the 500 GA generations. Such a large number of generations, however, impacts the computational load of these techniques severely. Therefore, to reduce this computational burden in both hybrid methods, the GA portion of the search was limited to 100 generations. Then the resulting GA solution was handed over as the initial starting point for the simplex or Powell algorithms.

The objective of this recipe synthesis procedure was to find the optimal deposition recipes for the following individual novel film characteristics: 100% thickness uniformity, low permittivity, zero residual stress, and low impurity concentration in the silicon dioxide film. Clearly, it is desirable to grow films with the best combination of all the desired qualities (i.e., 100% uniformity, low permittivity, low stress, and low impurity content). This involves processing tradeoffs, and the challenge, therefore, is to devise a means for designating the importance of a given response variable in determining the optimal recipe.

These multiple objectives were accomplished simultaneously by applying the fitness function in Eq. (8–14) with a specific set of K_r coefficients chosen for growing films optimized for electronics packaging applications. Because one of the most important qualities in SiO_2 films for this application is permittivity, Han set the weight of permittivity equal to 100. Weights for both water and silanol concentration were set to 50, and the weights for both uniformity and stress were set to 1. Table 8.20 shows the measured results of the five different synthesis procedures for optimizing multiple outputs. Overall, given the response weighting selected, the genetic algorithm and the hybrid GA/Powell algorithm provide the best compromise among the multiple objectives.

SUMMARY

In this chapter, we have provided an overview of how data derived from designed experiments can be used to construct process models of various types, including conventional regression models, as well as more contemporary artificial intelligence–based techniques. The models so derived are used to analyze, visualize, and predict semiconductor process behavior. We have also discussed how these models can be used "in reverse" to synthesize optimal process recipes. In the next chapter, we will discuss how data these models are applied for advanced process control.

PROBLEMS

8.1. Derive Eqs. (8.24).

8.2. Fit the yield data in Example 8.2 to a linear model using regression techniques, and perform analysis of variance to evaluate the quality of your model. Is the linear model sufficient?

8.3. A dry-etch step is used to etch 1.0 μm of polysilicon. A sample of five wafers is measured each hour. It is expected that over time the lower electrode of the etcher will become contaminated and ohmic contact with the wafer will therefore deteriorate, decreasing the etch rate. The electrode is cleaned before the first and seventh samples. The measured etch rate is given below. Draw a 2σ regression chart for this process.

$\bar{x}$	1.03	1.01	1.02	1.01	0.98	0.99	1.05	1.03	1.04	1.00	0.95	0.94
R	0.006	0.005	0.007	0.006	0.009	0.008	0.007	0.005	0.008	0.005	0.006	0.005

8.4. Consider a 3^3 factorial experiment conducted in a semiconductor manufacturing process with three normalized factors ($x_1, x_2,$ and x_3) and two responses (y_1 and y_2) [3]. The following data were collected:

x_1	x_2	x_3	y_1	y_2
−1	−1	−1	24.00	12.49
0	−1	−1	120.33	8.39
1	−1	−1	213.67	42.83
−1	0	−1	86.00	3.46
0	0	−1	136.63	80.41
1	0	−1	340.67	16.17
−1	1	−1	112.33	27.57
0	1	−1	256.33	4.62
1	1	−1	271.67	23.63
−1	−1	0	81.00	0.00
0	−1	0	101.67	17.67
1	−1	0	357.00	32.91
−1	0	0	171.33	15.01
0	0	0	372.00	0.00
1	0	0	501.67	92.50
−1	1	0	264.00	63.50
0	1	0	427.00	88.61
1	1	0	730.67	21.08
−1	−1	1	220.67	133.82

x_1	x_2	x_3	y_1	y_2
0	−1	1	239.67	23.46
1	−1	1	422.00	18.52
−1	0	1	199.00	29.44
0	0	1	485.33	44.67
1	0	1	673.67	158.21
−1	1	1	176.67	55.51
0	1	1	501.00	138.94
1	1	1	1010.00	142.45

(a) The response y_1 is the average of 3 resistivity readings (in Ω-cm) for a single wafer. Fit a quadratic model to this response.

(b) The response y_2 is the standard deviation of the three resistivity measurements. Fit a first-order model to this response.

(c) Where should x_1, x_2, and x_3 be set if the objective is to hold the mean resistivity at 500 Ω-cm and minimize the standard deviation?

8.5. The yield from the first four cycles of a chemical process is shown below, with the following variables: (1)% concentration (X_1) at levels 30 (L), 31 (M), and 32 (H) and (2) temperature (X_2) at 140 (L), 142 (M), and 144 (H) degrees. Analyze the data using EVOP.

Cycle	(1 M–M)	(2 L–L)	(3 H–H)	(4 H–L)	(5 L–H)
1	60.7	69.8	60.2	64.2	57.5
2	69.1	62.8	62.5	64.6	58.3
3	66.6	69.1	69.0	62.3	61.1
4	60.5	69.8	64.5	61.0	60.1

8.6. Suppose that data are collected on deposition rate (x_1) and uniformity (x_2) for a CVD process. The covariance matrix is

$$\Sigma = \begin{pmatrix} 80 & 44 \\ 44 & 80 \end{pmatrix}$$

(a) What is the first principal component, and what percentage of the total variation does it account for?

(b) Repeat (a) given

$$\Sigma = \begin{pmatrix} 8000 & 440 \\ 440 & 80 \end{pmatrix}$$

8.7. Derive Eq. (8.83).

8.8. The data tabulated below are deposition rates collected from a designed experiment to characterize a plasma-enhanced CVD process as a function of five input factors: SiH_4 flowrate, N_2O flowrate, temperature, pressure, and power. Using the backpropagation algorithm, design a neural network to approximate the deposition rate.

SiH$_4$ (sccm)	N$_2$O (sccm)	Temperature ($^\circ$C)	Pressure (Torr)	Power (W)	Deposition Rate (Å/min)
200	400	400	0.25	20	123
400	400	400	1.80	20	460
400	400	400	0.25	150	454
300	650	300	1.025	85	433
300	650	300	1.025	85	432
400	900	200	0.25	150	362
200	400	400	1.80	150	173
200	400	200	1.80	20	304
400	900	200	1.80	20	242
300	650	300	1.025	85	428
400	900	400	1.80	150	426
400	400	200	1.80	150	352
200	900	200	0.25	20	101
200	900	400	0.25	150	222
200	900	400	1.80	20	268
400	900	400	0.25	20	56
200	400	200	0.25	150	287
400	400	200	0.25	20	139
200	900	200	1.80	150	196

8.9. Suppose that we are evaluating the particle content of two adjacent areas (A and B) in a class 10 cleanroom. The number of particles/ft^3 in each area are described by the fuzzy membership functions m_A and m_B, where

$$m_A \in \Im(x) = \{(\text{particle count}) \text{ close to } 7 \text{ ft}^{-3}\}$$
$$m_B \in \Im(x) = \{(\text{particle count}) \text{ close to } 3 \text{ ft}^{-3}\}$$

These membership functions are shown in Figure P8.9.
Determine the following:

(a) The extent to which area B has counts nearly 7 ft^{-3}.

(b) The extent to which area B has counts NOT nearly 7 ft^{-3}.

(c) The extent to which a measurement of 6 ft^{-3} is nearly 7 ft^{-3}.

(d) The extent to which a measurement of 6 ft^{-3} is nearly 3 ft^{-3}.

(e) The extent to which a measurement of 6 ft^{-3} is nearly (3 AND 7) ft^{-3}.

(f) The extent to which a measurement of 6 ft^{-3} is nearly (3 OR 7) ft^{-3}.

8.10. Consider the plasma etching example in Section 8.2.2. Use Powell's method to optimize Eq. (8.61) to identify a recipe that achieves the highest etch rate possible.

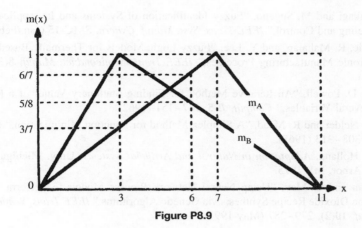

Figure P8.9

REFERENCES

1. G. Box, W. Hunter, and J. Hunter, *Statistics for Experimenters*, Wiley, New York, 1978.
2. *RS/Explore Primer*, BBN Software Products, Cambridge, MA, 1992.
3. D. Montgomery, *Introduction to Statistical Quality Control*, Wiley, New York, 1993.
4. G. May, J. Huang, and C. Spanos, "Statistical experimental Design in Plasma Etch Modeling," *IEEE Trans. Semiconduct. Manuf.* **4**(2) (May 1991).
5. *RS/Discover User's Guide*, BBN Software Products, Cambridge, MA, 1992.
6. D. Montgomery, *Design and Analysis of Experiments*, 3rd ed., Wiley, New York, 1991.
7. I. Joliffe, *Principal Component Analysis*, Springer-Verlag, New York, 1986.
8. S. Hong, G. May, and D. Park, "Neural Network Modeling of Reactive Ion Etching Using Optical Emission Spectroscopy Data," *IEEE Trans. Semiconduct. Manuf.* **16**(4) (Nov. 2003).
9. G. May, "Computational Intelligence in Microelectronics Manufacturing," in *Handbook of Computational Intelligence in Design and Manufacturing*, J. Wang and A. Kusiak, eds., CRC Press, Boca Raton, FL, 2001, Chapter 13.
10. J. Hopfield and D. Tank, "Neural Computation of Decisions in Optimization Problems," *Biol. Cybern.* **52** (1985).
11. G. May, "Manufacturing IC's the Neural Way," *IEEE Spectrum.* **31**(9) (Sept. 1994).
12. B. Irie and S. Miyake, "Capabilities of Three-Layered Perceptrons," *Proc. IEEE Intl. Conf. Neural Networks*, 1988.
13. C. Himmel and G. May, "Advantages of Plasma Etch Modeling Using Neural Networks over Statistical Techniques," *IEEE Trans. Semiconduct. Manuf.* **6**(2) (May 1993).
14. L. Zadeh, *Inform. Control.* **8** (1965).
15. J. Bezdek, "A Review of Probabilistic, Fuzzy, and Neural Models for Pattern Recognition," in *Fuzzy Logic and Neural Network Handbook*, C. H. Chen, ed., McGraw-Hill, New York, 1996, Chapter 2.

16. T. Takagi and M. Sugeno, "Fuzzy Identification of Systems and Its Applications to Modeling and Control," *IEEE Trans. Syst. Manuf. Cybern.* **SMC-15** (Jan./Feb. 1985).

17. H. Xie, R. Mahajan, and Y. Lee, "Fuzzy Logic Models for Thermally Based Microelectronic Manufacturing Processes," *IEEE Trans. Semiconduct. Manuf.* **8**(3) (Aug. 1995).

18. M. J. D. Powell, "An Iterative Method for Finding Stationary Values of a Function of Several Variables," *Comput. J.* **5**, 147–151 (1962).

19. J. A. Nelder and R. Mead, "A Simplex Method for Function Minimization," *Comput. J.* **7**, 308–313 (1965).

20. J. H. Holland, *Adaptation in Natural and Artificial Systems*, Univ. Michigan Press, Ann Arbor, MI, 1975.

21. S. Han and G. May, "Using Neural Network Process Models to Perform PECVD Silicon Dioxide Recipe Synthesis via Genetic Algorithms," *IEEE Trans. Semiconduct. Manuf.* **10**(2), 279–287 (May 1997).

9

ADVANCED PROCESS CONTROL

OBJECTIVES

- Provide an overview of various techniques for univariate and multivariate run-by-run control.
- Introduce and explore the concept of supervisory control.

INTRODUCTION

In Chapter 6, the basic concepts of statistical process control (SPC) were presented. Although SPC is indeed a powerful technique for monitoring reducing variation in semiconductor manufacturing processes, it is limited. The underlying assumption on which SPC is based is that the observations collected and plotted on control charts represent a random sample from a stable probability distribution. However, this assumption does not hold for many commonly encountered scenarios. Primary examples are processes that have undergone an abrupt shift or gradual drifts. Another limitation of SPC is that it is usually applied offline. As a result, corrective actions suggested by SPC alarms typically occur too long after process shifts, potentially leading to significant misprocessing.

One solution to this dilemma is *run-by-run* (RbR) control. The RbR approach is a discrete form of feedback control in which process recipes are modified between runs to minimize shifts, drifts, and other forms of process variability. A "run" can be a single wafer, a lot, a batch, or any other grouping of semiconductor

Fundamentals of Semiconductor Manufacturing and Process Control,
By Gary S. May and Costas J. Spanos
Copyright © 2006 John Wiley & Sons, Inc.

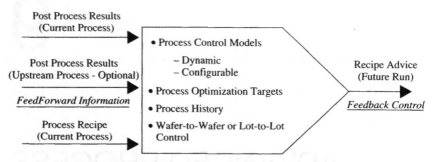

Figure 9.1. Block diagram of run-by-run control system [1].

products undergoing the same set of process conditions. RbR control is event-driven, since control actions are initiated by characterizing pre- and postprocess, as well as in situ, metrology data. These metrology data are compared to predictions generated from process models. When model predictions differ significantly from measurements, corrective action is initiated for the next run. The input/output structure of a typical RbR control system is shown in Figure 9.1. RbR control has, in recent years, become a proven and viable technology for process and equipment control.

A run-by-run control system that involves both feedforward and feedback control actions is known as a *supervisory* control system. Control of semiconductor processes can be examined at several levels (see Figure 9.2). Real-time control is at the lowest level of the hierarchy. In this case, adjustments are made to process variables during a run to maintain setpoints. A common example of the real-time level is the control loop used by mass flow controllers to regulate gas flow in a process chamber. The next level of the hierarchy is RbR control that adjusts process conditions between runs. Supervisory control highest level of the hierarchy. At this level, the progression of a wafer is tracked from unit process to unit process, and adjustments can be made to subsequent steps to account for variation in preceding steps.

This chapter explores both RbR and supervisory process control strategies. Such advanced process control techniques are required more and more for increasingly sophisticated modern semiconductor manufacturing applications. The chapter

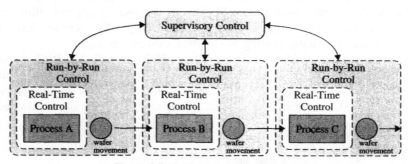

Figure 9.2. Process control hierarchy.

addresses control in two main categories. In this first, we deal with controllers that utilize polynomial process models, but limit their adaptation to the constant term of the model. Afterward, we discuss controllers that can adapt the entire model.

9.1. RUN-BY-RUN CONTROL WITH CONSTANT TERM ADAPTATION

Although a wide range of RbR control scenarios exist, there are three basic characteristics that are common to all RbR systems. First, some form of in situ (or *online*) or postprocess (*inline*) measurement is made, and the data from such measurements are used to trigger RbR control actions. Second, a dynamic model of the process undergoing RbR control must be established and maintained to relate tunable "recipe" inputs to the measurable process responses. Finally, process improvement is facilitated by control actions (i.e., adjustments to the tunable inputs) that occur between process runs.

Beyond these three common characteristics, RbR controllers may be categorized by the type of measurement data used to drive control actions, the type of control algorithm employed, or any number of other features. Here, we discuss RbR controllers as either single-variable or multivariate systems, with the common characteristic that they adapt only the constant term of a linear process model.

9.1.1. Single-Variable Methods

Some of the first research performed to establish RbR control as a viable technique in semiconductor manufacturing was conducted by Sachs et al. [2]. This RbR control architecture is depicted in Figure 9.3. This controller has two modes of operation: *rapid* and *gradual*. The rapid mode adapts to abrupt process shifts, such as those caused by maintenance operations. In this case, the control action must be decisive. The gradual mode, on the other hand, responds to drifts that occur over time, such as those caused by aging equipment. In these situations, the control action should also be gradual, and care must be taken to avoid overcontrol. The choice between modes is regulated by a *generalized SPC* approach, which allows SPC to be applied while the process is being tuned.

Consider a batch process in which runs are identified by a discrete time index t, the controllable input variables are denoted by x_t, and the output response y_t is a function of these inputs. In other words

$$y_t = \alpha_t + \beta_t x_t + \varepsilon_t \tag{9.1}$$

where the coefficients α_t and β_t are random variables that may change with time, and $\varepsilon_t \sim N(0, \sigma^2)$ is process noise. On the basis of this relationship, the appropriate prediction equation is

$$\hat{y}_t = a_{t-1} + b_{t-1} x_t \tag{9.2}$$

where a_{t-1} and b_{t-1} are estimates of the parameters α_t and β_t. The prediction equation [Eq. (9.2)] is used to select a "recipe" x_t for the next run at which the process output is likely to be close to some target value.

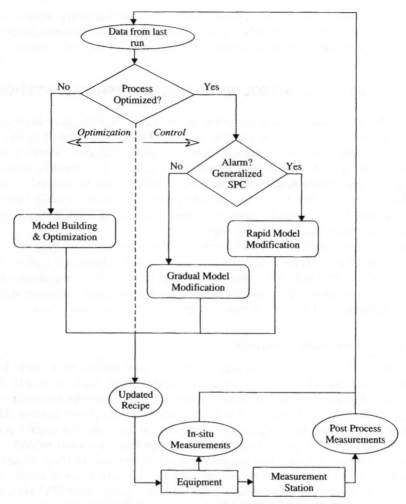

Figure 9.3. Flowchart depicting the Sachs RbR control architecture [2].

The output y_t is measured for every run to decide whether the process is behaving in a manner consistent with that of the predictive model. In its simplest implementation, an RbR controller assumes that the process sensitivities represented by b_{t-1} remain constant over time, and only the estimated intercept term a_{t-1} is updated after each output measurement. As long as no radical departure from predicted behavior occurs, the process model is updated gradually. However, if the last few output measurements disagree significantly with the predicted values, the model is updated rapidly to return the output to its target value.

The method used to evaluate the agreement of the output predicted values is "generalized" SPC. In this approach, the model residuals (which are given by $\varepsilon_t = y_t - \hat{y}_t$) are plotted on a standard Shewhart (i.e., $\bar{x}$), cusum (cumulative sum), EWMA (exponentially weighted moving-average), or other type of control chart (see Chapter 6). The control chart is used to distinguish between rapid shifts and slow drifts. A Shewhart chart, for example, would be more appropriate for identifying the former, whereas a cusum or EWMA chart would be more effective in detecting the latter. In the RbR controller, the gradual mode, which is designed to remove the effect of small process changes, is triggered by the detection of drifts. On the other hand, the rapid mode is engaged to compensate for larger variations when an abrupt shift is detected. These operations are described in more detail next.

9.1.1.1. Gradual Drift

To illustrate gradual mode operation, consider the hypothetical single-variable process shown in Figure 9.4. Assume that the estimated intercept a_{t-1} and slope b have been computed from runs 1 to $t - 1$. For simplicity, the slope is assumed to be constant. Process changes are accounted for by revising the intercept term whenever a new output measurement becomes available. The process drift can be visualized by letting the shaded line in Figure 9.4 (which represents the process) change its vertical position between successive runs.

To control the process, the current process model is used to predict the output of the next run as a function of the input variable. The RbR controller then selects an input value for which the predicted output matches the target specifications. This is illustrated in Figure 9.5. To update the constant term, the current intercept is calculated as

$$a_1 = y_1 - bx_1 \tag{9.3}$$

Since the process is subject to noise, it is reasonable to compute a_t as a weighted average of the past differences $(y_1 - bx_1, y_2 - bx_2, \ldots, y_t - bx_t)$. The EWMA approach has the advantage of allowing the weight (λ) of a given estimate to

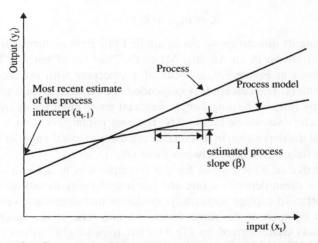

Figure 9.4. True process (shaded line) and process model (solid line) [2].

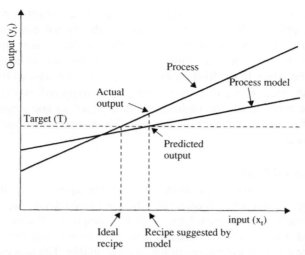

Figure 9.5. True process (shaded line), process model (solid line), and horizontal dashed line that intercepts the output axis at the target value [2].

decay gradually over time. The relationship is expressed in recursive form as

$$a_t = \lambda(y_t - bx_t) + (1 - \lambda)a_{t-1} \qquad (9.4)$$

Gradual mode operation is thus defined by this relationship along with

$$x_t = \frac{(T - a_{t-1})}{b} \qquad (9.5)$$

where T is the target response value. Equation (9.5) determines how the recipe for the tth run is selected, and Eq. (9.4) describes how the intercept is estimated after the process output is measured. Combining these two equations gives the following expression for updating the intercept term:

$$a_t = a_{t-1} + \lambda(y_t - T) \qquad (9.6)$$

To illustrate its effectiveness, Sachs applied this RbR controller to the control of silicon deposition in an Applied Materials 7800 barrel reactor. This reactor, which is shown in Figure 9.6, consists of a susceptor with six faces that each hold three wafers. The susceptor is suspended and rotated inside a quartz bell jar surrounded by infrared heating lamps. Reactant gases enter the chamber through the two injectors shown in Figure 9.6. The two parameters used to control the uniformity of the deposition in this reactor are the horizontal angle of the injectors (J_x) and the balance of flow between them (B_{mv}).

The objective of RbR control for this example was to achieve simultaneous control of the mean deposition rate and batch uniformity by adjusting these two input variables. To capture uniformity, thickness measurements were performed at five sites on each of the three wafers on one face of the susceptor. Batch uniformity was characterized by (1) the "left minus right" difference ($L - R$), which is the difference between the average of the left measurement sites of the

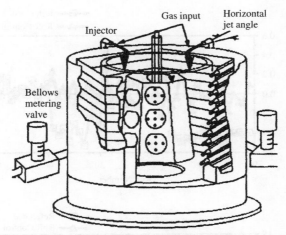

Figure 9.6. Schematic of Applied Materials 7800 reactor [2].

three wafers on a face and those on the right; and (2) the "top minus bottom" difference $(T - B)$, which is the difference between the top measurement site on the uppermost wafer and the bottom measurements site on the lower wafer. Least-squares regression models for the two uniformity metrics were derived from a 2^2 factorial experiment with two replicates. These models (with process parameters normalized over the range $[-1, 1]$ and standard errors of the coefficients in parentheses) are

$$L - R = -0.0173(\pm 0.0046) - 0.00313(\pm 0.0049)J_x + 0.118(\pm 0.0049)B_{mv}$$
$$(9.7)$$

$$T - B = -0.138(\pm 0.0061) - 0.174(\pm 0.0065)J_x + 0.0223(\pm 0.0065)B_{mv}$$
$$(9.8)$$

Gradual-mode RbR control was implemented using these models for generalized SPC using an EWMA control chart with a weight of 0.1. Fifty runs were performed over the course of three weeks. Changes in input settings using RbR control resulted in an improvement in the normalized process standard deviation in the $L - R$ response from 0.259 to 0.103 and in a similar improvement in the normalized $T - B$ standard deviation from 0.239 to 0.123. Figure 9.7 shows the $L - R$ response with and without RbR control, as well as the corresponding adjustments in the B_{mv} parameter in each situation. Note that when operated without RbR control, the process operator made a single adjustment that was too late and too large. The RbR controller made smaller and more frequent adjustments. Figure 9.8 is a similar plot for $T - B$ and J_x, respectively. Again, it is clear that the response stayed closer to the target under RbR control.

9.1.1.2. Abrupt Shifts
Rapid-mode control presents a different set of challenges than does gradual mode. When generalized SPC detects a shift of sufficient magnitude to trigger an alarm, in rapid mode, the RbR controller must estimate the magnitude of the disturbance,

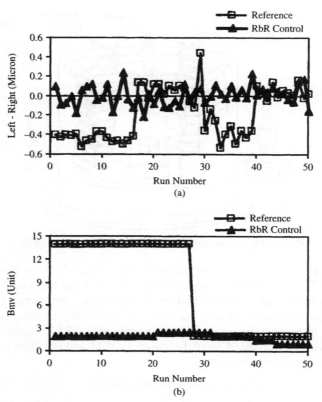

Figure 9.7. (a) $L - R$ response with and without RbR control; (b) corresponding adjustments in B_{mv} parameter [2].

assess the probability that a shift of that magnitude has taken place, and prescribe control actions. The strategy employed to accomplish these objectives begins with an assumption that the disturbance detected takes the form of a step function, and least-squares estimates of the magnitude and location of the step are then computed before control actions are prescribed.

The estimation procedure for the magnitude and location of a shift is illustrated in Figure 9.9. In this figure, the data points represent the intercept values (i.e., $y_t - bx_t$) computed for the last few runs. The levels of the two horizontal lines, as well as the position of the breakpoint in time, are fit to these data to minimize the sum of squared deviations from the lines. Let $z_t = y_t - bx_t$. Assuming the drift over the last k runs to be negligible, the intercept term changes very little during this period, and

$$z_t \approx a + e_t \tag{9.9}$$

where $a \approx a_i$ for $i = t - k + 1, \ldots, t$, and $e_t = y_t - T$. If the gradual mode of the RbR controller is performing correctly, then $e_t \sim N(0, \sigma^2)$. If a step of

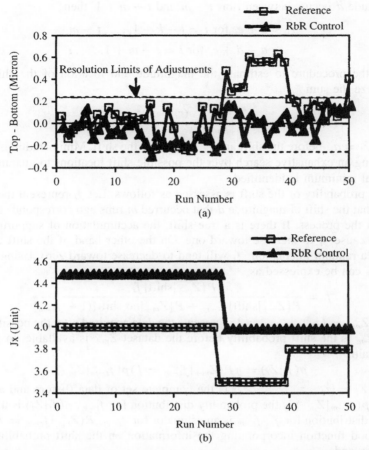

Figure 9.8. (a) $T - B$ response with and without RbR control; (b) corresponding adjustments in J_x parameter [2].

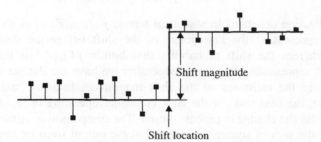

Figure 9.9. Illustration of least-squares procedure to estimate shift magnitude and location [2].

magnitude d occurs between runs $t - m$ and $t - m + 1$, then

$$z_i \approx a + e_i \text{ for } i = t - k + 1, \ldots, t - m \tag{9.10}$$
$$\approx a + d + e_i \text{ for } i = t - m + 1, \ldots, t$$

Thus, the procedure to estimate the magnitude and location of the shift is to minimize the sum

$$\sum_{i=t-k+1}^{t} (z_i - \hat{z}_t)^2 \tag{9.11}$$

where $\hat{z}_t = a_-$ before the shift and $\hat{z}_t = a_+$ after the shift. Minimization is carried out using an exhaustive search over the possible shift locations to guarantee that a global minimum is identified.

The probability of the shift is assessed as follows. Let f_t represent the probability that the shift of magnitude d that occurred m runs ago corresponds to a true shift in the process. If there is a true shift, the accumulation of supporting data should cause f_t to increase toward one. On the other hand, if the shift alarm is due to a random fluctuation, f_t will tend to decrease toward zero. Using Bayes' rule, f_t can be expressed as

$$f_t = \frac{P\{Z_{m+}|\text{shift}\} f_{t-m}}{P\{Z_{m+}|\text{shift}\} f_{t-m} + P\{Z_{m+}|\text{no shift}\}(1 - f_{t-m})} \tag{9.12}$$

where $Z_{m+} = \{z_{t-m+1}, \ldots, z_t\}$ represents the data acquired after the possible shift and f_{t-m} is the shift probability before the dataset Z_{m+} is available. In general, we obtain

$$p(f_t|Z_t) = h P(Z_{m+}|f_{t-m} = f) p(f_{t-m}|Z_{m-}) \tag{9.13}$$

where $Z_t = \{z_{1,\ldots,}z_{t-m}, \ldots, z_t\}$ is the complete set of data (before and after the shift), $p(f_{t-m}|Z_{m-})$ is the probability distribution for f_{t-m}, $p(f_t|Z_t)$ is the probability distribution for f_t, f_{t-m} is an estimator for f_{t-m}, $P(Z_{m+}|f_{t-m} = f)$ is the likelihood function incorporating the information on the shift probability from the data, and

$$h = \frac{1}{\displaystyle\int_0^1 P(Z_{m+}|f_{t-m} = f) p(f|Z_{m-}) df} \tag{9.14}$$

is a normalization constant. In statistical terms, $p(f_{t-m}|Z_{m-})$ is the "prior distribution" representing the knowledge of the shift before the dataset Z_{m+} is acquired, whereas the shift probability distribution $p(f_t|Z_t)$ is the "posterior distribution" representing that knowledge after we have the dataset Z_{m+}.

As soon as the estimates of the shift magnitude, location, and probability are available, the next task for the RbR controller operating in rapid mode is to compensate for the change in process output. The compensation criterion involves minimizing the sum of squared deviations of the output from its target. In other words, the quantity to be minimized is

$$E[(y_{t+1} - T)^2 Z_t] \tag{9.15}$$

where E represents the expectation. Assuming again that only the intercept term changes, the amount by which the intercept must be adjusted is $E[f_t]d = f_t d$.

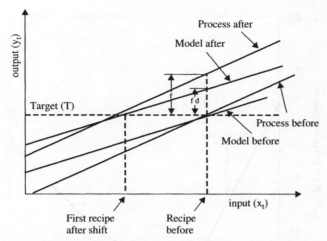

Figure 9.10. Illustration of RbR compensation in rapid mode [2].

This is illustrated in Figure 9.10. The adjustments required in the process input is then determined by solving for the new model input subject to the adjusted intercept.

To test rapid-mode operation of the RbR controller, control of silicon deposition in an Applied Materials 7800 barrel reactor was once again evaluated (see previous section). In one test, a disturbance in film thickness uniformity was induced by changing the operating point for its bellows metering valves. These valves control the overall flow characteristics in the reactor. The rapid-mode adjustments were able to recover control of the uniformity within three runs. This is illustrated in Figures 9.11 and 9.12.

9.1.2. Multivariate Techniques

9.1.2.1. Exponentially Weighted Moving-Average (EWMA) Gradual Model

The EWMA approach proposed by Sachs for RbR control can also be extended to the simultaneous control of multiple variables. If there are multiple inputs (i.e., if $\mathbf{b}$ and $\mathbf{x}_t$ in Eq. (9.5) are column vectors), then the form of the update relation for the intercept term changes only in the sense that the term bx_t is now interpreted as a the inner product $\mathbf{b}^T\mathbf{x}_t$. However, when b and x_t are scalars, Eq. (9.6) has a unique solution. This is no longer the case when these parameters are column vectors. In the latter case, assuming $\mathbf{b} \neq 0$, there are solutions for all points that satisfy $\mathbf{T} = \mathbf{a}_{t-1} + \mathbf{b}^T\mathbf{x}_t$. In this case, the recipe $\mathbf{x}_t$ is chosen as the point that minimizes the distance from the previous recipe $\mathbf{x}_{t-1}$. This point turns out to be

$$\mathbf{x}_t = \frac{\mathbf{T} - \mathbf{a}_{t-1}}{\mathbf{b}^T\mathbf{b}}\mathbf{b} + \left(I - \frac{\mathbf{b}\mathbf{b}^T}{\mathbf{b}^T\mathbf{b}}\right)\mathbf{x}_{t-1} \qquad (9.16)$$

9.1.2.2. Predictor–Corrector Control

Butler and Stefani proposed the use of in situ ellipsometry to drive a RbR controller, called a *predictor–corrector controller* (PCC), to alleviate the effect

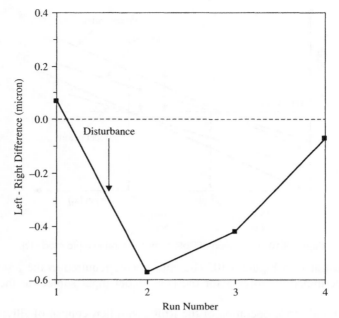

Figure 9.11. Recovery of $L - R$ in rapid mode [2].

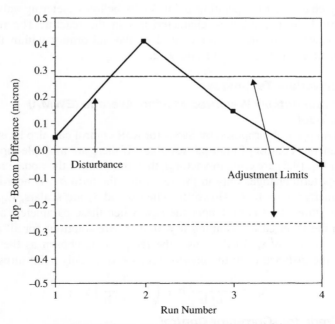

Figure 9.12. Recovery of $T - B$ in rapid mode [2].

of machine and process drift in reactive ion etching [3]. The process under investigation was the etching of polysilicon gates in a CMOS manufacturing line. This etching process determines the critical dimension, and thus the performance limits of the ICs produced. However, the process was known to drift because of aging of the reactor.

The response surface modeling technique was used to predict mean etch rate (MER) and uniformity from the ellipsometry data. The process conditions which served as inputs to these models were RF power, chamber pressure and total gas flow (HCl + HBr) rate. Model coefficients were obtained from a central composite experimental design that required 21 trials. The etch uniformity was estimated by deriving relationships between the etch rate at the center of each wafer (as measured by ellipsometry) and at each of 10 other specific sites on the wafer.

The predictive RSM models were employed by the PCC controller to generate optimal recipe settings to achieve etch rate and uniformity targets. The control system objectives were (1) target tracking without lag, (2) disturbance compensation, and (3) noise rejection. The key component of the PCC is the *double-exponential forecasting filter* (DEFF). A forecast filter such as this smooths current data (i.e., reduces noise) and provides a forecast. The DEFF consists of a filter to estimate the output and another filter to estimate its trend. In other words

$$\text{Current smoothed output} = (1 - \alpha)(\text{current actual output})$$

$$+ \alpha(\text{previous estimate}) \tag{9.17}$$

$$\text{Current smoothed trend} = (1 - \beta)(\text{trend estimate})$$

$$+ \beta(\text{previous trend}) \tag{9.18}$$

$$\text{Forecast} = (\text{current smoothed output})$$

$$+ (\text{current smoothed trend}) \tag{9.19}$$

where α and β are tuning constants. The output data to be filtered were the RSM model residuals (i.e., measurements minus predictions). The equations for the DEFF that correspond to Eqs. (9.17)–(9.19) are

$$\text{Fdelta}_t = (1 - \alpha)(\text{delta})_t + \alpha^* \text{Fdelta}_{t-1} \tag{9.20}$$

$$\text{PE}_t = (\text{delta})_t - \text{Fdelta}_{t-1} \tag{9.21}$$

$$\text{FPE}_t = (1 - \beta)\text{PE}_t + \beta^* \text{FPE}_{t-1} \tag{9.22}$$

$$\text{Prediction}_t = \hat{y}_t + \text{Fdelta}_t + \text{FPE}_{t-1} \tag{9.23}$$

where Fdelta_t is the filtered model error at time t, $(\text{delta})_t$ is the unfiltered model error at time t, PE_t is the unfiltered prediction error at time t (which serves as the trend estimate), FPE_t is the filtered prediction at time t, and $\hat{y}_t$ is the RSM model prediction at time t. Figure 9.13 is a block diagram of the PCC system.

The "controller" block in Figure 9.13 represents the commercial nonlinear optimization package *NPSOL* [4]. This package was used to solve the RSM equations "in reverse" to determine the optimal process recipe corresponding

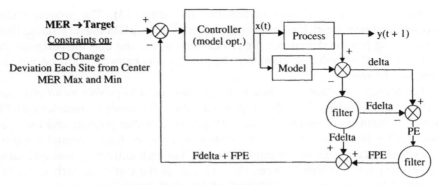

Figure 9.13. Block diagram of PCC system [3].

to the desired targets. Since multiple solutions are often possible, the controller chooses the solution closest to the current operating point. Set points for the last wafer are used if they produce predicted responses within one standard deviation of the target. If no solution is possible, the most recent recipe is repeated or the system quits so that the problem may be diagnosed.

Implementation of the PCC initially occurred in a 200-wafer demonstration experiment in which half of the wafers used a standard recipe and the other half used PCC-generated optimal recipes. During this demonstration, two equipment faults were simulated: (1) a miscalibrated power supply and (2) neglecting the prior wafer cleaning step. The controlled and uncontrolled measurement residuals for the process etch rate are shown in Figure 9.14. Overall, PCC resulted in a 36% decrease in the standard deviation from target for the mean etch rate, and similar results were achieved for uniformity. In addition, the natural variance of the process did not increase when PCC was used, indicating that continuous run-by-run control did not cause unnecessary control actions.

9.1.3. Practical Considerations

The methods discussed in the previous sections provide mathematically correct RbR control solutions. However, in practical industrial applications, there are several issues that arise that are not directly addressed by ideal theoretical RbR control algorithms. The additional constraints imposed by these issues must be considered before useful solutions can be found. Although the incorporation of strategies to address such constraints can complicate otherwise simple RbR control approaches, they provide a valuable complement to theoretical control solutions.

9.1.3.1. Input Bounds

The adjustments in process conditions suggested by RbR controller must account for limitations in the possible ranges of settings that a given input parameter might have. In other words, computed input conditions are constrained by actual equipment capabilities, and the control system must avoid recommending recipe

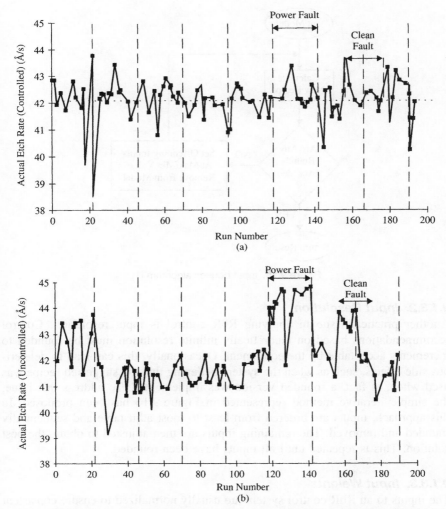

Figure 9.14. Actual etch rate (dotted line is the target): (a) controlled; (b) uncontrolled [3].

conditions that are nonphysical (such as a negative pressure) or beyond the ability of the equipment to reach. One way to address such input constraints is to simply determine optimal recipes without input bounds and then set all RbR control recommendations that exceed these bounds to the closest realizable setting. However, this simple approach can generally lead to less than optimal equipment settings.

Another approach to meet input constraints is to modify the RbR algorithm to use the iterative approach shown in Figure 9.15. In this scheme, after process variables have been modified with respect to their maximum ranges, they are removed from the system and the process is repeated. This method does not guarantee an optimal solution, but is computationally inexpensive and reduces the possibility of suboptimal results.

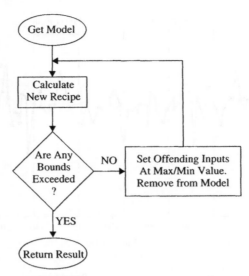

Figure 9.15. Input bounds algorithm [1].

9.1.3.2. Input Resolution

Another practical issue in applying RbR control is input resolution. Control recommendations based on theoretically infinite resolution must be rounded to increments acceptable by the equipment. Occasionally, this can lead to deleterious side effects, such as when the system believing that the suggested recipe was used when, in fact, a rounded version was implemented. To address this issue, the simple iterative method represented in Figure 9.16 has been proposed. In this approach, inputs are ordered from least to most adjustable and sequentially rounded and removed. The remaining inputs are then adjusted to obtain the best solution. This is repeated until all inputs have been rounded.

9.1.3.3. Input Weights

The inputs to an RbR control system are usually normalized to ensure consistent operation on a common scale. However, on some occasions, some inputs may be of greater importance to the user than others. In such cases, weights may be applied to the more critical inputs to provide another level of adjustability. In this way, more heavily weighted input variables can be modified with greater magnitude relative to lightly weighted inputs.

One weighting scheme that achieves the desired objectives adjust the normalized input variables so that the least-squared distance between the new control advice (x_t) and the previous input value (x_{t-1}) is modified by the input weight. In underdetermined systems in which the model provides a set of solutions where all output requirements are met, the new input recipe is identified using the added constraint that it is as close to the previous recipe as possible. This constraint can be biased by the relative weighting of the inputs. Heavily weighted inputs are forced to be the least adjustable as a result of their larger effect on the error

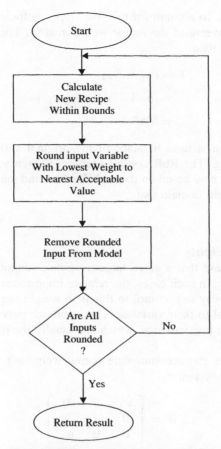

Figure 9.16. Input resolution algorithm [1].

calculation for the recipe (i.e., the difference between the target response and the value predicted by the model using the recommended recipe).

The matrix

$$V = \begin{bmatrix} v_1 & 0 & 0 \\ 0 & \cdots & 0 \\ 0 & 0 & v_n \end{bmatrix}$$ (9.24)

is used to apply the input weighting, where $v_1, \ldots, v_n$ are the weights for the n inputs. Note that input weighting has no effect on overdetermined problems or those with exact solutions. In those situations, the inputs do not affect the calculation of the error of the final solution, so their relative magnitude is irrelevant.

In order to achieve a correct solution, the weight must be applied to both the recipe adjustment and the first-order (slope) term in the process model. This is because the application of V to x changes the least-squared error generated by x when determining the closest solution. The side effect of input weighting is that the new output generated by these inputs is inconsistent with the original problem formulation. To account for this, the slope coefficient [b in Eq. (9.5)] is weighted with the inverse of the recipe weight matrix. The resulting system of equations to solve is then

$$T = bx_t + a_{t-1}$$
$$= (b \cdot V^{-1})(Vx_t) + a_{t-1} \qquad (9.25)$$
$$= b^*x^* + a_{t-1}$$

This new formulation is used in place of the original variables to incorporate the desired weighting. The RbR control problem is then solved as before. The solution, however, is now based on the scaled values and must be converted back to the original problem domain using

$$x = V^{-1}x^* \qquad (9.26)$$

9.1.3.4. Output Weights

It is also often the case that a given target response cannot be reached because of system constraints. In such cases, the relative importance of each output must be ascertained. One way to accomplish this is to weight the outputs in a manner inversely proportional to their variance. This approach puts greater emphasis on output variables with low variance, which are usually the ones that can be most accurately controlled.

An RbR controller can accommodate output weighting by applying the following matrix to the system

$$W = \begin{bmatrix} w_1 & 0 & 0 \\ 0 & \cdots & 0 \\ 0 & 0 & w_m \end{bmatrix} \qquad (9.27)$$

where $w_1, \ldots, w_m$ are the weights for the m inputs. The resulting system of equations is then

$$WT = Wbx_t + Wa_{t-1}$$
$$W(T - a_{t-1}) = Wbx_t$$
$$(Wb)^T W(T - a_{t-1}) = (Wb)^T Wbx_t \qquad (9.28)$$
$$(b^T W^T Wb)^{-1} b^T W^T W(T - a_{t-1}) = x_t$$

This scheme works by biasing the magnitude of selected outputs such that when a least-squares solution is calculated, those outputs with higher weights contribute a greater penalty to the solution if they are off target. In this way, outputs with higher weights are set closer to their targets than are those with lower weights. The application of output weights to an exact or underdetermined system has no effect.

Other bias terms related to output weighting are the model update weights. These coefficients (such as λ for EWMA control, or α and β for PCC control) also play a key role in determining the aggressiveness of the RbR controller. These parameters can also be used to minimize the impact of noisy outputs on the model update. The result is a system that can rapidly adapt to changing process conditions while resisting responding to process noise.

9.2. MULTIVARIATE CONTROL WITH COMPLETE MODEL ADAPTATION

Among the many challenges that impede the effectiveness of various process controllers, two stand out. The first has to do with the true nature of a process drift over time. The controllers discussed thus far make the general assumption that a process can be described by a linear model that has a gain factor (linking the process input to the process output), and a constant term. Further, they assume that as the process changes over time, it is only the constant term that needs to be adjusted. This works well in many cases, but in reality, the *gain* relationship between the input parameter and the process output (which is captured by the slope coefficient in the linear model) also changes. This presents a problem when such a model is used for feedback control, since the erroneous gain value will diminish the effectiveness of the controller. The problem might not be so obvious to the user, since the controller can compensate for the error in the gain value by adjusting the constant term accordingly. Using the wrong value of the gain term, however, has much more serious implications when the model is used for feedforward control. In this case, an erroneous prediction results, and the controller fails. In summary, while many feedback operations are robust to errors in the gain parameter of the model, feedforward applications are not robust to such errors, and depend on accurate process models for their operation.

The second challenge has to do with the necessary economy of control actions. This means that a run-to-run controller should take action only when needed. Further, the nature of the needed action may depend on the situation. For example, at one instant, the most appropriate action might be the gradual reestimation of the constant term of the model, while at another instant, it might be necessary to reevaluate the gain term of the model. In other cases the most appropriate action might be the termination of any automatic adaptation and the involvement of a human operator that might be better suited to resolve an emerging problem.

In this section, we present a run-to-run control approach that addresses these issues. This approach does so by including a more complete set of possible responses—such as various modes of model adaptation, feedback and feedforward calculation, and malfunction declarations. Each of these responses is gated by the outcome of a series of statistical tests. This approach is particularly suitable for controlling a sequence of process steps, where feedback and feedforward control actions must be coordinated. That type of multistep (supervisory) control operation is described in Section 9.3.

The demonstration vehicle for this approach is the photolithography process sequence. Here, the role of the controller is to provide an intelligent system for generating initial process recipes, correcting process drifts, and detecting equipment or process malfunctions on a run-by-run basis. Subsequently, on the detection of a process drift or malfunction, a diagnostic system (see Chapter 10) linked to the controller offers an educated guess of the cause of the problem.

We present two process control approaches for multiprocess sequences. The first keeps tight control over each machine in the sequence. When the outputs of a machine drift, the controller generates a new recipe to bring them back on target. The second approach, on the other hand, keeps the final target of the process sequence fixed, while intermediate targets are subject to dynamic adjustments. Before describing the details of the control actions, we first outline the conditions and statistical tests that trigger them.

9.2.1. Detection of Process Disturbances via Model-Based SPC

The default state of the complete-model adaptation controller is dormant, until a "disturbance" is detected. There are two types of disturbance. The first manifests itself through sudden, statistically significant changes in the process output. This indicates the presence of a malfunction that needs to be addressed by a human operator. This type of disturbance triggers a *malfunction alarm*. The second type of disturbance manifests itself as a systematic process drift that can be corrected by the control system. This type of disturbance triggers a *control alarm*, which in turn triggers the control system.

9.2.1.1. Malfunction Alarms

Malfunction alarms identify conditions that require operator attention. These are cases where the variation of a monitored parameter increases, or when sudden changes that are not consistent enough to be compensated by recipe adjustments are encountered. A malfunction alarm is also generated if the change cannot be compensated unless one (or more) of the controlling parameters moves beyond its acceptable range.

These conditions can be identified with the application of a special SPC scheme that can accommodate multiple parameters. This scheme must be able to ignore intentional changes in equipment settings such as those that might occur due to control actions. Such a scheme has been developed using an extension of the regression chart (see Section 8.1.5) [5] and Hotelling's T^2 statistic (see Section 6.5) [6].

Using this scheme, malfunction alarms are generated in two stages. First, the controller uses response surface models to predict new measurements. Then, it plots the difference between the readings and the model predictions. When the process is under statistical control, this difference is a random number with a known mean and variance. The method is described for univariate regression models by Mandel [5], and it has been generalized for multivariable response surface models by Lee [7]. A brief description of this method follows.

Let $\bar{\mathbf{y}}$ be the $p \times 1$ vector corresponding to p equipment outputs, where each element is the average reading of n samples. Let $\hat{\mathbf{y}}$ be the $p \times 1$ vector predicted by the equipment models. If the process is under control, the residual vector $(\bar{\mathbf{y}} - \hat{\mathbf{y}})$ has a multivariate normal distribution with mean zero and variance–covariance matrix S. Once estimates of these parameters have been computed, the multiple responses are merged together using the T^2 statistic

$$T^2 = n(\bar{\mathbf{y}} - \hat{\mathbf{y}})^T \mathbf{S}^{-1}(\bar{\mathbf{y}} - \hat{\mathbf{y}}) \qquad (9.29)$$

where n is the sample size and S is the estimated covariance matrix of a process assumed to be in statistical control. Even when a process is in statistical control, it can yield noisy *estimates* of S when the sample size is small. Therefore, it is not advisable to use these run-by-run estimates, but rather depend on the original estimate of S that was obtained when the models of the process where created, and presumably, when the process was in control. This estimate is calculated using the methodology described in Section 6.5. Once the T^2 statistic is calculated, it is plotted on a single-sided control chart whose upper control limit (UCL) can be formally set at the desired probability of erroneously stopping a good process, by using the F distribution, or

$$\text{UCL} = \frac{p \cdot (N-1) \cdot F_{p,N-p}}{N-p} \qquad (9.30)$$

where N is the sample size used in estimating S. Note that the sample size n used to calculate S is *different* from the sample size N used to determine the UCL. When the UCL is exceeded, the automated control system stops, and a human operator investigates the malfunction in the same way that a traditional SPC out-of-control condition is investigated.

9.2.1.2. Alarms for Feedback Control

Control alarms identify process drifts and then trigger the feedback control system. These drifts and disturbances are detected using a cusum scheme (see Section 6.4.5) that is very efficient at identifying small, consistent changes, while ignoring outliers that are not useful for feedback corrections. The controller will compensate this type of disturbance by appropriate model adaptation, followed by a recipe change.

The alarms are generated by the multivariate cusum scheme described by Crosier [8]. Crosier's scheme forms a cusum vector directly from the residuals between the experimental data $(\mathbf{y}_j)$ and their respective model predictions, after shrinking them by a factor $[1 - (k/C_j)]$. In other words

$$\mathbf{s}_j = 0 \quad \text{if} \quad C_j < k \qquad (9.31)$$

$$\mathbf{s}_j = (\mathbf{s}_{j-1} + \mathbf{y}_j - \hat{\mathbf{y}}) \left(1 - \frac{k}{C_j}\right) \quad \text{if} \quad C_j \geq k \qquad (9.32)$$

where C_j is the variance-normalized length of the residual cusum vector $(\mathbf{s}_{j-1} + \mathbf{y}_j - \hat{\mathbf{y}})$, that is

$$C_j = \sqrt{[(\mathbf{s}_{j-1} + \mathbf{y}_j - \hat{\mathbf{y}})^T \cdot \mathbf{S}^{-1} \cdot (\mathbf{s}_{j-1} + \mathbf{y}_j - \hat{\mathbf{y}})]} \qquad (9.33)$$

where S is the same estimate of the covariance matrix used for generating malfunction alarms, and is obtained from the designed experiments used to create the process models (when the process is in control). Typically, we want a process to return to its original target. Sometimes, this is not possible because the multiple outputs are not completely independent of each other. A corollary is that measurements should not be compared against fixed targets, which are sometimes unattainable, since this would generate control alarms too often. Comparison of the experimental data to the model predictions, on the other hand, properly generates an alarm only if the updated models do not represent the experimental data well. So, the control alarm is generated only when the model represents the data inadequately. This scheme yields an alarm when the variance-normalized length of the residual cusum vector s_n is greater than a constant η:

$$Y_j = \sqrt{[s_j^T \cdot S^{-1} \cdot s_j]} > \eta \tag{9.34}$$

The sensitivity of the alarm depends on the number of output parameters p and the constants k and η, which can be adjusted for the desired probability of stopping erroneously a good process. Equivalently, we can adjust the average runlength (ARL) between false alarms when the process is in control, also called *on-target ARL*.

9.2.2. Full Model Adaptation

The goal of feedback control is to ensure that the distribution of the process outputs stays centered on target. Triggered by control alarms that detect output drifts, the feedback controller first updates the equipment models of the machine and then finds a new recipe to bring the machine's outputs back on target. If the machine has multiple outputs that cannot be simultaneously brought back on target by a new recipe, because of correlation among outputs, a *compromise* recipe, which brings all the outputs as close as possible back on target, is generated.

The model update algorithm uses stepwise regression, which depends on matrix computations. The $k \times q$ *input setting matrix* X contains the q input settings of the k process runs, which are fed into a $k \times t$ *model term* matrix T, which stores the input settings as model terms. The number of terms inside the model is t, which can also be interpreted as the number of coefficients in the model.

As an example, assume that two wafers are processed by a photolithography wafer track. The first wafer undergoes a spin speed (SPS) of 4600 rpm, a baking time (BTI) of 60 s, and a baking temperature (BTE) of 90°C, or (SPS, BTI, BTE) = $(4600, 60, 90)$. The second wafer is processed by the recipe (SPS, BTI, BTE) = $(4800, 65, 90)$. It is known that the resist thickness is approximately proportional to time and temperature, and inversely proportional to the square root of the spin speed [9]. Therefore, the resist thickness model has the following terms: $1/\sqrt{SPS}$, BTI, and BTE. That information is stored as

$$X = \begin{bmatrix} 4600 & 60 & 90 \\ 4800 & 65 & 90 \end{bmatrix} \quad \text{and} \quad T = \begin{bmatrix} 1/\sqrt{4600} & 60 & 90 \\ 1/\sqrt{4800} & 65 & 90 \end{bmatrix} \tag{9.35}$$

Note that X and T do not necessarily have the same number of columns. If the resist thickness model also contained the SPS term, T would have four columns: SPS, $1/\sqrt{\text{SPS}}$, BTI, and BTE, or

$$T = \begin{bmatrix} 4600 & 1/\sqrt{4600} & 60 & 90 \\ 4800 & 1/\sqrt{4800} & 65 & 90 \end{bmatrix} \tag{9.36}$$

Next, the algorithm applies two transformations to T to prevent it from being ill-conditioned. First, it centers the resulting matrix, by subtracting the average of each column, and then divides it by a *range matrix* D, so that the variances of each term are of comparable magnitudes. D is a $t \times t$ diagonal matrix that contains the experimental range of each model term. This results in a *standardized matrix* Y, which is composed of unitless numbers with comparable magnitudes, or

$$Y = (T - T_{\text{av}}) \cdot D^{-1} \tag{9.37}$$

The second transformation that the algorithm applies on matrix Y is a principal-component transformation (see Section 8.4) to ensure that each column of Y is mutually orthogonal. This is necessary in order to apply stepwise regression. Since Y has been standardized, principal-component transformation on the covariance matrix is numerically stable.

The first step of the model update algorithm consists of entering all the machine settings into X. Since the performance of the machine changes with time, the performance obtained from older settings should not be weighted as much as that obtained from newer settings. Therefore, a *forgetting factor* w_{kk} is applied to the input settings, emphasizing the more recent ones. (The parameter k corresponds to the number of sets of input settings). Thus

$$X' = W \cdot X \tag{9.38}$$

where W is a diagonal matrix containing the forgetting factor w_{kk} of each set of input settings. In a typical implementation, the number of sets of input settings is also limited to a specific number, called the *window size*, based on how often the machine performance drifts with time. A good way to choose the weighting factors depends on the autocorrelation structure of the measurements. Lacking such information, one can use an empirically chosen exponential weighting factor. Depending on the application, the effective memory of such a weighting function should be on the order of 10–20 runs.

Next, the input setting matrix is transformed into T, which is then transformed into a unitless matrix Y using Eq. (9.37). A principal-component transformation on Y ensures that each column is orthogonal. This sets the stage for the stepwise regression that follows. Next, the difference between the measurements and the current model predictions, defined as a $k \times p$ *output discrepancy matrix* Δz, is calculated. As before, p is the number of output variables, and k is the number of sets of input settings, i.e., the number of wafers in the window. The output discrepancy matrix is computed as follows for each output variable i, $i = 1, \ldots, p$

$$\Delta z_i = z_{i,\text{meas}} - z_{i,\text{model}} = z_{i,\text{meas}} - (Y_{pc}^T \cdot \gamma + c_0) \qquad (9.39)$$

where $\gamma = \boldsymbol{B}^T \cdot \boldsymbol{D} \cdot c$ represents the vector of term coefficients of the model, transformed into the principal-component space, c is a $t \times 1$ vector containing all the model coefficients, c_0 is its constant term, and $\boldsymbol{D}$ is the range matrix.

Stepwise regression is performed considering each PC separately, in order to obtain a *vector of correction term coefficients* $\Delta\gamma$. The statistical significance based on the t distribution of each correction coefficient $\Delta\gamma_j (j = 1, \ldots, t)$ is calculated. If the significance is greater than a certain threshold, the correction coefficient is updated to $\Delta\gamma_j$; otherwise, it is set to zero. Then, Y_{pc} is multiplied by the updated set of new coefficients $\Delta\gamma$ and subtracted from the output discrepancy vector $\Delta\mathbf{z}$. If the resulting constant term Δc_0 is significant, it is also updated. Finally, the modified correction coefficients $\Delta\gamma$ are transformed back to their original space (resulting in a set of correction coefficients Δc) and added to the current model coefficients c, to result in a newly updated set of coefficients c_{updated}, or

$$c_{\text{updated}} = c + \Delta c = c + D^{-1} \cdot B \cdot \Delta\gamma \qquad (9.40)$$

$$c_{0,\text{updated}} = c_0 + \Delta c_0 \qquad (9.41)$$

This concludes the model update procedure. The next step for the feedback controller is to find a new recipe that will bring the machine's outputs back on target.

9.2.3. Automated Recipe Generation

Since empirical equipment models are relatively simple, routine nonlinear transformation can be used for recipe generation, eliminating the need for more complex geometric centering techniques. Given a set of input settings x, a machine output vector $f(x)$, and the desired output from the machine, the recipe generation problem is mathematically formulated as follows. First, we solve for x, such that

$$\min \sum_{i=1}^{p} W_i \cdot (f_i(x) - \hat{z}_i)^2 \qquad (9.42)$$

subject to the constraint

$$e_m \leq x \leq e_M \qquad (9.43)$$

where e_m is a vector of minimum input settings, e_M is the vector of maximum input settings, W_i is the weight of the ith output variable and p is the number of output variables. This is a typical optimization problem, which can be solved in many different ways. For this application, the iterative Gauss–Seidel algorithm [10, 11] was chosen.

The weights in Eq. (9.42) are needed because some output variables must be better controlled than some others. These weights could be derived from the specification limits of the output variables. Specifically

$$\Delta \hat{z}' = W_z^{-1} \Delta \hat{z} = W_z^{-1}[f(x_i) - \hat{z}] \tag{9.44}$$

where $W_z = 2 \times \min(\text{USL} - \hat{z}, \hat{z} - \text{LSL})$. However, a better weighting scheme is based on the sensitivity of the final process output relative to the intermediate output variables. For example, if the critical dimension, which is the final lithography process output, is as sensitive to a 3% change in the amount of photoactive compound (PAC) remaining in the photoresist as it is to a 100-Å change in resist thickness (which are both intermediate process outputs), then $W_z = \begin{bmatrix} 0.03 \\ 100 \end{bmatrix}$. More formally, the output weights are chosen using

$$W_z = \begin{bmatrix} 1 \Big/ \dfrac{\partial z_{\text{final}}}{\partial z_1} \\ \dots \\ 1 \Big/ \dfrac{\partial z_{\text{final}}}{\partial z_p} \end{bmatrix} \tag{9.45}$$

When solving for a new recipe, one must also apply weights for input variables (as discussed in Section 9.1.3.3), since some input settings have a wider range of operation than others do. Weights are also used to favor changing the input settings that would cause fewer side effects. For example, changing the spin speed is often preferable to changing the bake temperature. The entire feedback algorithm with full model adaptation is depicted in Figure 9.17.

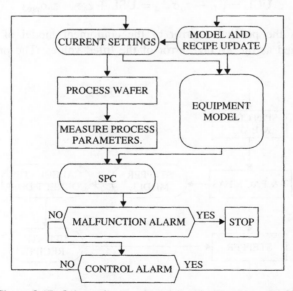

Figure 9.17. Schematic representation of feedback procedure.

9.2.4. Feedforward Control

The primary task of the feedforward control mechanism is to adjust any down-stream process steps in order to compensate for the variability of the present step. A feedforward controller complements a feedback controller, which centers the process on the target by reducing the process variability. Before sending the wafer on to the next process step, the outputs of the current step are analyzed to see if they are likely to produce a wafer within specifications after the next step, assuming normal settings. If the analysis is positive, no feedforward control is performed. However, if the analysis shows that the wafer is unlikely to meet specifications, a feedforward alarm is triggered and activates a feedforward controller, which then finds a corrective recipe for the next machine, using the same recipe generation scheme described earlier and depicted in Figure 9.18.

In highly controllable process steps, a feedforward control system can even compensate for inherent variability, thereby increasing the overall process capability. Currently, however, feedforward control mechanisms are not well accepted in the semiconductor industry because of the high stakes involved. A corrective action that worsens a process cannot be tolerated. Therefore, feedforward control should be activated only when the problem is clearly confirmed. Like feedback control, this mechanism is also activated by a formal statistical test.

The feedforward control alarm is a variant of the acceptance chart (see Section 6.4.4). Given specification limits for a fraction of nonconforming wafers of at most δ and a specified type I error of α, the upper and lower control limits are

$$\text{LCL} = \mu_L - z_\alpha \sigma_{\text{pred}} = \text{LSL} + z_\delta \sigma - z_\alpha \sigma_{\text{pred}} \tag{9.46}$$

$$\text{UCL} = \mu_U - z_\alpha \sigma_{\text{pred}} = \text{USL} + z_\delta \sigma - z_\alpha \sigma_{\text{pred}} \tag{9.47}$$

where σ_{pred} is the prediction error of the equipment model of the machine, which is defined as the average error of the fitted values. The prediction error

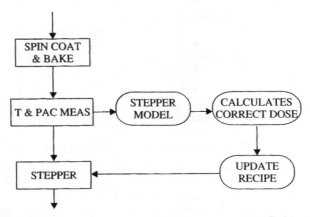

Figure 9.18. Example of the feedforward control procedure applied to a stepper.

is calculated from the standard error between the modeled data y_i and its fitted values, which is

$$\sigma_{model} = \sqrt{\frac{1}{m-1} \sum_{i=1}^{m} (y_i - \hat{y}_i)^2} \qquad (9.48)$$

$$\sigma_{pred} = \sqrt{\frac{t}{m}} \sigma_{model} \qquad (9.49)$$

where $i = 1, \ldots, m$, m is the number of wafers used in building the equipment model, and t is the number of parameters in the model.

When the predicted output falls between the lower and upper control limits, no feedforward action is taken. On the other hand, when a prediction falls outside the control limits, an alarm signals the feedforward control system to generate new recipe(s) for the next machine(s) in the sequence in order to prevent the final process output from drifting outside the specification limits. Although the recipe generator should always try to correct the error using only the next stage, its success is not guaranteed and may require considering several subsequent steps. If the situation cannot be corrected by any means, the feedforward controller sends the wafer to be reworked.

The combination of the full-model adaptation feedback and feedforward approach leads to a robust local control system that is capable of reducing process variability of a process step, and centering the process mean on target by applying SPC to accurate equipment models.

9.3. SUPERVISORY CONTROL

As discussed in the introduction to this chapter, control of semiconductor processes can be examined at several levels (refer to Figure 9.2). Supervisory control is the highest level of the control hierarchy. At this level, the progression of a wafer is tracked from unit process to unit process, and adjustments can be made to subsequent steps to account for variation in preceding steps. Both feedback and feedforward adjustments are made in a supervisory control system.

9.3.1. Supervisory Control Using Complete Model Adaptation

Although the local controllers have been shown to significantly improve the overall capability of a process sequence, they have one major caveat. When intermediate machines with multiple outputs drift, it is not always possible to bring the process back to its original point because the outputs are correlated. This will cause an inherent deviation in the output of the next machine downstream, and ultimately, in the final outputs of the process. The source of this problem is the inflexible specifications of each machine. In actuality, only the final specification needs to be kept on target. Those of the intermediate machines are flexible and should be changed, if they prevent downstream processes from keeping the final parameter on target.

The solution is to link the local controllers and integrate them into one global controller, which fixes the specification of only the last machine of the process sequence, and sets optimal specifications for the other machines upstream in the sequence, so that control of the final parameter of interest is optimized. This specification propagation concept leads to a significant improvement of the overall capability of the process sequence. Furthermore, it also results in a more controllable process, which reacts effectively to specification changes or synthesizes a solution to a new process faster.

The global supervisory control algorithm shares the same control algorithms as a local controller, but is improved by specification propagation. Consider two machines linked together in a sequence. The first part of the specification propagation algorithm determines the region of acceptable input settings of the downstream machine (and therefore, the region of acceptable outputs of the upstream machine) through Monte Carlo simulation of the downstream process. The acceptable input setting region is defined as a region of settings that would keep the process output within specifications. Mathematically, each input setting is tagged with a cost, which quantifies how close the resulting outputs are to their targets. If the process outputs are independent of each other, the total cost is

$$\text{Cost} = \sum_{i=1}^{p} k_i (y_i - y_{i,\text{target}})^2 \tag{9.50}$$

where y_i is process output i and p is the number of outputs. If the process outputs are dependent of each other, the total cost associated with each input setting is

$$\text{Cost} = \sum_{j=1}^{p} \sum_{i=1}^{p} k_{ij} (y_i - y_{i,\text{target}})(y_j - y_{j,\text{target}}) \tag{9.51}$$

The scaling coefficients k_i are chosen so that the cost equals one when the process output equals its specification limit, with the other process outputs being on target (see Figure 9.19). Typically, though, process outputs are not independent, and a principal-component transformation is applied to the raw process outputs to obtain independent output variables. The coefficients k_{ij} are then determined in a similar fashion as k_i, except that a coordinate transformation is involved. Once done, the new specifications of the upstream process are determined from the geometric center of the acceptable input settings region of the downstream process. This procedure is repeated for upstream processes, updating along the specifications of each machine, so that the final process outputs are on target. After the specifications have been set, then the cost function is used during recipe generation to center the process into the desirable region of operation.

The stability of this scheme is ensured by a multitude of mechanisms. In addition to the statistically driven alarms (whose sensitivity can be set to eliminate control oscillations due to noise), stability is also guaranteed by implementing hard limits for the specifications of each measurable process parameter used by the supervisory controller.

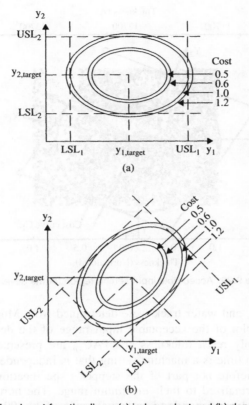

Figure 9.19. Equal cost function lines: (a) independent and (b) dependent outputs.

9.3.1.1. Acceptable Input Ranges of Photolithographic Machines

This supervisory controller has been tested on three photolithography steps: the wafer track, the stepper, and the developing station. These three steps process the wafer sequentially. The wafer track spin-coats the wafer with photoresist and bakes it. The stepper exposes the wafer through a patterned mask. The developer develops the photoresist pattern.

The final output of the process sequence is the linewidth of the photoresist pattern or critical dimension (CD). The inputs of the wafer track model are the input settings of the machine, which include spin speed, spin time, baking time, and baking temperature. The outputs are resist thickness and photoactive compound (PAC) concentration. The inputs to the stepper model are dose, PAC, and resist thickness, while its output is PAC concentration. The postexposure PAC concentration is differentiated from its preexposure value and denoted as PAC_{xp}. Finally, the inputs of the developer model are resist thickness, PAC, and develop time, and its output is the CD.

Given a CD target of 1.72 ± 0.06 μm (the average CD $\pm 1\sigma$ process noise when the process is in control), the acceptable range of input settings for the

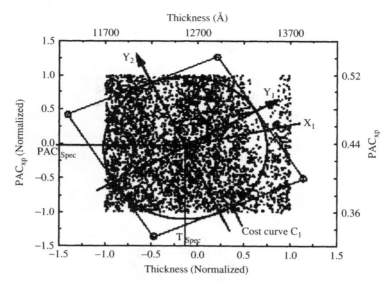

Figure 9.20. Acceptable input setting region for the developer.

developer, stepper, and wafer track were determined from Monte Carlo simulations. The scatterplot of the acceptable input range of the developer is shown in Figure 9.20. Only resist thickness and PAC_{xp} are presented in the scatterplot, since develop time is a machine setting that is independent from upstream processes and therefore not part of the stepper's specifications. Both parameters have been normalized to their maximum range. The normalized thickness range from -1 to 1 corresponds to an actual range of $11,700–13,700$ Å, while the normalized PAC_{xp} range from -1 to 1 corresponds to an actual range of $0.36–0.52$. Principal-component analysis revealed that PAC_{xp} and resist thickness were slightly correlated. In order to use this region of acceptable developer inputs to find the region of acceptable stepper inputs, the thickness and PAC_{xp} must be transformed into their principal components, Y_1 and Y_2, so that they are independent. The region is not necessarily convex. However, when the developer's input settings are allowed to change, an acceptable convex input settings region can be approximated by deriving the PC ellipse for the data that meets the specifications.

Next, the region of acceptable input settings for the stepper is determined. The specification for the stepper is taken from the coordinates of the centroid X_1 of the previously determined region (Figure 9.20), and the cost tagged to each input setting is calculated using the principal components Y_1 and Y_2. For example, cost curve C_1, which represents the maximum acceptable stepper outputs (i.e., where the cost equals 1.0), is shown in Figure 9.20. Using that cost, the region of acceptable input setting of the stepper is determined and shown in Figure 9.21. Again, only PAC and thickness are shown, since the third input of the stepper model is dose, which is a setting not included among the wafer track outputs. This acceptable input setting region of the stepper assumes that the dose can be set to any value within the range of the machine. As before, both PAC and resist

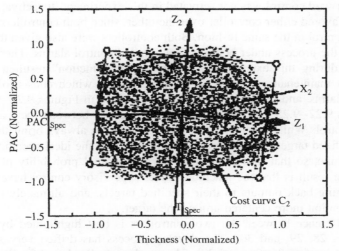

Figure 9.21. Acceptable input setting region for the stepper.

thickness are normalized to their maximum range. In order to use this region to quantify the cost of each input setting combination of the wafer track, PAC and resist thickness are transformed into independent principal components Z_1 and Z_2. The specifications for the wafer track are given by the coordinates of the centroid of the acceptable input settings region of the stepper, X_2, and the cost tagged to each input setting of the wafer track is derived using the principal components Z_1 and Z_2. For example, the cost C_2 that represents the maximum acceptable wafer track outputs is also shown in Figure 9.21.

Finally, the region of acceptable input settings of the wafer track is given by the range of input settings of the machine itself. Since the wafer track is the first machine of the process sequence, there is no process upstream whose specification needs to be determined from the region of acceptable inputs of the wafer track. The supervisory controller is in many ways superior to a simple sequence of fixed run-by-run controllers, since it has the ability to manipulate collectively several process steps in a synergistic fashion.

9.3.1.2. Experimental Examples

Both local and supervisory controllers have been implemented on photolithography equipment at the University of California at Berkeley [12], where an experiment was run to test the capabilities of both controllers and compare them to an uncontrolled process. The experiment consisted of processing 4-in. p-type silicon wafers coated with 1000 Å of oxide through the photolithography sequence. Control was applied on a lot-by-lot basis instead of on a run-by-run basis, with each lot consisting of three wafers. Each wafer is sampled 4 times, and the average reading is recorded. Three groups of 10 lots were processed in an alternating fashion during the experiment. Every 2 days, three lots of wafers were processed: an uncontrolled baseline lot, a lot subject to local controllers, and a lot subject to global supervisory control.

Feedforward control was not activated in this experiment. Its activation would not have favored either controller over the other, since both controllers use feed-forward control in the same fashion. Both controllers were also given the latitude to correct the process under both malfunction and control alarms. There were no instances during this experiment in which a "malfunction" resulted in actual equipment maintenance. Details of the experiments, which consist of machine outputs, alarms, and recipe changes, are summarized in Figures 9.22–9.25.

Figures 9.22–9.25 explicitly show the differences between local and super-visory control. While the local controller attempts to always bring the outputs back to a fixed target, the supervisory controller finds the ideal specifications for each machine, so that the final output has the optimal probability of being on target. The result is that the machines under supervisory control have an easier time bringing back outputs to their specified targets, and ultimately result in a CD distribution more centered around the target.

The difference between the two controllers is best highlighted by the 10th lot (wafers 28, 29, and 30). The develop process has drifted very sharply in the 9th lot (wafers 25, 26, and 27). The local controller has difficulty bringing the CD back on target at the 10th lot, because it only tried to correct for the drift through a new develop time, whereas the supervisory controller involved a change in exposure dose in addition to a change in develop time, and therefore was able to bring the CD much closer to target. This difference in corrective action was due to the fact that although a malfunction alarm was triggered on the stepper during the 10th lot under local control, the operator chose not to take any corrective action because the measurements were still within specifications. When the bad lot was processed by the developer, the local controller could use only the feedback control mechanism of the developer to compensate for the process shift, whereas the supervisory controller used the control mechanisms of all the machines to compensate for the process shift by changing the targets of the previous machines. The final comparison between both control mechanisms is presented in Figure 9.26. Here the CD distributions of wafers processed under both controllers are compared. Although the supervisory controller has a clear advantage over the local controller, the latter is still a significant improvement over an uncontrolled process.

An additional benefit of the supervisory controller is that it can be used to synthesize new process recipes. Because of the broadly applicable equipment models, the supervisory controller can start with the final specifications and syn-thesize not just optimal recipes for all the modeled process steps but also optimum intermediate specifications as well.

9.3.2. Intelligent Supervisory Control

Chapter 8 (in Section 8.5) introduced the concept of intelligent modeling techniques such as neural networks. These techniques can also be applied to supervisory control systems. As an example of such a system, Kim has developed a model-based supervisory control algorithm based on computational intelligence techniques and applied this approach to reduce undesirable behavior

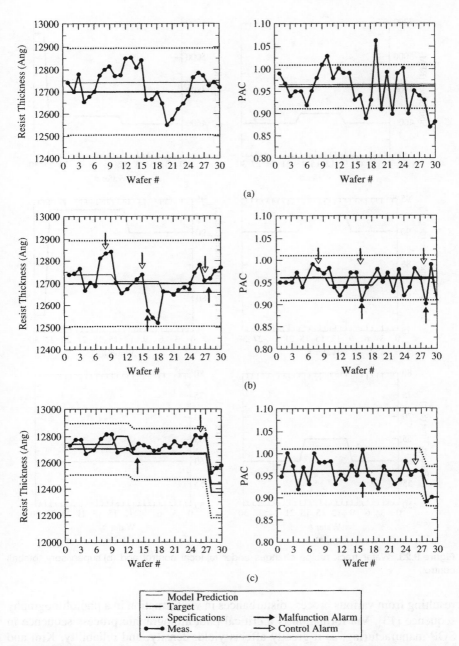

Figure 9.22. Wafer track outputs under (a) no control (baseline process), (b) local control, and (c) supervisory (global) control.

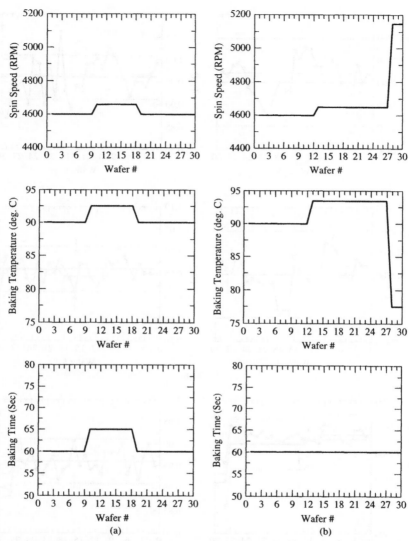

Figure 9.23. Wafer track recipe changes under (a) local control and (b) supervisory (global) control.

resulting from various process disturbances in via formation in a photolithography sequence [13]. Via formation is a critical photolithographic process sequence in SOP manufacturing, as it greatly affects yield, density, and reliability. Kim and May [14] presented a modeling approach for via formation in dielectric layers composed of photosensitive benzocyclobutene (BCB) based on the mapping capabilities of neural networks. Photosensitive BCB is a negative imaging material (i.e., unexposed areas are removed during development), and it is sensitive to 365 nm radiation (I-line and/or broadband exposure). The basic unit

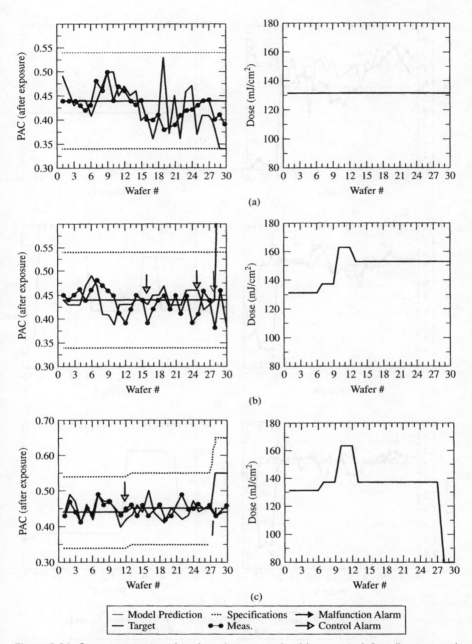

Figure 9.24. Stepper output and recipe changes under (a) no control (baseline process), (b) local control, and (c) supervisory (global) control.

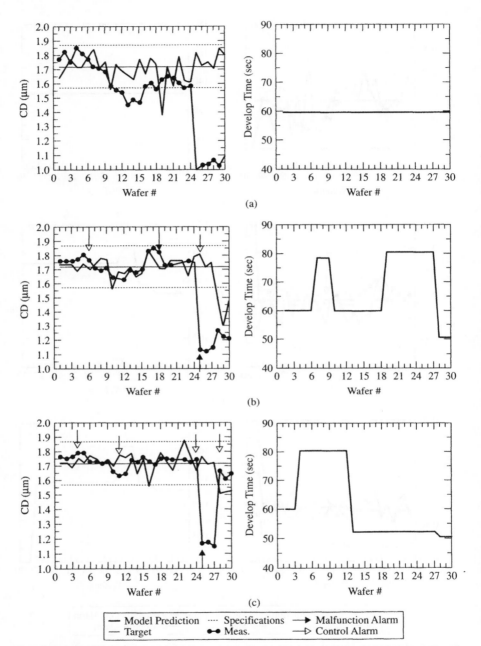

Figure 9.25. Developer output and recipe changes under (a) no control (baseline process), (b) local control, and (c) global control.

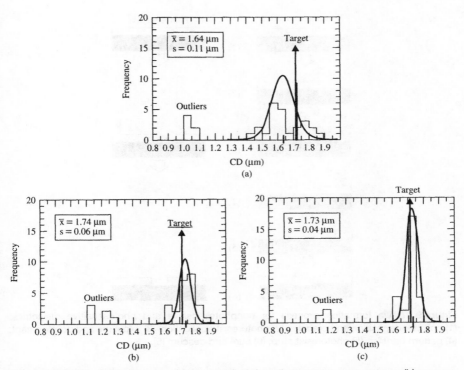

Figure 9.26. CD distribution for (a) uncontrolled baseline process sequence, (b) process sequence under local control, and (c) process sequence under global control.

process steps in via formation are polymer deposition, prebaking, pattern transfer (exposure and development), curing, and plasma descumming. This process sequence is illustrated in Figure 9.27, which compares process sequences for via formation using photosensitive and nonphotosensitive dielectric materials. As shown in the figure, the use of photosensitive material allows reduction of the number of process steps. This allows a reduction in the process cycle time, saves material and labor, and lowers cost.

A series of designed experiments were performed to characterize the complete via formation workcell (which consists of the spin coat, soft bake, expose, develop, cure, and plasma descum unit process steps). The output characteristics considered were film thickness, refractive index, uniformity, film retention, and via yield. To reduce the number of experimental trials required, the entire via formation process was divided into four subprocesses (spin-prebake, exposure–development, cure, and descum). Each subprocess was then modeled individually, and each subprocess model was linked to previous subprocess outputs and subsequent subprocess inputs.

Using a unique sequential scheme, each workcell subprocess is modeled individually, and each subprocess model is linked to previous subprocess outputs and subsequent subprocess inputs (see Figure 9.28). The sequential neural network process models were used for system identification, and genetic algorithms were

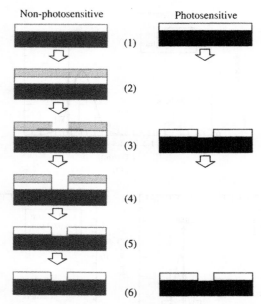

Figure 9.27. Via formation process for nonphotosensitive and photosensitive dielectrics: (1) polymer deposition and prebake; (2) photoresist application; (3) exposure and development; (4) pattern transfer; (5) photoresist strip; (6) cure and descum [6].

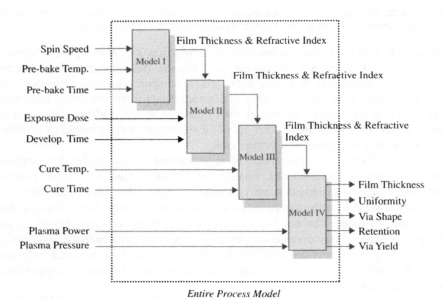

Entire Process Model

Figure 9.28. Block diagram of sequential modeling scheme [14].

applied to synthesize process optimal recipes (see Chapter 8). Afterward, the neural process models were used for optimal recipe generation using genetic algorithms. Recipe generation using this approach may be viewed as an example of offline process control where the objective is to estimate optimal operating points.

The goal in this study was to develop a supervisory process control system for via formation to maintain system reliability in the face of process disturbances. Supervisory control can reduce variability in two ways. The first involves reducing the variability of each contributing step by feedback control. The second requires accounting for the variation of consecutive steps so that their deviations cancel each other by feedforward control. In this RbR system, dielectric film thickness and refractive index were used as process monitors for each subprocess, and via yield, film retention, and film nonuniformity were added as the final response characteristics to be controlled. Based on appropriate decision criteria, model and recipe updates for consecutive subprocess were determined.

Figure 9.29 shows the general structure of the proposed supervisory control scheme. Nine neural networks were required: one global process model for optimal process recipe synthesis, four models for each subprocess model, and four for recipe updates to realize the supervisory algorithm. To construct the process supervisor, recipe update modules were developed individually for each subprocess. The neural networks for the recipe update modules are trained offline and are updated online as necessary. As illustrated in Figure 9.28, the outputs of these modules are the next subprocess' inputs. The inputs consist of the previous

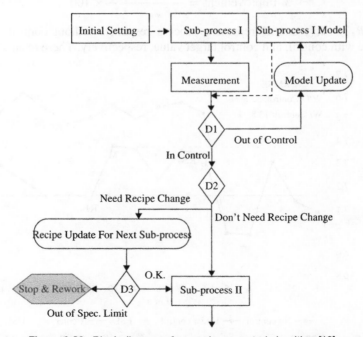

Figure 9.29. Block diagram of supervisory control algorithm [13].

subprocess' measured outputs, the desired final responses (final film thickness, via yield, film retention, and film nonuniformity), and the next subprocess's desired outputs. Based on the neural networks used for recipe updates, genetic algorithms generate optimal process recipes for the next subprocess.

The basic algorithm for supervisory control is as follows. The process was initiated using predetermined optimal recipes based on the operator's requirements. During the process, based on the results of two different decisions (D_1 and D_2 in Figure 9.29), the control system updated recipes to achieve the desired system outputs. These decisions were required for each subprocess. After the completion of a subprocess, film thickness and refractive index were measured. Generally, unpredicted outputs are the result of either system changes or process noise. In the case of system changes, appropriate control methods are required, and the models need to be updated. For such system changes, accompanying changes in the mean and/or variance of the outputs are also expected. Therefore, to differentiate system changes from noise, Shewhart control charts employing the Western Electric rules were applied (see Chapter 7).

When the supervisory control algorithm was applied to a real via formation process, experimental results showed significant improvement in film thickness and via yield control, as compared to open-loop operation. Figure 9.30 illustrates the performance of the system for controlling a shift in film thickness. Table 9.1 compares the final responses of the process with and without control. The "% improvement" column in this table is calculated using

$$\% \text{ Improvement} = \frac{(R_{\text{woc}} - R_{\text{wc}})}{(R_{\text{woc}} - T)} \times 100 \tag{9.52}$$

where R_{woc}, R_{wc}, and T represent process response without control, process response with control, and control target value, respectively. These results showed

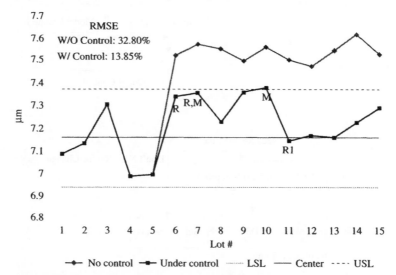

Figure 9.30. $\bar{x}$ chart showing control actions for a shift in film thickness [5].

Table 9.1. Supervisory control results.

Response	Without Control	With Control	% Improvement
Thickness (μm)	7.53	7.19	64.4
Yield (%)	84.7	97.3	82.6
Nonuniformity (%)	1.93	1.60	17.3
Film retention (%)	73.30	72.89	−1.5

that the supervisory control system significantly increased via yield, and the final film thickness was very close to the control target compared to the result of the experiment without control.

SUMMARY

In this chapter, we have introduced two key concepts in advanced process control: run-by-run and supervisory control. These advanced process control techniques can be used to supplement traditional statistical process control methods to provide enhanced variation reduction and responses to disturbances in sophisticated modern semiconductor manufacturing processes and equipment. In the next chapter, we introduce other advanced techniques for automated diagnosis of semiconductor manufacturing equipment.

PROBLEMS

9.1. A manufacturer desires to improve the performance of a three-zone induction heating furnace used to produce ceramic material for superconducting wire using RbR control [1]. The furnace is shown in Figure P9.1.

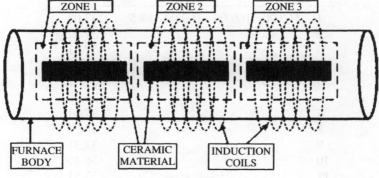

Figure P9.1

Each zone in the furnace is heated directly with its own set of induction coils. Each coil can be ramped to full power by an independently controlled ramp gradient R_i. The controllable input factors in this process are the ramp rates in each zone (R_1, R_2, R_3) and a trace element additive (E). The response variable to be controlled is the ductility of the wire produced (Y). A 2^{4-1} fractional factorial experiment with four centerpoints for this process yielded the following results:

Run	R_1	R_2	R_3	E	Y
1	160	160	160	18.1	37.3
2	159	161	161	16.3	34.8
3	161	159	161	16.3	34.7
4	160	160	160	18.1	37.4
5	159	159	159	16.3	34.7
6	161	161	159	16.3	34.8
7	159	161	159	20.0	40.1
8	160	160	160	18.1	37.4
9	161	161	161	20.0	40.0
10	159	159	161	20.0	39.9
11	161	159	159	20.0	40.0
12	160	160	160	18.1	37.3

(a) Derive a regression model using the data from this experiment. Evaluate the quality of the model using standard techniques.

(b) Sixty furnace runs are subsequently performed. The initial process recipe and data for these 60 runs are tabulated below. Given a target ductility of 37.3, develop and apply an EWMA-based RbR control system for this process to meet this target.

Run	R_1	R_2	R_3	E	Uncontrolled Y
1	160.00	160.00	160.00	17.5	37.08
2					37.82
3					37.01
4					36.43
5					37.72
6					36.40
7					37.41
8					36.29
9					37.31
10					36.40
11					37.51
12					37.20
13					37.17

Run	R_1	R_2	R_3	E	Uncontrolled Y
14					36.63
15					36.04
16					36.56
17					37.06
18					36.99
19					35.90
20					35.91
21					36.02
22					35.84
23					35.95
24					35.93
25					35.73
26					36.21
27					35.81
28					35.77
29					35.74
30					36.57
31					35.67
32					35.49
33					35.45
34					35.42
35					36.67
36					36.40
37					35.82
38					35.79
39					36.26
40					35.37
41					35.17
42					35.28
43					36.12
44					36.12
45					35.54
46					35.51
47					36.01
48					34.93
49					34.90
50					35.91
51					35.84
52					34.75
53					34.71
54					35.73
55					35.73

					Uncontrolled
Run	R_1	R_2	R_3	E	Y
56					34.79
57					35.90
58					35.59
59					35.83
60					35.79

9.2. Suppose that for the same three-zone induction heating furnace described in Problem 9.1, we are also interested in controlling the superconductivity in zone 1 (Z_1). The 2^{4-1} fractional factorial experiment for this response yielded the following results:

Run	R_1	R_2	R_3	E	Z_1
1	160	160	160	18.1	10.39
2	159	161	161	16.3	10.30
3	161	159	161	16.3	10.49
4	160	160	160	18.1	10.42
5	159	159	159	16.3	10.35
6	161	161	159	16.3	10.49
7	159	161	159	20.0	10.34
8	160	160	160	18.1	10.41
9	161	161	161	20.0	10.48
10	159	159	161	20.0	10.34
11	161	159	159	20.0	10.45
12	160	160	160	18.1	10.43

Data for the 60 subsequent furnace runs appears below. If the target superconductivity is 10.5, repeat Problem 1 to meet this target and the target ductility simultaneously.

					Uncontrolled
Run	R_1	R_2	R_3	E	Z_1
1	160.00	160.00	160.00	17.5	10.39
2					10.50
3					10.39
4					10.12
5					10.50
6					11.72
7					10.56
8					10.01
9					10.45

Run	R_1	R_2	R_3	E	Uncontrolled Z_1
10					10.34
11					10.51
12					10.45
13					10.45
14					11.06
15					10.40
16					11.06
17					10.57
18					10.46
19					10.19
20					10.30
21					10.35
22					10.30
23					10.35
24					10.69
25					10.30
26					10.41
27					10.36
28					10.36
29					10.36
30					10.47
31					10.36
32					10.31
33					10.31
34					10.31
35					10.53
36					10.59
37					10.42
38					10.42
39					10.48
40					10.70
41					10.32
42					10.37
43					10.48
44					10.60
45					10.43
46					10.43
47					10.60
48					10.32
49					10.77
50					10.60
51					10.49

Run	R_1	R_2	R_3	E	Uncontrolled Z_1
52					10.11
53					10.16
54					10.50
55					10.61
56					10.39
57					10.55
58					10.50
59					10.56
60					10.56

REFERENCES

1. J. Moyne, E. del Castillo, and A. Hurwitz, *Run-to-Run Control in Semiconductor Manufacturing*, CRC Press, Boca Raton, FL, 2001.

2. E. Sachs, A. Hu, and A. Ingolfsson, "Run by Run Process Control: Combining SPC and Feedback Control," *IEEE Trans. Semiconduct. Manuf.* **8**(1) (Feb. 1995).

3. S. Butler and J. Stefani, "Supervisory Run-to-Run Control of Polysilicon Gate Etch Using In-Situ Ellipsometry," *IEEE Trans. Semiconduct. Manuf.* **7**(2) (May 1994).

4. P. Gill, W. Murray, M. Saunders, and M. Wright, *User's Guide for NPSOL (Smoothed Nonlinear Constrained Optimization) (Version 4.0): A Fortran Package for Nonlinear Programming*, Stanford Univ., Stanford, CA, 1986.

5. B. Mandel, "The Regression Control Chart," *J. Quality Technol.* **1**(1), 1–9 (Jan. 1969).

6. R. Harris, *A Primer of Multivariate Statistics*, Academic Press, New York, 1975.

7. S. Lee, "A Strategy for Adaptive Regression Modeling of LPCVD Reactors," *Special Issues in Semiconductor Manufacturing*, 69–80 (Univ. California, Berkeley/ERL M90/8) (Jan. 1990).

8. R. Crosier, "Multivariate Generalizations of Cumulative Sum Quality-Control Schemes," *Technometrics* **30**(3) (Aug. 1988).

9. A. Emslie, F. Bonner, and L. Peck, "Flow of a Viscous Liquid on a Rotating Disk," *J. Appl. Phys.* **29**, 858 (1958).

10. B. Bombay, *The BCAM Control and Monitoring Environment*, Univ. California Berkeley, M.S. thesis, Memorandum UCB/ERL M92/113, 1992.

11. G. Golub and C. Van Loan, *Matrix Computations*, 2nd ed., John Hopkins Univ. Press, Baltimore, 1989.

12. S. Leang and C. Spanos, "A General Equipment Diagnostic System and Its Application on Photolithographic Sequence," *IEEE Trans. Semiconduct. Manuf.* **10**(3) (1997).

13. T. Kim and G. May, "Intelligent Control of Via Formation by Photosensitive BCB for MCM-D Applications," *IEEE Trans. Semiconduct. Manuf.* **12**(4), 503–515 (Nov. 1999).

14. T. Kim and G. May, "Sequential Modeling of Via Formation in Photosensitive Dielectric Materials for MCM-D Applications," *IEEE Trans. Semiconduct. Manuf.* **12**(3), 345–352 (Aug. 1999).

10

PROCESS AND EQUIPMENT DIAGNOSIS

OBJECTIVES

- Survey various techniques for automated diagnosis of semiconductor manufacturing processes and equipment.
- Compare and contrast these techniques.

INTRODUCTION

As we have established throughout this book, maintaining product quality throughout a semiconductor manufacturing facility requires the strict control of literally thousands of process variables. These variables serve as input and output parameters for hundreds of distinct process steps. Individual process steps are conducted by sophisticated and expensive fabrication equipment. A certain amount of inherent variability exists in this equipment regardless of how well the machine is designed or maintained. This variation is the result of numerous small and essentially uncontrollable causes. However, when this variability becomes large compared to background noise, significant performance shifts may occur.

As an example of such a shift, consider the standard Shewhart control chart shown in Figure 10.1. This figure depicts a shift in the thickness of a particular thin film as integrated circuit wafers are processed in a fabrication line. Such shifts are often indicative of equipment malfunctions. When unreliable equipment performance causes operating conditions to vary beyond an acceptable level,

Fundamentals of Semiconductor Manufacturing and Process Control,
By Gary S. May and Costas J. Spanos
Copyright © 2006 John Wiley & Sons, Inc.

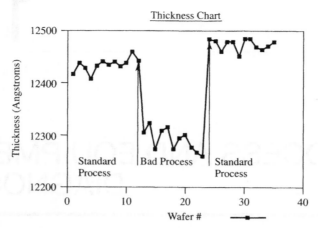

Figure 10.1. Shewhart control chart illustrating a process shift [1].

overall product quality is jeopardized. Consequently, fast and accurate equipment malfunction *diagnosis* is essential to the success of the semiconductor production process. This diagnosis involves determining the *assignable causes* for the equipment malfunctions and correcting them quickly to prevent the continued occurrence of expensive misprocessing. Fortunately, with the advent of proficient sensors to monitor process conditions in situ (see Chapter 3), it has become feasible to perform malfunction diagnosis on a real-time basis.

Several diagnostic systems have had the objective of performing automated diagnosis of faults in both manufacturing processes and equipment. *Algorithmic systems* have been developed to identify process faults from statistical inference procedures and electrical measurements performed on finished IC wafers. Although this technique makes good use of quantitative models of process behavior, it can arrive at useful diagnostic conclusions only in the limited regions of operation over which these models are valid. When catastrophic faults destroy circuit functionality, these models can no longer adequately describe the failure mechanism. Moreover, in some process steps (such as plasma etching), the theoretical basis for determining causal relationships is not well understood, thereby limiting the usefulness of physical models.

When attempting to diagnose unstructured problems that lack a solid conceptual foundation for reasoning, some success has been attained by approaches based on quantifying expert knowledge. *Expert systems* are designed to draw on experiential knowledge to develop qualitative models of process behavior. In this way, they are able to circumvent the difficulties encountered by algorithmic systems when quantitative relationships break down. Yet a purely knowledge-based approach often lacks the precision inherent in the deep-level physical models, and is thus incapable of deriving solutions for unanticipated situations from the underlying principles surrounding the process. Another shortcoming of purely expert diagnosis is its inability to identify concurrent multiple faults.

Neural networks have also emerged as an effective tool for equipment diagnosis. Diagnostic problem solving using neural networks requires the association of input patterns representing quantitative and qualitative process behavior to fault identification. Robustness to noisy sensor data and high-speed parallel computation make neural networks an attractive alternative for real-time diagnosis. However, the pattern recognition-based neural network approach is not without limitations. First, a complete set of fault signatures is hard to obtain, and the representational inadequacy of a limited number of datasets can induce network overtraining, thus increasing the misclassification (or "false alarm") rate. Also, pattern matching approaches in which diagnostic actions take place following a sequence of several processing steps are suboptimal since evidence pertaining to potential equipment malfunctions accumulates at irregular intervals throughout the process sequence. At the end of a sequence, significant misprocessing and yield loss may have already taken place, making postprocess diagnosis alone economically undesirable.

This chapter presents several approaches for the automated malfunction diagnosis of IC fabrication equipment. The methodologies discussed include quantitative algorithmic diagnosis, qualitative experiential, and pattern recognition-based neural network approaches. The use of process and equipment diagnosis can contribute to maintaining consistent manufacturing processes, increasing the probability of identifying faults caused by equipment malfunctions, and ultimately leading to improved process yields.

10.1. ALGORITHMIC METHODS

10.1.1. Hippocrates

Hippocrates is a system developed by Spanos in 1986 designed for the statistical diagnosis of noncatastrophic IC process faults [2]. Diagnosis in Hippocrates is based on the automatic selection of the minimum required set of electrical measurements and the subsequent solution of a sequence of nonlinear minimization problems that yield information regarding a process fault. This methodology is best suited to postprocess fault identification for finished IC wafers.

The statistical variations of an IC process are due primarily to the existence of a set of low-level, nonmeasurable, noncontrollable, independently varying physical quantities called *process disturbances*. A few examples of process disturbances include dopant diffusivity fluctuations and mask misalignments. These disturbances result in *process faults*. As discussed in Chapter 5, there are two categories of faults. *"Hard" faults* are manufacturing defects that destroy the functionality of a circuit. These faults are not addressed by Hippocrates. *"Soft" faults* are departures in circuit performance that do not lead to a loss in functionality. Such faults are observed through *symptoms*, which are defined as abnormal changes in the value of the statistics of one or more measurable parameters that relate to circuit performance. *Process diagnosis* is thus the inference of causes of changes in performance statistics resulting from process disturbances.

10.1.1.1. Measurement Plan

Diagnosis begins with the selection of a *measurement plan*, which is the data acquisition approach used to derive a *symptom vector*. The selection of an appropriate set of measurements whose deviations from expectations are used for diagnosis is critical. The cost, speed, and accuracy of data acquisition are obviously important. Furthermore, even a set of accurate but poorly selected measurements can be of limited use if the *observability* (i.e., the maximum number of independent process faults that can be identified) of the measurement plan is insufficient.

Assuming an initial measurement plan with m elements, the first step is to create a fault matrix F, where

$$F_{ij} = \frac{(s_i^s)_j - s_i^s}{s_i^s} \qquad i = 1, \ldots, m \quad j = 1, \ldots, n \qquad (10.1)$$

where s_i^s is the regular $(s_i^s)_j$ is the faulty value of the ith element of the symptom vector due to the occurrence of the jth fault. These quantities are evaluated by means of measurement (or simulation) of the circuit response due to process disturbances $d_1^0, d_2^0, \ldots, d_n^0$. Thus, F_{ij} is the normalized change in the symptom vector due to the jth fault (i.e., the addition of a perturbation δd_j to the normal value of the jth disturbance).

The next step is to cross-correlate the columns of the fault matrix to create an $m \times m$ *symptom correlation matrix* (SCM). In this step, all entries whose value is less than a specified threshold $C_{\max s}$ (where $C_{\max s} = 0.95$ is a typical initial selection) are discarded. In other words

$$\text{SCM}_{ij} = \begin{cases} |\rho_s| \equiv \text{cor} (\{F_{i1}, F_{i2}, \ldots, F_{in}\}, \{F_{j1}, F_{j2}, \ldots, F_{jn}\}) \\ \qquad\qquad\qquad\qquad \text{for } |\rho_s| \geq C_{\max s} \\ |\rho_s| \equiv 0 \qquad\qquad\qquad \text{for } |\rho_s| < C_{\max s} \end{cases} \qquad (10.2)$$

The symptoms are then grouped into m' high correlation groups by pivoting the truncated SCM so that it assumes a block diagonal form. The most sensitive symptom (i.e., the one that reflects the maximum sum of normalized, absolute changes due to all simulated faults) is then selected from each group. In so doing, a reduced measurement plan with $m' < m$ elements is obtained.

Next, fault observability is tested by cross-correlating the reduced symptom vectors for all simulated faults, thereby creating an $n \times n$ *fault correlation matrix* (FCM). Once again, entries whose value is less than a specified threshold $C_{\max f}$ (where $C_{\max f} = 0.8$ is typical) are discarded. Then

$$\text{FCM}_{ij} = \begin{cases} |\rho_f| \equiv \text{cor} (\{F_{1i}, F_{2i}, \ldots, F_{m'i}\}, \{F_{1j}, F_{2j}, \ldots, F_{m'j}\}) \\ \qquad\qquad\qquad\qquad \text{for } |\rho_f| \geq C_{\max f} \\ |\rho_f| \equiv 0 \qquad\qquad\qquad \text{for } |\rho_f| < C_{\max f} \end{cases} \qquad (10.3)$$

The FCM is likewise pivoted to a block diagonal form. One drawback of using this approach is that faults that end up in the same group are indistinguishable. If desired, the process can be repeated with different values of $C_{\max s}$ and $C_{\max f}$.

10.1.1.2. Fault Diagnosis

After a suitable reduced measurement plan has been identified, a symptom vector is generated. This is accomplished using a statistical process simulator that is capable of mapping faults to symptoms. The FABRICS simulator discussed in Chapter 5 (see Section 5.5.2) is an example of such a program. Next, nonlinear regression models are constructed to relate process disturbances to developed symptoms. The result is a *cost function* that serves as an analytical approximation of the "distance" between the measured and nominal (as provided by the simulation) statistics of the process. In other words

$$\text{Cost}(d_1, \ldots, d_n, s_1, \ldots, s'_m) \cong \sum_{i=1}^{m'} [s_i - \text{sim}_i(d_1, \ldots, d_n)]^2 \qquad (10.4)$$

where sim_i is the ith simulation.

Once the cost function is established, a sequence of minimization problems are solved to diagnose the fault. Hippocrates initially assumes that single faults occur independently. However, if Hippocrates fails to infer a single fault that explains the symptom vector, the system looks for increasingly complex combinations of multiple faults. Faults are identified by solving the minimization problem

$$\min_{d_1, \ldots, d_n} \text{cost}(d_1, \ldots, d_n, s_1, \ldots, s'_m) + \pi_k(d_1, \ldots, d_n) \qquad (10.5)$$

$$\text{such that } a_j < d_j < b_j \qquad j = 1, \ldots, n$$

where a_j and b_j are constraints that represent regions of validity for the cost function and π_k is a penalty supplement for a multiple fault that explains k disturbances. For example, the penalty supplements for $k = 2, 3$ are

$$\pi_2 = \sum_{i \neq j \neq k} (d_i - d_i^0)^2 (d_j - d_j^0)^2 (d_k - d_k^0)^2 \qquad (10.6)$$

$$\pi_3 = \sum_{i \neq j \neq k \neq l} (d_i - d_i^0)^2 (d_j - d_j^0)^2 (d_k - d_k^0)^2 (d_l - d_l^0)^2 \qquad (10.7)$$

Before an inferred fault is accepted, it is tested by producing a simulated symptom vector and comparing that simulated symptom vector to the measured symptom vector.

10.1.1.3. Example

Spanos demonstrated the effectiveness of Hippocrates using measurements collected from an NMOS fabrication line [2]. A lot consisting of 15 wafers containing 72 test die each was used for reference purposes to establish baseline (nominal) behavior. Eleven on each of approximately 500 test dies provided current–voltage data for characterizing process disturbance statistics. The test die contained enhancement and depletion mode MOSFETs of various sizes.

The initial measurement plan called for measurements on 11 devices ranging in width/length from 3 μm/4 μm to 25 μm/25 μm. The plan also called for

Table 10.1. Identifiable independent faults.

No.	Disturbance	Effect
1	Linewidth variation of nitride	Channel width
4	Linewidth variation of polysilicon	Channel length
14	Diffusivity of boron	Enhancement region
20	Diffusivity of arsenic	Depletion region
30	Linear rate of gate oxidation	Gate oxide thickness
29	Parabolic rate of gate oxidation	Gate oxide thickness
35	Etch rate of gate oxide	Gate oxide thickness

gate voltages to be swept in 1-V increments from -2 to 2 V for the depletion-mode devices and 2 to 6 V for the enhancement-mode devices. In each case, the gate voltage was swept for drain voltages of 2, 4, and 6 V, and for substrate voltages of 0, -3, and -7 V. This resulted in a total of 495 measurements. However, the application of a reduced measurement plan derived using the algorithm described in Section 10.1.1.1 required only 54 bias points to be measured. Testing the observability of this reduced plan resulted in the potential independent identification of the faults listed in Table 10.1.

After an approximation to the cost function was derived using a statistical modeling package, diagnosis using *Hippocrates* ensued. The successfully diagnosed faults correlated well with the history of the faulty wafer. A shift in channel length reduction (disturbance 4) was already known and properly diagnosed by the system. In addition, a shift in the arsenic profiles was traced to a 10% increase in the arsenic implantation dose between the reference and faulty lots.

10.1.2. MERLIN

The *me*asurement *rel*ational *in*terpreter (MERLIN) is another approach to computer-aided diagnosis of IC parametric test data that was developed by Freeman [3]. In contrast with a purely rule-based expert systems approach (see Section 10.2), the foundation of the knowledge base used by MERLIN is library of analytical device equations. This flexible approach can readily adapt to changing measurement data availability, as well as provide guidance in the selection of additional measurements to clarify diagnostic ambiguities.

Diagnosis is typically performed in three stages: *device diagnosis, process diagnosis*, and *root-cause diagnosis*. In the device stage, which is the focus of MERLIN, electrical measurements are analyzed to ascertain any potential physical anomalies in device structure. In this stage, fundamental first-principles knowledge embedded in equations describing device behavior plays a key role in the interpretation of measurement data. The challenge here is in mapping such knowledge into an adequate machine representation. However, since the knowledge is centered around a relatively small number of well-defined equations, the acquisition, maintenance, and modification of the overall knowledge base is simplified significantly. The result is that this knowledge base can be generated with less effort and is less susceptible to errors introduced during regular maintenance.

At the heart of MERLIN is a symbolic representation that encodes device model equations. This representation is complemented by a diagnostic inference mechanism that is capable of explaining abnormal test data by reasoning from the models.

10.1.2.1. Knowledge Representation

MERLIN's analytical reasoning capability is enabled by its internal model representation, which allows users to manipulate device models through a graphical user interface. MERLIN can also dynamically adapt to any combination of available measurements and is capable of incorporating experiential knowledge to complement its analytical models.

As an object-oriented package, MERLIN defines four classes of objects—variables, constants, equations, and device–components—as basic data types in its knowledge base. Specialized versions of these objects represent portions of equations. For example, an object called **VT** is a type of *variable* representing the threshold voltage in a MOS transistor. Various attributes, including a description, a list of equations that refer to it, and its approximate values, are associated with such an object. An object editor (see Figure 10.2) is used to view or modify these attributes. *Constant* objects are similarly defined, except since their values are independent of any set of measurements, their value is specified along with the model.

Equation objects describe a particular relationship among variables and constants. Within an *equation*, this relationship is encoded as a symbolic expression with references to objects representing other relevant quantities (such as *variables* and *constants*). For example, an object called **VT-EQN** is shown in Figure 10.3. To simplify the representation, all equations are stored in a an "equal-to-zero" format (i.e., $\langle expression \rangle = 0$). If there is a dependent variable in the expression, it is listed as such in the attribute section of the equation object. Since many devices are described by different equations in different regions of operation, MERLIN provides a special "PIECEWISE-CONDITION" attribute that allows the representation to be described by a list of equations, with each equation tagged by a condition under which it applies.

Device–component objects are defined to relate variables to the physical structures they represent. Examples of such structures are gate oxides, source–drain

```
OBJECT: VT
SYNONYMS:
GENERALIZATIONS: VARIABLE
GROUPS:
TYPE: CLASS

UNITS: V
APPROX-VALUE: 0.7
DESCRIPTION: Threshold Voltage
PART-OF-EQN: I_D-EQN, VT-EQN, and MU-EQN
DEPENDENT-VARIABLE-IN-EQN: VT-EQN
```

Figure 10.2. MERLIN's object editor [3].

```
OBJECT: VT-EQN
SYNONYMS:
GENERALIZATIONS: EQUATION
GROUPS:
TYPE: CLASS

  ACTUAL-EQUATION: ((- VT ...) (- VT ...))
  PIECEWISE-CONDITION: ((>= LGD ...) (< LGD ...))
  DEPENDENT-VARIABLE: VT
  DESCRIPTION: Threshold voltage, with terms for substrate bias, threshold implant, and
DIBL
    REFERENCE: SZE, p442 eq30, p442 eq32, p458 eq58, p475 eq94 - simplified to
eliminate Rj
```

Figure 10.3. Object editor for equation class [3].

```
OBJECT: NMOS-CAPACITOR
SYNONYMS:
GENERALIZATIONS: DEVICE-STRUCTURE
GROUPS:
TYPE: CLASS

  CONTAINS-DEVICE-STRUCTURE-ELEMENTS: N-DOPED-POLY-GATE, NMOS-
GATE-OXIDE, NMOS-THRESHOLD-IMPLANT-ON-SUBSTRATE, NMOS-CAPACITOR-
INVERSION-LAYER, and NMOS-CAPACITOR-DEPLETION-LAYER
  VARIABLES-DESCRIBING-DEVICE: WM, PHI_B, PHI_M, PHI_MS, VFB, COX, and
V_BI
```

Figure 10.4. Object editor for device–component class [3].

diffusion regions, or even entire transistors. MERLIN provides a representation for describing these structures through two attributes of this class: CONTAINS-DEVICE-STRUCTURE-ELEMENTS and PARAMETERS-DESCRIBING-DEVICE-COMPONENT. An example of a device–component object is shown in Figure 10.4.

To capture experiential knowledge, MERLIN allows expert users to complement an equation model with heuristic information, such as the abnormal likelihood of a variable or the accuracy of an equation. Such information is represented as an attribute. For variables, this can be a number between 0 and 1 (with 1 representing the most likely case), or as a function that when evaluated, returns a likelihood value between 0 and 1. When performing diagnosis, use of these heuristics is optional.

Measurements are entered into MERLIN as special instances of *variable* objects. For example, a threshold voltage measurement **VT-1** created under the class **VT** with attributes like ACTUAL-VALUE and UNITS is shown in Figure 10.5. To denote the origin of the measurement, lot, wafer, and site labels are listed in the GROUP attribute. Other attributes are inherited from the parent **VT** class. The QUALIFIERS attribute describes the device layout parameters and test conditions of the measurement. Measurements and their qualifiers are used to create a consistent set of related equations. MERLIN determines which equations are shared by those measurements, resulting in a unique set of equations that are

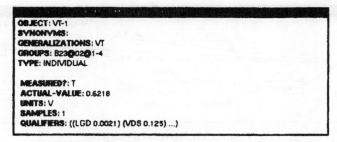

Figure 10.5. Object editor for VT-1 instance of VT class [3].

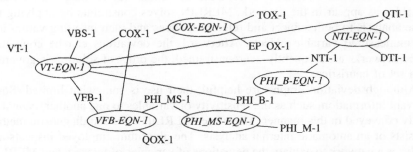

Figure 10.6. Network of equations, variables, and constants [3].

associated with quantities that they reference. This is illustrated schematically in Figure 10.6.

10.1.2.2. Inference Mechanism

MERLIN views device equations as constraints, where, if values are known for all except one of the variables, the value of the remaining variable can be computed. These values may be associated with individual measurements, means, medians, defect densities, or yields. Collectively, a set of measurements can characterize a single site, a wafer, or a lot. In analyzing a set of measurements, both *subject* and *reference* values are specified. The reference values represent the nominal conditions to which subject measurements are compared. The reference values are defined in the same manner as the subjects, except that the GROUP and ACTUAL-VALUE attributes differ.

Once validated, measurements are analyzed for the purpose of diagnosis. Two diagnostic methods are available. The first is visual inspection of a graphical representations of the data. This visualization is accomplished by means of a *deviation graph* (see Figure 10.7). These graphs contain a summary of all parameter information and display the percent deviation of subject values compared to reference values. To differentiate measurements from unmeasured (computed) values, boxes appear around variables that have been measured. Each deviation graph also contains *dependence links*, which provide structure to the graph and identify which variables are most likely causes of deviations.

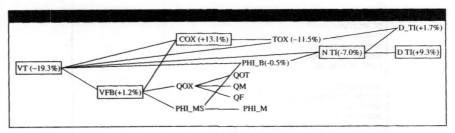

Figure 10.7. Deviation graph [3].

A user can ask MERLIN to compute deviations for as many unmeasured variables as appear in the network. MERLIN solves constraints by applying the same analysis to both subject and reference values and then inserting values into the unmeasured variable objects. Afterward, the deviations can be computed. Since networks are frequently overconstrained, the order of solution is governed by a set of heuristics.

Although deviation graphs are helpful, their use is somewhat limited. Some relevant information, such as the sensitivity of variables to one another, cannot be easily conveyed in this manner. Therefore, MERLIN's second diagnostic method consists of an automatic internal analysis. The algorithm employed inspects and analyzes a network to explain the deviations of variables of interest, and MERLIN returns the variable that it believes to be the most likely cause of the problem and why. Such an inquiry may be triggered interactively by the user or automatically when any measurement is found to be outside of its specification limits.

After identifying such variables, hypotheses must be generated to determine the causes of the deviations. An example of such a hypothesis might be "**TOX** is 20 angstroms too low." In the graph in Figure 10.7, deviations exist along the path between **VT** and **TOX**. This seems to be the path through which **VT** may have been affected, leading to the conclusion that **TOX** is a possible cause of the deviation in **VT**. In finding such paths, MERLIN computes a *measure of association* for each path. The measure of association is a value between -1 and 1 that indicates how well the deviations along a path predict the deviation in the variable of interest.

In addition, on a given path, each variable with a known deviation is assigned a level of *predictability*, which is another value between -1 and 1 that indicates how well that specific deviation predicts the deviation in the variable of interest. To do so, through a propagation algorithm, MERLIN then computes a predicted deviation ($\Delta V_{0,\text{pred}}$) in the variable of interest (V_0) based on the known variation (ΔV_i) in the variable along the path. The predictability of variable i is determined by applying a heuristic that compares to $\Delta V_{0,\text{pred}}$ the actual deviation (ΔV_0). This is illustrated in Figure 10.8. A value of -1 indicates poor predictability, and a value of 1 represents an exact match. The measure of association is an average of the predictabilities over a given path.

To verify a hypothesis, MERLIN performs a reverse traversal to look for evidence in the network that supports it. Through inspection of the deviations of

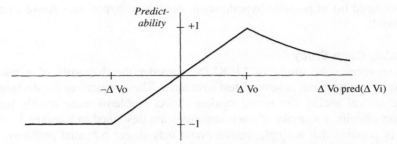

Figure 10.8. Representation of heuristic measure of how well a predicted deviation reflects a known deviation [3].

the affected variables encountered during this traversal, MERLIN calculates two measures of verification: a value measure and a correlation measure. The *value measure* reflects the known deviation (subject minus reference) in dependent variables to deviation values that would be expected if the hypothesis were true. For example, if the hypothesis were a 50-Å decrease in **TOX**, then one would expect a 350-pF increase in **COX**. MERLIN automatically computes these expected deviations, and for each affected variable, a comparison between the predicted and known deviations is summarized using the same predictability measure employed for computing the measure of association. The value measure of verification is simply the average of these predictabilities.

The correlation measure of verification, on the other hand, reflects an attempt to compare trends between different measurements of the same variable. These measurements may represent the same test conditions applied to several different devices (such as devices with different dimensions) or several different test conditions applied to the same device. For each hypothesis, MERLIN finds those sets of dependent variables for which multiple instances are available and computes the fractional deviation for each variable in that set that would be expected if the hypothesis were true. The set of fractional deviations is

$$X = \frac{\text{subject} - \text{reference}}{\text{reference}} \qquad (10.8)$$

Let Y be the known values of the variable set. The correlation measure of verification (ρ) is simply the correlation coefficient between X and Y, or

$$\rho = \frac{\text{cov}(X, Y)}{\sqrt{[\text{var}(X)\text{var}(Y)]}} \qquad (10.9)$$

where cov is the covariance and var is the variance of these variables. A correlation measure of $+1$ indicates that the trend predicted by the hypothesis matches the measured trend perfectly, and -1 reflects the opposite observed trend.

To determine an overall diagnosis, MERLIN takes into account the measures of association, value verification, correlation verification, and any experiential knowledge provided by expert users. The system is capable of producing a rank-ordered list of possible hypotheses or only those hypotheses above a certain threshold.

10.1.2.3. Case Study

To demonstrate the utility of MERLIN, consider the diagnosis of a junction leakage yield problem in several test structures. These structures are designed so that electrical testing can reveal random defect problems more readily than in product circuits. Examples of such structures are described in Chapter 5.

It is possible that a single measurement may detect potential problems with more than one structural component. For example, a junction leakage measurement on the same structure may reflect structural issues indicative of random defects at the interface between the junction and field isolation or the area under the contacts. MERLIN is capable of determining which structural components are affected the most by the random defects by encoding the critical areas as control variables within its models. Test data, along with associated critical areas, are passed to MERLIN at the beginning of an analysis, and MERLIN examines the corresponding sensitivities to perform a diagnosis.

In this example, test data were obtained through the measurement of five defect monitors—four of which emphasize individual structural features of n^+-NMOS source–drain junctions (i.e., junction area, junction edge adjacent to field oxide, junction edge under the gate, and junction under a metal contact), and one that combines these components as part of a contact string. The physical relationships between these components are captured by the expressions

$$\text{SDJ-D} = \text{SDJ-A-DD} \times \text{JA} + \text{SDJ-FE-DD} \times \text{JFE} + \text{SDJ-C-DD}$$
$$\times \text{NC} + \text{SDJ-GE-DD} \times \text{GE} \tag{10.10}$$

$$\text{SDJ-Y} = e^{-\text{SDJ-D}} \tag{10.11}$$

where SDJ-A-DD, SDJ-FE-DD, SDJ-C-DD, and SDJ-GE-DD are the defect densities for the area, field edge, contact, and gate edge components of the junction, respectively. SDJ-D is the average number of defects, and SDJ-Y is the yield of a given structure with junction area JA, field edge JFE, number of contacts NC, and gate edge GE. The latter four parameters are controllable and specified along with measurements as *qualifiers*. The contact string is described by similar equations.

The yield of each of the test structures was extracted, and the qualifiers specified along with the yield measurements were determined. When this information is provided to MERLIN, it constructs a network relating the measurements, computes the average defects per structure, and solves simultaneous equations to approximate the defect densities. The resulting deviation graph is shown in Figure 10.9, and the associated diagnosis of the CONTACT-STR-Y region appears in Figure 10.10. In this report, the evidence is strongly in favor of contact-related defects as the primary detractor from the contact string yield. The rationale for this conclusion is that since the contact string contained no gate structural features, which was ruled out as a possibility. Comparison of the

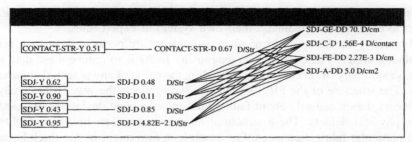

Figure 10.9. Deviation graph for source-drain junction leakage test structure case study [3].

	H-Dev	Assc	VTyV	VTyC	Likhd
SDJ-C-D	+5.91 E–5	0.54	0.19	0.89	0.50
SDJ-A-DD	+2.81 E+2	0.36	0.18	–0.42	0.50
SDJ-FE-DD	+4.07 E–2	0.38	0.12	–0.48	0.50
SDJ-GE-DO		–1.00	–1.00	–1.00	0.50

	H-Dev	Assc	VTyV	VTyC	Likhd
SDJ-C-D	+5.91 E–5	0.54	0.19	0.89	0.50

Figure 10.10. Analysis of CONTACT-STR-Y component in Figure 10.9 [3].

remaining three structural features revealed that the association evidence and value verification evidence were too close, and therefore inconclusive. However, the correlation measure of verification was strongly in favor of the contacts as the likely source of the problem.

10.2. EXPERT SYSTEMS

Algorithmic diagnostic systems identify process faults from electrical measurements, statistical inference procedures, and quantitative process models. These systems are somewhat limited in the sense that they can only arrive at diagnostic conclusions in the regions of operation in which these models are valid. In some process steps, however, the theoretical basis for establishing quantitative models that determine causal relationships is not well understood. *Expert systems* have been designed to draw on experiential knowledge to develop qualitative models of process behavior. This rule-based approach has attained some success in attempting to diagnose unstructured problems that lack a solid conceptual foundation for reasoning.

10.2.1. PIES

The parametric interpretation expert system (PIES) is a knowledge-based methodology for making inferences about parametric test data collected during semiconductor manufacturing [4]. PIES transforms voluminous measurement data into a concise statement of the "health" of a manufacturing process and the nature and probable cause of any anomalies. The structure of the PIES knowledge base mimics the rationale used by process engineers in diagnosing problems and allows users to construct and maintain their own system of expert rules.

Typically, hundreds of electrical measurements are performed on each semiconductor wafer. The challenge in diagnosing faults is to reduce these data to a concise summary of process status, including the cause of any potential abnormalities. The structure of the PIES knowledge base reflects the way failure analysis engineers reason causally about faults. First measurement deviations are used to infer physical defects. These structural anomalies are then linked to problems with particular fabrication steps. For example, a film might be too thick because a wafer was left in an oven too long. Ultimately, such problems are traced to root causes (i.e., the wafer was in the oven too long because a timer broke). The multilevel nature of the PIES knowledge base allows process engineers to codify their experiential knowledge using causal links that associate evidence at each level with hypotheses at the next. A knowledge editor supports this conceptual structure.

The usual duties of a failure analysis engineer involve diagnosing process faults and recommending corrective action. PIES enhances the efficiency of this process by reducing the volume of raw test data that must be analyzed and ensuring objective assessment and analysis of this data. This is accomplished by representing the diagnostic domain in terms of multiple causal levels. Figure 10.11 shows the causal chain PIES uses to represent the origination and

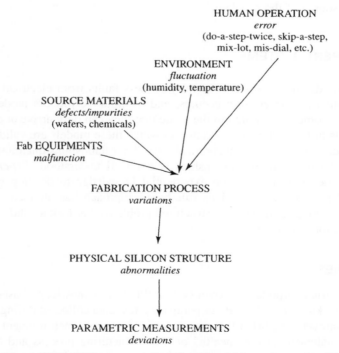

Figure 10.11. PIES representation of multilevel failure propagation [4].

propagation of fabrication failures. The root cause of these failures is either an equipment malfunction, contamination in source materials or the environment, or human error.

The diagnostic approach involves isolating the possible causes of observed symptoms by "reversing" the causal chain using the following sequence: measurement deviations $\rightarrow$ physical structure $\rightarrow$ abnormalities $\rightarrow$ process variations $\rightarrow$ root causes. At the physical structure level, failure modes consist of incorrect film thicknesses, doping densities, etc. At the process level, they include incorrect temperatures, pressures, or gas flows during particular steps. Rules provided by process engineers link failure modes at adjacent levels. For example, EPI-THICKNESS-HIGH might be associated with an abnormally high temperature during epitaxy.

Thus, associated with each structural anomaly are a set of observable symptoms and a corresponding set of possible causes. Diagnosis proceeds as a multilevel hypothesis verification exercise. Parametric measurements are first transformed from numeric values into quantitative ranges (normal, high, low, etc.). Each abnormal measurement implicates one or more structural problems. The symptoms associated with each hypothesized structural issue are compared with the complete set of abnormal measurements. A score is assigned based on how well the expected symptoms match those that have been observed, and a hypothesis verification process is used to select the most probable failure(s). Finally, the root causes are selected that best explain the highest likelihood failures.

10.2.1.1. Knowledge Base

The PIES knowledge base is organized according to the four levels described in Figure 10.11. The causal sequence among this hierarchy is described by a set of symbolic links, which are used by both a *knowledge editor* and a *diagnostic reasoner*. At each causal level, the knowledge base is decomposed into structures called *failure cases* that encode information about the type of failure at that level. Examples of such information are the popular name used by experts to refer to the case, comments from process engineers about the case, and *associational links* that describe how the case is related to other types of failures. A link can either be the *causes* or *caused-by* type, and it can be either *intralevel* or *interlevel*. Each link also has an *associational strength*, which is a heuristic estimate of the strength of the causal relationship. Strengths have five possible states: *must, very likely, likely, probably,* or *maybe*. An example of an associational link is shown in Figure 10.12, which is the PIES representation of a BASE-DISTRIBUTION-deep fault in a bipolar process.

A knowledge editor allows a process engineer serving as a domain expert to build and maintain the knowledge base. The primary function of the PIES knowledge editor is to guide the domain expert in codifying knowledge in a form syntactically and semantically consistent with the PIES knowledge base. The knowledge editor allows the addition, deletion, revision, or replacement of associational links, as necessary.

Possible effects at measurement level --
1: ((parametric-measurement WE10BETA low) very-likely)
2: ((parametric-measurement RB1 low) probably)
3: ((parametric-measurement RB2 low) very-likely)
4: ((parametric-measurement WE10-CBO low) probably)
5: ((parametric-measurement SOT2-CBO low) probably)
6: ((parametric-measurement SOT-B-SU very-low) probably)
7: ((parametric-measurement SOTBETAF low) probably)

Possible causes at process level --
1: ((BASE-IMPLANT ENERGY high) likely)
2: ((BASE-DRIVE FURNACE-TEMPERATURE high) likely)
3: ((BASE-DRIVE DIFFUSION-TIME long) likely)

Possible causes at SAME physical-structure level --
1: ((BASE-OXIDE THICKNESS low) likely)

Figure 10.12. PIES representation of bipolar BASE-DISTRIBUTION-deep structural defect [4].

10.2.1.2. Diagnostic Reasoning

The diagnostic reasoning mechanism in PIES uses the multiple causal level structure shown in Figure 10.11 to identify the root causes of failures from a set of parametric test data. Before doing so, symbolic symptoms must be abstracted from raw test data. This occurs in two steps: (1) any noisy data are removed statistically and (2) the average and standard deviation for each measurement over all wafers in a given lot are computed for the remaining data. The averages and standard deviations are then compared with limits provided by experts to produce a qualitative estimate that describes the measurement (such as EPI-R-very-low in the experiment described in Section 10.2.1.3). These estimates form the initial symptom set.

The diagnostic process then proceeds by progressing through each causal level by a sequence of hypotheses and confirmations. At each level, a set of possible failures is filtered from hypotheses suggested by likely faults isolated at the previous level. At each stage, isolation is achieved in four steps: hypothesization, implication, confirmation, and thresholding. This process continues until a final diagnostic conclusion is reached.

The objective of the *hypothesization* step is to heuristically retrieve a "suspect set" that includes those failures that are reasonably implicated by the symptoms, given some adjustable threshold for inclusion. *Implication*, the next reasoning step, expands the suspect set by including additional hypotheses that are implicated by any failure case already included. This implication step is based on the intralevel causalities coded in the knowledge base. In the *confirmation* step, the expected symptoms for each failure case in the expanded suspect set are matched against the hypotheses concluded by the diagnostic process thus far, and a "score" is computed for each case. The score indicates how close the symptoms and derived conclusions match. Following confirmation, the failures in the suspect set are sorted according to these scores. *Thresholding* excludes those cases with relatively low scores.

10.2.1.3. Examples

As a typical example of a PIES application, consider the measurement of the resistance of an epitaxial layer in a bipolar process. EPI-R is the nomenclature used for this measurement from a test structure. One possible explanation for a low observed EPI-R value is an epitaxial layer that is too thick. A thick layer can result from an abnormally high temperature during the epitaxial growth process. The objective of PIES is to determine the root cause of such a failure, which might be a faulty thermostat or another equipment failure.

Another example from the same process involves a physical structure failure (BASE-OXIDE-THICKNESS-low, shown in Figure 10.12) that might cause another similar failure (BASE-DISTRIBUTION-deep). This types of failure, as illustrated in Figure 10.13, requires the PIES knowledge base to make use of *intralevel causalities* during the implication phase of diagnosis.

PIES was initially implemented in 1986 for the ISO-Z bipolar process at Fairchild Semiconductor, which is the process used in the examples above. In the knowledge base for this process, a total of 342 types of failure cases were identified: 101 at the measurement level, 82 at the physical structure level, and 159 at the process level. The knowledge base encoded approximately 600 associational links among these cases. After initial system tuning and enhancement of the PIES knowledge base, PIES achieved the correct diagnosis nearly 100% of the time in cases for which complete knowledge was available.

10.2.2. PEDX

The plasma etch diagnosis expert (system) (PEDX) is a tool that automatically interprets endpoint traces generated in real time by optical emission spectroscopy [5]. This system combines signal-to-symbol transformations for data abstraction and rule-based reasoning to detect and classify process faults.

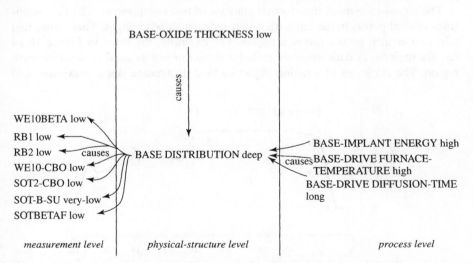

Figure 10.13. Concepts causally related to BASE-DISTRIBUTION-deep structural defect [4].

An experienced plasma process engineer can detect problems by interpreting OES endpoint traces, which contain information regarding the amount of particular chemicals produced as material is removed from a wafer. When examining a trace, engineers usually make the following assumptions:

- Approximately horizontal regions correspond to periods of etching through a single layer of material.
- Sharply ascending regions reflect periods when power has just been turned on or when a new layer of material has been reached.
- Sharply descending regions correspond to periods when the power has just been turned off or when a new layer of material has been reached.

As an example, consider the trace for the polysilicon and silicon nitride etch depicted in Figure 10.14. Initially, this trace rises sharply when the power is turned on. Polysilicon is removed in the flat region 2, and a nitride layer is removed in the two regions from point D to point F. Process parameters are adjusted for the nitride etch between points C and D, and the last sharp decline reflects power shutting down.

A faulty etch can be identified by comparing nominal traces such as this to abnormal traces. For example, too long a flat region may be caused by a previous process depositing too much polysilicon. PEDX was developed to automatically interpret traces to infer such problems.

10.2.2.1. Architecture

The architecture of PEDX is illustrated in Figure 10.15. A *signal-to-symbol* transformer takes an OES trace and creates and symbolic representation. A set of rules operating on this symbolic description is then executed to detect and diagnose problems.

The signal-to-symbol transformer consists of two components. The first intensifies critical points in the input trace that indicate slope changes. These lines that join two critical points forms a *region*. For example, the trace in Figure 10.14 has six regions. A data structure called a *region object* is used to describe each region. The attributes of a region object include its average slope, maximum and

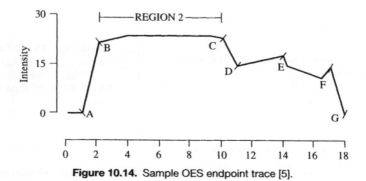

Figure 10.14. Sample OES endpoint trace [5].

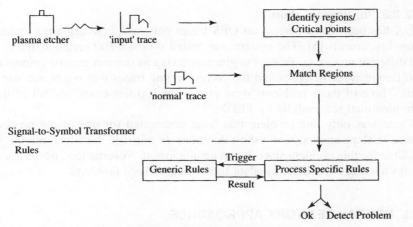

Figure 10.15. PEDX architecture [5].

minimum intensity values, material etched, starting and ending times, and times and intensities of the critical points. The second component of the transformer is the *matching algorithm.* This algorithm compares regions of an input trace with corresponding regions from a normal trace. Discrepancies between the two are accounted for by the PEDX rule-based reasoning system.

10.2.2.2. Rule-Based Reasoning

PEDX uses two categories of rules to operate on the output of the signal-to-symbol transformer: generic and process-specific. *Generic* rules compare regions of an input trace to the corresponding regions of a normal trace to identify any abnormalities. They compare intensities, slopes, and times of critical points to provide symbolic conclusions such as "too early," "too late," or "no problem." An example of a *generic* rule is

If (time for a normal ending critical point - time for input ending critical point) < threshold
then conclude "too late"

Process-specific rules identify the causes of problems. These rules are divided into different groups that represent knowledge about different regions. Process-specific rules are developed by domain experts. An example of such a rule is

If the generic rule for testing ending time determines "too late" for region 1 then conclude "thin material on poly" (polysilicon)

After a wafer is etched, process-specific rules for each region are executed. In this way, PEDX detects abnormalities if the size of differences between input and normal traces is outside an acceptable range. In cases where a problem is detected, PEDX can shut down an etcher and report its diagnosis to a technician for corrective action.

10.2.2.3. Implementation

PEDX has been demonstrated on OES traces collected from a plasma etcher at Texas Instruments [5]. The system was tested on over 200 endpoint traces for two different processes. Process engineers serving as domain experts enumerated 13 different types of faults and the corresponding traces that might account for them. Three of these problems were present in 100 test cases, and all of these were identified successfully by PEDX.

There was only one problem that went undetected for one of the processes studied (a slope anomaly for a short duration). In this case, however, no rule was available for this problem since it was unanticipated. Nevertheless, new rules can always be added to PEDX to adapt to such unforeseen problems.

10.3. NEURAL NETWORK APPROACHES

Although very useful and powerful, both the algorithmic and expert diagnostic approaches discussed in Sections 10.1 and 10.2, respectively, have limitations. Algorithmic systems make good use of quantitative models of process behavior, but can arrive at diagnostic conclusions only in the limited regions of operation over which the analytical models on which they depend are valid. Expert systems, on the other hand, draw on experiential knowledge to develop qualitative process models, but purely knowledge-based techniques lack the precision inherent in analytical models, and are therefore incapable of deriving solutions for previously unanticipated situations.

Neural networks (see Chapter 8) have emerged as another effective tool for malfunction diagnosis in semiconductor processes [6–9]. Diagnostic problem solving using neural networks requires the association of input patterns representing quantitative and qualitative process behavior to fault identification. Robustness to noisy sensor data and high-speed parallel computation make neural networks an attractive alternative for fault diagnosis.

10.3.1. Process Control Neural Network

For diagnosis at the integrated circuit level, Plummer developed a process control neural network (PCNN) to identify faults in bipolar operational amplifiers ("op-amps") based on electrical test data [6]. The PCNN exploits the capability of neural nets to interpret multidimensional data and identify clusters of performance within such a dataset. This provides enhanced sensitivity to sources of variation that are not distinguishable from observing traditional single-variable control charts. Given a vector of electrical test results as input, the PCNN can evaluate the probability of membership in each set of clusters, which represent different categories of circuit faults. The network can then report the various fault probabilities or select the most likely fault category.

Representing one of the few cases in semiconductor manufacturing in which backpropagation networks are not employed, the PCNN is formed by replacing the output layer of a probabilistic neural network with a Grossberg layer

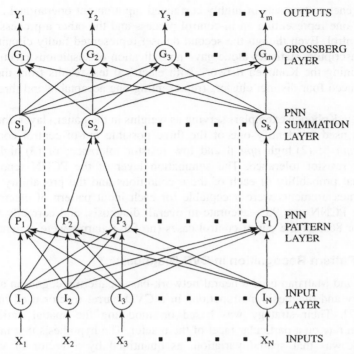

Figure 10.16. Process control neural network formed by replacing the output layer of a probabilistic neural network with a Grossberg layer whose outputs reflect probabilities that constitute a Pareto distribution of possible causes for a given input vector [6].

(Figure 10.16). In the probabilistic network, input data are fed to a set of pattern nodes. The pattern layer is trained using weights developed with a Kohonen self-organizing network. Each pattern node contains an exemplar vector of values corresponding to an input variable typical of the category it represents. If more than one exemplar represents a single category, the number of exemplars reflects the probability that a randomly selected pattern is included in that category. The proximity of each input vector to each pattern is computed, and the results are analyzed in the summation layer.

The Grossberg layer functions as a lookup table. Each node in this layer contains a weight corresponding to each category defined by the probabilistic network. These weights reflect the conditional probability of a cause belonging to the corresponding category. Then outputs from the Grossberg layer reflect the products of the conditional probabilities. Together, these probabilities constitute a Pareto distribution of possible causes for a given test result (which is represented in the PCNN input vector). The Grossberg layer is trained in a supervised manner, which requires that the cause for each instance of membership in a fault category be recorded beforehand.

Despite its somewhat misleading name, Plummer applied the PCNN in a diagnostic (as opposed to a control) application. The SPICE circuit simulator was used to generate two sets of highly correlated input/output operational amplifier test data, one representing an in-control process and the other a process grossly out of control. Even though the second dataset represented faulty circuit behavior, its descriptive statistics alone gave no indication of suspicious electrical test data. Training the Kohonen network with electrical test results from these data sets produced four distinct clusters (representing one acceptable and three faulty states).

With the Kohonen exemplars serving as weights in the pattern layer, the PCNN then was used to identify one of the three possible out-of-control conditions: (1) low *npn* β; (2) high *npn* β and low resistor tolerance; or (3) high *npn* β and high resistor tolerance. The summation layer of the PCNN reported the conditional probability of each of these conditions and the probability that the op-amp measurements were acceptable for each input pattern of electrical test data. The PCNN was 93% accurate in overall diagnosis, and correctly sounded alarms for 86% of the out-of-control cases (no false alarms were generated).

10.3.2. Pattern Recognition in CVD Diagnosis

Bhatikar and Mahajan used a neural-network-based pattern recognition approach to identify and diagnosis malfunctions in a CVD barrel reactor used in silicon epitaxy [7]. Their strategy was based on modeling the spatial variation of deposition rate on a particular facet of the reactor. The hypothesis that motivated this work was that spatial variation, as quantified by a vector of variously measured standard deviations, encoded a pattern reflecting the state of the reactor. Thus, faults could be diagnosed by decoding this pattern using neural networks.

Figure 10.17 shows a schematic diagram of the CVD reactor. In this reactor, silicon wafers are positioned in shallow pockets of a heated graphite susceptor.

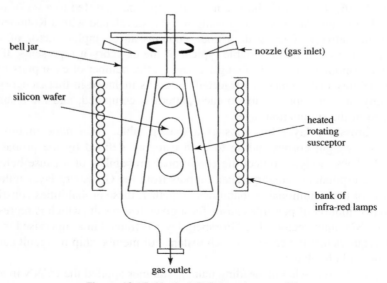

Figure 10.17. Vertical CVD barrel reactor [7].

Reactive gases are introduced into the reactor through nozzles at the top of the chamber and exit from the outlet at the bottom. The six controllable reactor settings include flow velocity at the left and right nozzles, the settings of the nozzles in the horizontal and vertical planes, the main flow valve reading, and the rotational flow valve reading.

Bhatikar and Mahajan chose the uniformity of the deposition rate as the response variable to optimize. Each side of the susceptor held three wafers, and deposition rate measurements were performed on five sites on each wafer. Afterward, a polynomial regression model that described the film thickness at each of the five measurement locations for each wafer as a function of the six reactor settings was developed. Next, backpropagation neural networks were trained as event classifiers to detect significant deviations from the target uniformity. Eight specific distributions of thickness measurements were computed. These are depicted in Figure 10.18. As a group, these eight standard deviations constituted a process signature. Patterns associated with normal and specific types of abnormal behavior were captured in these signatures.

Three disparate events were then simulated to represent deviations from normal equipment settings: (1) a mismatch between the left and right nozzles, (2) a horizontal nozzle offset, and (3) a vertical nozzle offset. The first event was simulated with a 5% mismatch, and offsets from 0% to 20% were simulated for both the vertical and horizontal directions. A neural network was then trained to match these events with their process signatures as quantified by the vector of eight standard deviations. The network had eight input neurons and three outputs (one for each event). The number of hidden layer neurons was varied from five to seven, with six providing the best performance. Each output was a binary response, with one or zero representing the presence or absence of a given event. The threshold for a binary "high" was set at 0.5. Training consisted of exposing

standard deviation from top to bottom, from left to right   standard deviation from top to bottom, from right to left

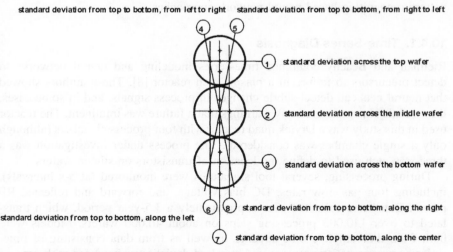

Figure 10.18. Vectors that characterize spatial variation of thickness distribution [7].

the network to an equal number of representative signatures for each event. When tested on twelve signatures not seen during training (four for each event), the network was able to discriminate between the three faults with 100% accuracy.

This scheme was then applied to a fault detection task (as opposed to fault classification). This required the addition of a "nonevent" representing normal equipment operation. Since there was only one signature corresponding to the nonevent, this signature was replicated in the training data with the addition of white noise to the optimal equipment settings to simulate typical random process variation. The network used for detection had the same structure as that used for classification, with the exception of having seven hidden layer neurons rather than six. After an adjustment of the "high" threshold to a value of 0.78, 100% classification accuracy was again achieved.

10.4. HYBRID METHODS

Even the pattern-recognition-based neural network approach has limitations. First, a complete set of fault signatures is hard to obtain, and the representational inadequacy of a limited number of datasets can induce network overtraining, thus increasing the misclassification or false-alarm rate. Also, pattern matching approaches in which diagnostic actions take place following a sequence of several processing steps are suboptimal since evidence pertaining to potential equipment malfunctions accumulates at irregular intervals throughout the process sequence. At the end of a sequence, significant misprocessing and yield loss may have already taken place, making postprocess diagnosis alone economically undesirable.

To address these concerns, hybrid methods have emerged as an effective tool for process modeling and fault diagnosis [9]. These methods attempt to combine the best characteristics of quantitative algorithmic, qualitative experiential, and pattern-recognition-based neural network approaches.

10.4.1. Time-Series Diagnosis

Rietman and Beachy combined time-series modeling and neural networks to detect precursors to failure in a plasma etch reactor [8]. These authors showed that neural nets can detect subtle changes in process signals, and in some cases, these subtle changes were early warnings that a failure was imminent. The reactor used in this study was a Drytek quad reactor with four process chambers (although only a single chamber was considered). The process under investigation was a three-step etch used to define the location of transistors on silicon wafers.

During processing, several tool signatures were monitored (at 5-s intervals), including four gas flow rates, DC bias voltage, and forward and reflected RF power. Data were collected over approximately a 3.5-year period, which translated to over 140,000 processing steps on about 46,000 wafers. Models were built from the complete time streams, as well as from data consisting of time-series summary statistics (mean and standard deviation values) for each process signature for each wafer. Samples that deviated by more than four standard deviations from the mean for a given response variable were classified as failure events. According to this classification scheme, a failure occurred approximately every 9000 wafers.

Rietman focused on pressure for response modeling. The models constructed for summary statistical data had the advantage that the mean and the standard deviation of the time series could be expected to exhibit less noise than the raw data. For example, a model was derived from process signatures for 3000 wafers processed in sequence. The means and standard deviations for each step of the three-step process served as additional sources of data. A neural network with 21 inputs (etch end time, total etch time, step number, mean and standard deviations for four gases, RF applied and reflected, pressure, and DC bias), five hidden units, and a single output was used to predict pressure. The results of this prediction for one, 12, and 24 wafers in advance is shown in Figure 10.19.

To demonstrate malfunction prediction, Rietman again examined summary data, this time in the form of the standard deviation time streams. The assumption was that fluctuations in these signatures would be more indicative of precursors to equipment failure. For this part of the investigation, a neural time-series model with inputs consisting of five delay units, one current time unit, one recurrent time unit from the network output, and one bias unit was constructed. This network had five hidden units and a single output. Figure 10.20a shows the mean value of pressure at each of the three processing steps. This was the time stream to be modeled. A failure was observed at wafer 5770. Figure 10.20b shows the corresponding standard deviation time stream, with the failure at 5770 clearly observable, as well as precursors to failure beginning at 5710–5725. Figure 10.20c shows the RMS error of the network trained to predict the standard deviation signal as a function of the number of training iterations. Finally, Figure 10.20d compares the network response to the target values, clearly indicating that the network is able to detect the fluctuations in standard deviation indicative of the malfunction.

10.4.2. Hybrid Expert System

Kim employed a hybrid scheme that uses neural networks and traditional expert systems for real-time, automated malfunction diagnosis of reactive-ion etching equipment [9]. This system was implemented on a Plasma Therm 700 series RIE to outline a general diagnostic strategy applicable to other rapid single-wafer processes. Diagnostic systems that rely on postprocess measurements and electrical test data alone cannot rapidly detect process shifts and also identify process faults. Because unreliable equipment jeopardizes product quality, it is essential to diagnose the root causes for the malfunctions quickly and accurately.

Kim's approach integrates evidence from various sources using the *Dempster–Shafer theory* of evidential reasoning [10]. Diagnosis is conducted by this system in three chronological phases: the maintenance phase, the online phase, and the inline phase. Neural networks were used in the maintenance phase to approximate the functional form of the failure history distribution of each component in the RIE system. Predicted failure rates were subsequently converted to

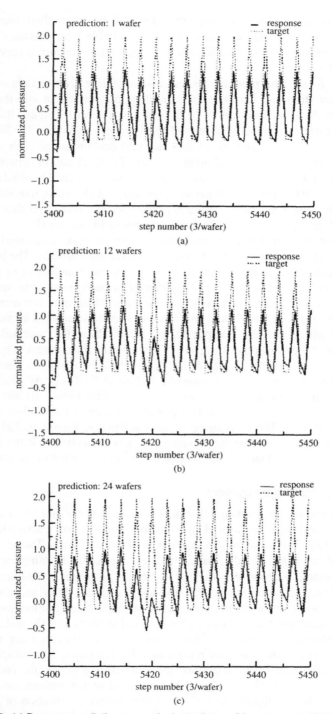

Figure 10.19. (a) Pressure prediction one wafer in the future; (b) pressure prediction 12 wafers in the future; and (c) pressure prediction 24 wafers in the future [8].

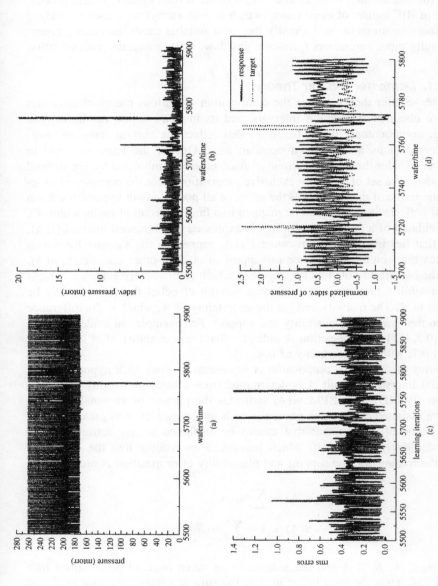

Figure 10.20. (a) Mean value of pressure between 5500 and 5900 samples. A failure can be seen, but no precursors to the failure are seen in the mean value data. (b) Standard deviation of pressure of the same time segment. Here, precursors are seen at about 5700 and the failure occurs at 5775. The precursors thus show up about 12 wafers prior to the actual failure. (c) Segment of neural network learning curve showing the detection of the precursors shown in (b). (d) Target and response curve for the same neural network predicting pressure [8].

belief levels. For online diagnosis of previously encountered faults, hypothesis testing on the statistical mean and variance of the sensor data was performed to search for similar data patterns and assign belief levels. Finally, neural process models of RIE figures of merit (such as etch or uniformity) were used to analyze the inline measurements and identify the most suitable candidate among potentially faulty input parameters (pressure, gas flow, etc.) to explain process shifts.

10.4.2.1. Dempster–Shafer Theory

Dempster–Shafer theory allows the combination of various pieces of uncertain evidence obtained at irregular intervals, and its implementation results in time varying, nonmonotonic belief functions that reflect the current status of diagnostic conclusions at any given point in time. One of the basic concepts in Dempster–Shafer theory is the *frame of discernment* (symbolized by Θ), defined as an exhaustive set of mutually exclusive propositions. For the purposes of diagnosis, the frame of discernment is the union of all possible fault hypotheses. Each piece of collected evidence can be mapped to a fault or group of faults within Θ. The likelihood of a fault proposition A is expressed as a bounded interval $[s(A), p(A)]$ that lies in $[0, 1]$. The parameter $s(A)$ represents the *support* for which measures the weight of evidence in support of A. The other parameter, $p(A)$, called the *plausibility* of A, is the degree to which contradictory evidence is lacking. Plausibility measures the maximum amount of belief that can possibly be assigned to A. The quantity $u(A)$ is the uncertainty of A, which is the difference between the evidential plausibility and support. For example, an evidence interval of $[0.3, 0.7]$ for proposition A indicates that the probability of A is between 0.3 and 0.7, with an uncertainty of 0.4.

In terms of diagnosis, proposition A represents a given fault hypothesis. An evidential interval for fault A is determined from a basic probability mass distribution (BPMD). The BPM $m\langle A\rangle$ indicates the portion of the total belief in evidence assigned exactly to a particular fault hypothesis set. Any residual belief in the frame of discernment that cannot be attributed to any subset of Θ is assigned directly to Θ itself, which introduces uncertainty into the diagnosis. Using the framework, the support and plausibility of proposition A are given by

$$s(A) = \sum m\langle A_i\rangle \tag{10.12}$$

$$p(A) = 1 - \sum m\langle B_i\rangle \tag{10.13}$$

where $A_i \subseteq A$, $B_i \subseteq \overline{A}$ and the summation is taken over all propositions in a given BPM. Thus the total belief in A is the sum of support ascribed to A and all subsets thereof.

Dempster's rules for evidence combination provide a deterministic and unambiguous method of combining BPMDs from separate and distinct sources of evidence contributing varying degrees of belief to several propositions under a common frame of discernment. The rule for combing the observed BPMs of two arbitrary and independent knowledge sources m_1 and m_2 into a third (m_3) is

$$m_3 = \frac{\sum m_1 \langle X_i \rangle \cdot m_2 \langle Y_j \rangle}{1 - k} \qquad (10.14)$$

where $Z = X_i \cap Y_j$ and

$$k = \sum m_1 \langle X_i \rangle \cdot m_2 \langle Y_j \rangle \qquad (10.15)$$

where $X_i \cap Y_j = \emptyset$. Here X_i and Y_j represent various propositions which consist of fault hypotheses and disjunctions thereof. Thus, the BPM of the intersection of X_i and Y_j is the product of the individual BPMs of X_i and Y_j. The factor $(1 - k)$ is a normalization constant that prevents the total belief from exceeding unity due to attributing portions of belief to the empty set.

To illustrate, consider the combination of m_1 and m_2 when each contains different evidence concerning the diagnosis of a malfunction in a reactive-ion etcher. Such evidence could result from two different sensor readings. In particular, suppose that the sensors have observed that the flow of one of the etch gases into the process chamber is too low. Let the frame of discernment $\Theta = \{A, B, C, D\}$, where $A, \ldots, D$ symbolically represent the following mutually exclusive equipment faults:

A = mass flow controller miscalibration

B = gas line leak

C = throttle valve malfunction

D = incorrect sensor signal

These components are illustrated graphically in the gas flow system shown in Figure 10.21.

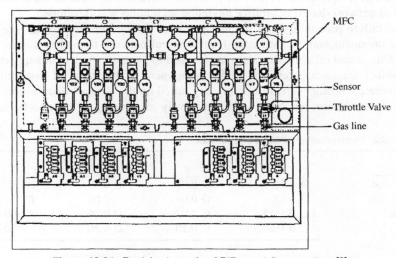

Figure 10.21. Partial schematic of RIE gas delivery system [9].

Suppose that belief in this frame of discernment is distributed according to the BPMDs:

$$m_1\langle A \cup C, B \cup D, \Theta \rangle = \langle 0.4, 0.3, 0.3 \rangle$$

$$m_2\langle A \cup B, C, D, \Theta \rangle = \langle 0.5, 0.1, 0.2, 0.2 \rangle$$

The calculation of the combined BPMD (m_3) is shown in Table 10.2. Each cell of the table contains the intersection of the corresponding propositions from m_1 and m_2, along with the product of their individual beliefs. Note that the intersection of any proposition with Θ is the original proposition. The BPM attributed to the empty set, k, which originates from the presence of various propositions in m_1 and m_2 whose intersection is empty, is 0.11. By applying Eq. (10.14), BPMs for the remaining propositions result in

$$m_3\langle A, A \cup C, A \cup B, B, B \cup D, C, D, \Theta \rangle$$

$$= \langle 0.225, 0.089, 0.169, 0.067, 0.079, 0.135, 0.067 \rangle$$

The plausibilities for propositions in the combined BPM are calculated by applying Eq. (10.13). The individual evidential intervals implied by m_3 are $A[0.225, 0.550]$, $B[0.169, 0.472]$, $C[0.079, 0.235]$, and $D[0.135, 0.269]$. Combining the evidence available from knowledge sources m_1 and m_2 thus leads to the conclusion that the most likely cause for the insufficient gas flow malfunction is a miscalibration of the mass flow controller (proposition A).

10.4.2.2. Maintenance Diagnosis

During maintenance diagnosis, the objective is to derive evidence of potential component failures based on historical performance. The available data consist of the number of failures a given component has experienced and the component age. To derive evidential support for potential malfunctions from this information, a neural network-based reliability modeling technique was developed.

The failure probability and the instantaneous failure rate (or *hazard rate*) for each component may be estimated from a neural network trained on failure history. This neural reliability model may be used to generate evidential support and plausibility for each potentially faulty component in the frame of discernment. To illustrate, consider reliability modeling based on the *Weibull* distribution. The Weibull distribution has been used extensively as a model of time-to-failure in

Table 10.2. Illustration of BPMD combination.

m_1				
$A \cup C$ 0.4	A 0.20	C 0.04	$\varnothing$ 0.08	$A \cup C$ 0.08
$B \cup D$ 0.3	B 0.15	$\varnothing$ 0.03	D 0.06	$B \cup D$ 0.06
Θ 0.3	$A \cup B$ 0.15	C 0.03	D 0.06	Θ 0.06
	$A \cup B$ 0.50	C 0.10	D 0.20	Θ 0.20
		m_2		

electrical and mechanical components and systems. When a system is composed of a number of components and failure is due to the most serious of a large number of possible faults, the Weibull distribution is a particularly accurate model.

The cumulative distribution function (which represents the failure probability of a component at time t) for the two-parameter Weibull distribution is given by

$$F(t) = 1 - \exp\left[-\left(\frac{t}{\alpha}\right)^{\beta}\right] \qquad (10.16)$$

where α and β are called "scale" and "shape" parameters, respectively. The hazard rate is given by

$$\lambda(t) = \frac{\beta t^{\beta-1}}{\alpha^{\beta}} \qquad (10.17)$$

The failure rate may be computed by plotting the number of failures of each component versus time and finding the slope of this curve at each timepoint. Following shape and scale parameter estimation, the evidential support for each component is obtained from the Weibull distribution function in Eq. (10.16). The corresponding plausibility is the *confidence level* (C) associated with this probability estimate, which is

$$C(t) = 1 - [1 - F(t)]^{n} \qquad (10.18)$$

where n is the total number of component failures that have been observed at time t. Applying this methodology to the Plasma Therm 700 series RIE yielded a ranked list of components faults similar to that shown in Table 10.3.

10.4.2.3. Online Diagnosis

In diagnosing previously encountered faults, neural network-based time-series (NTS) models are used to describe data indicating specific fault patterns [11]. The similarity between stored NTS fault models and the current sampled pattern is measured to ascertain their likelihood of resemblance. An underlying assumption is that malfunctions are triggered by inadvertent shifts in process settings.

Table 10.3. Fault ranking after maintenance diagnosis.

Component	Support	Plausibility
Capacitance manometer	0.353	0.508
Pressure switch	0.353	0.507
Electrode assembly	0.113	0.267
Exhaust valve controller	0.005	0.160
Throttle valve	0.003	0.159
Communication link	0.003	0.157
DC circuitry	0.003	0.157
Pressure transducer	0.003	0.157
Turbopump	0.003	0.157
Gas cylinder	0.002	0.157

This shift is assumed to be larger than the variability inherent in the processing equipment. To ascribe evidential support and plausibility to such a shift, statistical hypothesis tests are applied to sample means and variances of the time-series data. This requires the assumption that the notion of statistical *confidence* is analogous to the Dempster–Shafer concept of plausibility [1].

To compare two data patterns, it is assumed that if the two patterns are similar, then their means and variances are similar. Further, it is assumed that an equipment malfunction may cause either a shift in the mean or variance of a signal. The comparison begins by testing the hypothesis that the mean value of the current fault pattern ($\overline{x}_0$) equals the mean of previously stored fault patterns ($\overline{x}_i$). Letting s_0^2 and s_i^2 be the sample variances of current pattern and stored pattern, the appropriate test statistic is

$$t_0 = \frac{\overline{x}_0 - \overline{x}_i}{\sqrt{\dfrac{s_0^2}{n_0} + \dfrac{s_i^2}{n_i}}} \tag{10.19}$$

where n_0 and n_i are the sample sizes for the current and stored pattern, respectively. The statistical significance level for this hypothesis test (α_1) satisfies the relationship $t_0 = t_{\alpha, \nu_1}$, where ν is the number of degrees of freedom. A neural network that takes the role of a t-distribution "learner" can be used to predict α_1 based on the values of t_0 and ν. After the significance level has been computed, the probability that the mean values of the two data patterns are equal (β_1) is equal to $1 - \alpha_1$.

Next, the hypothesis that the variance of the current fault pattern (σ_0^2) equals the variance of each stored pattern (σ_i^2) is tested. The appropriate test statistic is

$$F_0 = \frac{s_0^2}{s_i^2} \tag{10.20}$$

The statistical significance for this hypothesis test (α_2) satisfies the relationship $F_0 = F_{\alpha_2, \nu_0, \nu_i}$, where ν_0 and ν_i are the degrees of freedom for s_0^2 and s_i^2. A neural network trained on the F distribution is used to predict α_2 using ν_0, ν_i, and F_0 as inputs. The resultant probability of equal variances is $\beta_2 = 1 - \alpha_2$. After completing the hypothesis tests for equal mean and variance, the support and plausibility that the current pattern is similar to a previously stored pattern are defined as

$$\text{Support} = \min(\beta_1, \beta_2) \tag{10.21}$$

$$\text{Plausibility} = \max(\beta_1, \beta_2)$$

Using the rules of evidence combination, the support and plausibility generated at each timepoint are continuously integrated with their prior values.

To demonstrate, data corresponding to the faulty CHF_3 flow in Figure 10.22 were used to derive an NTS model. The training set for the NTS model consisted of one out of every 10 data samples. The NTS fault model is stored in a database,

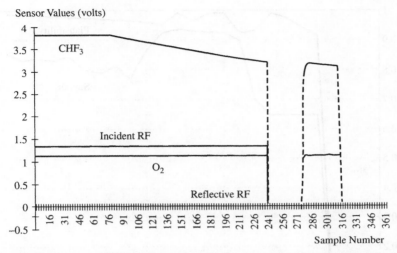

Figure 10.22. Data signatures for a malfunctioning chloroform mass flow controller [9].

from which it is compared to other patterns collected by sensors in real time so that the similarity of the sensor data to this stored pattern could be evaluated. In this example, the pattern of CHF_3 flow under consideration as a potential match to the stored fault pattern was sampled once for every 15 sensor data points. Following evaluation of the data, the evidential support and plausibility for pattern similarity are shown in Figure 10.23.

To identify malfunctions that have not been encountered previously, May established a technique based on the cusum control chart (see Chapter 6). The approach allows the detection of very small process shifts, which is critical for fabrication steps such as RIE, where slight equipment miscalibrations may have sufficient time to manifest themselves only as small shifts when the total processing time is on the order of minutes. In this application, the cusum chart monitors such shifts by comparing the cumulative sums of the deviations of the sample values from their targets.

Using this method to generate support requires the cumulative sums

$$S_H(i) = \max[0, x_i - (\mu_0 + K) + S_H(i - 1)] \qquad (10.22)$$

$$S_L(i) = \max[0, (\mu_0 - K) - x_i + S_L(i - 1)] \qquad (10.23)$$

where S_H is the sum used to detect positive process shifts, S_L is used to detect negative shifts, x_i is the mean value of the current sample, and μ_0 is the target value. In these equations, K is the *reference value*, which is chosen to be halfway between the target mean and the shifted mean to be detected (μ_1). If the shift is expressed in terms of the standard deviation as $\mu_1 = \mu_0 + \delta\sigma$, then K is

$$K = \frac{\delta}{2}\sigma = \frac{|\mu_1 - \mu_0|}{2} \qquad (10.24)$$

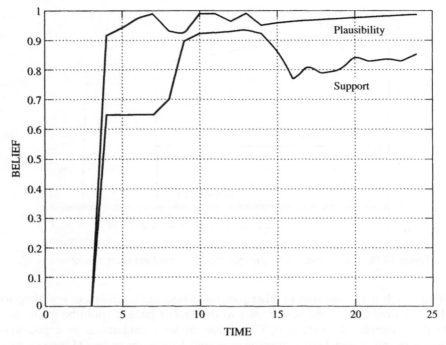

Figure 10.23. Plot of real-time support and plausibility for a recognized gas flow fault [9].

When either S_H or S_L exceeds the decision interval (H), this signals that the process has shifted out of statistical control. The decision interval may be used as the process tolerance limit, and the sums S_H and S_L are treated as measurement residuals. Support is derived from the cusum chart using

$$s(S_{H/L}) = \frac{1 - u}{1 + \exp\left[-\left(\dfrac{S_{H/L}}{H} - 1\right)\right]} \qquad (10.25)$$

where the uncertainty u is dictated by the measurement error of the sensor. As S_H or S_L become large compared to H, this function generates correspondingly larger support values.

To illustrate this technique, the faulty CHF_3 data pattern in Figure 10.22 is used again, this time under the assumption that no similar pattern exists in the database. The two parameters b and h vary continuously as the standard deviation of the monitored sensor data is changing. Equation (10.22) was used to calculate the accumulated deviations of CHF_3 flow. Each accumulated shift was then fed into the sigmoidal belief function in Eq. (10.25) to generate evidential support value. Figure 10.24 shows the incremental changes in the support values, clearly indicating the initial fault occurrence and the trend of process shifts.

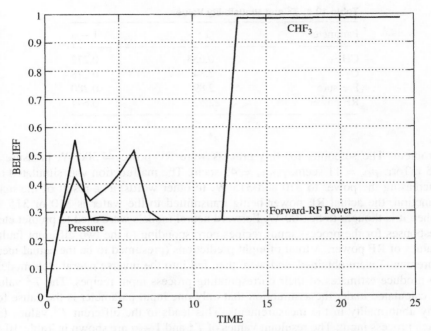

Figure 10.24. Support variations using cusum technique [9].

10.4.2.4. Inline Diagnosis

For inline diagnosis, measurements performed on processed wafers are used in conjunction with inverse neural process models. Inverse models are used to predict the etch recipe values (RF power, pressure, etc.) that reduce deviations in the measured etch responses. Since the setpoint recipes are different from those predicted by the inverse model, the vector of differences between them (Δx_0) can be used in a hypothesis test to determine the statistical significance of the deviations. That statistical significance can be calculated by testing the hypothesis that $\Delta x_0 = 0$.

Hotelling's T^2 statistic (see Chapter 6) is employed to obtain confidence intervals on the incremental changes in the input parameters. The value of the T^2 statistic is

$$T^2 = n \ \Delta x_0^T S^{-1} \Delta x_0 \qquad (10.26)$$

where n and S are the sample size and covariance matrix of the p process input parameters. Recall that the T^2 distribution is related to the F distribution by the relation

$$T_{\alpha,p,n-p}^2 = \frac{p(n-1)}{n-p} F_{\alpha,p,n-p} \qquad (10.27)$$

Plausibility values calculated for each input parameter are equal to $1 - \alpha$.

To illustrate, consider a fault scenario in which increased RF power was supplied to an RIE system during silicon dioxide etching due to an RF generator

Table 10.4. T^2 and Plausibility Values.

Parameter	T^2	$1 - \alpha$
CHF_3	0.053	0.272
O_2	2.84	0.278
Pressure	2.89	0.280
RF power	22.52	0.694

problem. The setpoints for this process were RF power = 300 W, pressure = 45 mTorr, p_{O_2} = 11 sccm, p_{CHF_3} = 45 sccm. The malfunction was simulated by increasing the power to 310 and 315 W. In other words, as a result of the malfunction, the actual RF power being transmitted to the wafer is 310 or 315 W when it is thought to be 300 W. Forward neural models were used to predict etch responses for the process input recipes corresponding to the two different faulty values of RF power. A total of eight predictions (presumed to be the actual measurements) were obtained, and were then fed into the inverse neural etch models to produce estimates of their corresponding process input recipes. The T^2 value is calculated under the assumption that only one input parameter is the cause for any abnormality in the measurements. This leads to the different T^2 values for each process input. The resultant values of T^2 and $1 - \alpha$ are shown in Table 10.4. As expected, RF power was the most significant input parameter since it has the highest plausibility value.

Hybrid neural expert systems offer the advantage of easier knowledge acquisition and maintenance and extracting implicit knowledge (through neural network learning) with the assistance of explicit expert rules. A disadvantage of such systems, however, is that, unlike other rule-based systems, the somewhat non-intuitive nature of neural networks makes it difficult to provide the user with explanations about how diagnostic conclusions are reached.

SUMMARY

This chapter has described a variety of methods for diagnosing problems in semiconductor manufacturing processes and equipment, including quantitative algorithmic techniques, qualitative experiential approaches, neural network-based pattern recognition methods, and hybrid combinations thereof. Such techniques are invaluable for reducing equipment downtime, limiting misprocessing, and enhancing manufacturing productivity and throughput.

PROBLEMS

10.1. Suggest an appropriate diagnostic situation for each of the process problems below. Justify your answers.

(a) A systematic photolithographic defect pattern is identified in a CMOS gate definition step.

(b) A PECVD system consistently produces films that are too thin.

(c) The threshold voltage in a batch of NMOS test transistors for a microprocessor line is out of specification.

10.2. Consider the combination of m_1 and m_2 when each contains different evidence concerning the diagnosis of a malfunction in the plasma etching application. Such evidence could result from two different sensor readings. In particular, suppose that the sensors have observed that the flow of one of the etchant gases into the process chamber is too low. Let the frame of discernment $\theta = \{A, B, C, D, E\}$, where $A, \ldots, E$ symbolically represent the following mutually exclusive equipment faults:

$$A = \text{mass flow controller miscalibration}$$

$$B = \text{gas line leak}$$

$$C = \text{throttle valve malfunction}$$

$$D = \text{incorrect sensor signal}$$

$$E = \text{the no-fault condition}$$

Suppose that belief in this frame of discernment is distributed according to the BPMDs:

$$m_1 \langle A \cup B, C, D, E, \Theta \rangle = \langle 0.48, 0.12, 0, 0.2, 0.2 \rangle$$
$$m_2 \langle B, A \cup C, D \cup E, \Theta \rangle = \langle 0, 0.7, 0.1, 0.2 \rangle$$

Use Dempster–Shafer theory to calculate a combined BPMD (m_3), as well as the individual evidential intervals implied by m_3.

REFERENCES

1. G. May and C. Spanos, "Automated Malfunction Diagnosis of Semiconductor Fabrication Equipment: A Plasma Etch Application," *IEEE Trans. Semiconduct. Manuf.* **6**(1), 28–40 (Feb. 1993).

2. C. Spanos, "Hippocrates: A Methodology for IC Process Diagnosis," *Proc. ICCAD*, 1986, pp. 513–516.

3. G. Freeman, W. Lukaszek, and J. Pan, "MERLIN: A Device Diagnosis System Based on Analytic Models," *IEEE Trans. Semiconduct. Manuf.* **6**(4), 306–317 (Nov. 1993).

4. J. Pan and J. Tenenbaum, "PIES: An Engineer's 'Do-it-Yourself' Knowledge System for Interpretation of Parametric Test Data," *Proc. 5th Nat. Conf. on AI*, 1986, pp. 836–843.

5. S. Dolins, A. Srivastava, and B. Flinchbaugh, "Monitoring and Diagnosis of Plasma Etch Processes," *IEEE Trans. Semiconduct. Manuf.* **1**(1), 23–27 (Feb. 1988).

6. J. Plummer, "Tighter Process Control with Neural Networks," *AI Expert* **10**, 49–55 (1993).

7. S. Bhatikar and R. Mahajan, "Artificial Neural Network Based Diagnosis of CVD Barrel Reactor," *IEEE Trans. Semiconduct. Manuf.* **15**(1), 71–78 (Feb. 2002).

8. E. Rietman and M. Beachy, "A Study on Failure Prediction in a Plasma Reactor," *IEEE Trans. Semiconduct. Manuf.* **11**(4), 670–680 (1998).

9. B. Kim and G. May, "Real-Time Diagnosis of Semiconductor Manufacturing Equipment Using Neural Networks," *IEEE Trans. Compon. Pack. Manuf. Technol. C* **20**(1), 39–47 (Jan. 1997).

10. G. Shafer, *A Mathematical Theory of Evidence*, Princeton Univ. Press, Princeton, NJ, 1976.

11. M. Baker, C. Himmel, and G. May, "Time Series Modeling of Reactive Ion Etching Using Neural Networks," *IEEE Trans. Semiconduct. Manuf.* **8**(1), 62–71 (Feb. 1995).

APPENDIX A

SOME PROPERTIES OF THE ERROR FUNCTION

w	$\text{erf}(w)$	w	$\text{erf}(w)$	w	$\text{erf}(w)$	w	$\text{erf}(w)$
0.00	0.000 000	0.19	0.211 840	0.38	0.409 009	0.57	0.579 816
0.01	0.011 283	0.20	0.222 703	0.39	0.418 739	0.58	0.587 923
0.02	0.022 565	0.21	0.233 522	0.40	0.428 392	0.59	0.595 936
0.03	0.033 841	0.22	0.244 296	0.41	0.437 969	0.60	0.603 856
0.04	0.045 111	0.23	0.255 023	0.42	0.447 468	0.61	0.611 681
0.05	0.056 372	0.24	0.265 700	0.43	0.456 887	0.62	0.619 411
0.06	0.067 622	0.25	0.276 326	0.44	0.466 225	0.63	0.627 046
0.07	0.078 858	0.26	0.286 900	0.45	0.475 482	0.64	0.634 586
0.08	0.090 078	0.27	0.297 418	0.46	0.484 655	0.65	0.642 029
0.09	0.101 281	0.28	0.307 880	0.47	0.493 745	0.66	0.649 377
0.10	0.112 463	0.29	0.318 283	0.48	0.502 750	0.67	0.656 628
0.11	0.123 623	0.30	0.328 627	0.49	0.511 668	0.68	0.663 782
0.12	0.134 758	0.31	0.338 908	0.50	0.520 500	0.69	0.670 840
0.13	0.145 867	0.32	0.349 126	0.51	0.529 244	0.70	0.677 801
0.14	0.156 947	0.33	0.359 279	0.52	0.537 899	0.71	0.684 666
0.15	0.167 996	0.34	0.369 365	0.53	0.546 464	0.72	0.691 433
0.16	0.179 012	0.35	0.379 382	0.54	0.554 939	0.73	0.698 104
0.17	0.189 992	0.36	0.389 330	0.55	0.563 323	0.74	0.704 678
0.18	0.200 936	0.37	0.399 206	0.56	0.571 616	0.75	0.711 156

Fundamentals of Semiconductor Manufacturing and Process Control,
By Gary S. May and Costas J. Spanos
Copyright © 2006 John Wiley & Sons, Inc.

w	erf(w)	w	erf(w)	w	erf(w)	w	erf(w)
0.76	0.717 537	1.16	0.899 096	1.56	0.972 628	1.97	0.994 664
0.77	0.723 822	1.17	0.902 000	1.57	0.973 603	1.98	0.994 892
0.78	0.730 010	1.18	0.904 837	1.58	0.974 547	1.99	0.995 111
0.79	0.736 103	1.19	0.907 608	1.59	0.975 462	2.00	0.995 322
0.80	0.742 101	1.20	0.910 314	1.60	0.976 348	2.01	0.995 525
0.81	0.748 003	1.21	0.912 956	1.61	0.977 207	2.02	0.995 719
0.82	0.753 811	1.22	0.915 534	1.62	0.978 038	2.03	0.995 906
0.83	0.759 524	1.23	0.918 050	1.63	0.978 843	2.04	0.996 086
0.84	0.765 143	1.24	0.920 505	1.64	0.979 622	2.05	0.996 258
0.85	0.770 668	1.25	0.922 900	1.65	0.980 376	2.06	0.996 423
0.86	0.776 110	1.26	0.925 236	1.66	0.981 105	2.07	0.996 582
0.87	0.781 440	1.27	0.927 514	1.67	0.981 810	2.08	0.996 734
0.88	0.786 687	1.28	0.929 734	1.68	0.982 493	2.09	0.996 880
0.89	0.719 843	1.29	0.931 899	1.69	0.983 153	2.10	0.997 021
0.90	0.796 908	1.30	0.934 008	1.70	0.983 790	2.11	0.997 155
0.91	0.801 883	1.31	0.936 063	1.71	0.984 407	2.12	0.997 284
0.92	0.806 768	1.32	0.938 065	1.72	0.985 003	2.13	0.997 407
0.93	0.811 564	1.33	0.940 015	1.73	0.985 578	2.14	0.997 525
0.94	0.816 271	1.34	0.941 914	1.74	0.986 135	2.15	0.997 639
0.95	0.820 891	1.35	0.943 762	1.75	0.986 672	2.16	0.997 747
0.96	0.825 424	1.36	0.945 561	1.76	0.987 190	2.17	0.997 851
0.97	0.829 870	1.37	0.947 312	1.77	0.987 691	2.18	0.997 951
0.98	0.834 232	1.38	0.949 016	1.79	0.988 641	2.19	0.998 046
0.99	0.838 508	1.39	0.950 673	1.80	0.989 091	2.20	0.998 137
1.00	0.842 701	1.40	0.952 285	1.81	0.989 525	2.21	0.998 224
1.01	0.846 810	1.41	0.953 852	1.82	0.989 943	2.22	0.998 308
1.02	0.850 838	1.42	0.955 376	1.83	0.990 347	2.23	0.998 388
1.03	0.854 784	1.43	0.956 857	1.84	0.990 736	2.24	0.998 464
1.04	0.858 650	1.44	0.958 297	1.85	0.991 111	2.25	0.998 537
1.05	0.862 436	1.45	0.959 695	1.86	0.991 472	2.26	0.998 607
1.06	0.866 144	1.46	0.961 054	1.87	0.991 821	2.27	0.998 674
1.07	0.869 773	1.47	0.962 373	1.88	0.992 156	2.28	0.998 738
1.08	0.873 326	1.48	0.963 654	1.89	0.992 479	2.29	0.998 799
1.09	0.876 803	1.49	0.964 898	1.90	0.992 790	2.30	0.998 857
1.10	0.880 205	1.50	0.966 105	1.91	0.993 090	2.31	0.998 912
1.11	0.883 533	1.51	0.967 277	1.92	0.993 378	2.32	0.998 966
1.12	0.886 788	1.52	0.968 413	1.93	0.993 656	2.33	0.999 016
1.13	0.889 971	1.53	0.969 516	1.94	0.993 923	2.34	0.999 065
1.14	0.893 082	1.54	0.970 586	1.95	0.994 179	2.35	0.999 111
1.15	0.896 124	1.55	0.971 623	1.96	0.994 426	2.36	0.999 155

w	erf(w)	w	erf(w)	w	erf(w)	w	erf(w)
2.37	0.999 197	2.78	0.999 916	3.20	0.999 993 97	3.61	0.999 999 670
2.38	0.999 237	2.79	0.999 920	3.21	0.999 994 36	3.62	0.999 999 694
2.39	0.999 275	2.80	0.999 925	3.22	0.999 994 73	3.63	0.999 999 716
2.40	0.999 311	2.81	0.999 929	3.23	0.999 995 07	3.64	0.999 999 736
2.41	0.999 346	2.82	0.999 933	3.24	0.999 995 40	3.65	0.999 999 756
2.42	0.999 379	2.83	0.999 937	3.25	0.999 995 70	3.66	0.999 999 773
2.43	0.999 411	2.85	0.999 944	3.26	0.999 995 98	3.67	0.999 999 790
2.44	0.999 441	2.86	0.999 948	3.27	0.999 996 24	3.68	0.999 999 805
2.45	0.999 469	2.87	0.999 951	3.28	0.999 996 49	3.69	0.999 999 820
2.46	0.999 497	2.88	0.999 954	3.29	0.999 996 72	3.70	0.999 999 833
2.47	0.999 523	2.89	0.999 956	3.30	0.999 996 94	3.71	0.999 999 845
2.48	0.999 547	2.90	0.999 959	3.31	0.999 997 15	3.72	0.999 999 857
2.49	0.999 571	2.91	0.999 961	3.32	0.999 997 34	3.73	0.999 999 867
2.50	0.999 593	2.92	0.999 964	3.33	0.999 997 51	3.74	0.999 999 877
2.51	0.999 614	2.93	0.999 966	3.34	0.999 997 68	3.75	0.999 999 886
2.52	0.999 634	2.94	0.999 968	3.35	0.999 997 838	3.76	0.999 999 895
2.53	0.999 654	2.95	0.999 970	3.36	0.999 997 983	3.77	0.999 999 903
2.54	0.999 672	2.96	0.999 972	3.37	0.999 998 120	3.78	0.999 999 910
2.55	0.999 689	2.97	0.999 973	3.38	0.999 998 247	3.79	0.999 999 917
2.56	0.999 706	2.98	0.999 975	3.39	0.999 998 367	3.80	0.999 999 923
2.57	0.999 722	2.99	0.999 976	3.40	0.999 998 478	3.81	0.999 999 929
2.58	0.999 736	3.00	0.999 977 91	3.41	0.999 998 582	3.82	0.999 999 934
2.59	0.999 751	3.01	0.999 979 26	3.42	0.999 998 679	3.83	0.999 999 939
2.60	0.999 764	3.02	0.999 980 53	3.43	0.999 998 770	3.84	0.999 999 944
2.61	0.999 777	3.03	0.999 981 73	3.44	0.999 998 855	3.85	0.999 999 948
2.62	0.999 789	3.04	0.999 982 86	3.45	0.999 998 934	3.86	0.999 999 952
2.63	0.999 800	3.05	0.999 983 92	3.46	0.999 999 008	3.87	0.999 999 956
2.64	0.999 811	3.06	0.999 984 92	3.47	0.999 999 077	3.88	0.999 999 959
2.65	0.999 822	3.07	0.999 985 86	3.48	0.999 999 141	3.89	0.999 999 962
2.66	0.999 831	3.08	0.999 986 74	3.49	0.999 999 201	3.90	0.999 999 965
2.67	0.999 841	3.09	0.999 987 57	3.50	0.999 999 257	3.91	0.999 999 968
2.68	0.999 849	3.10	0.999 988 35	3.51	0.999 999 309	3.92	0.999 999 970
2.69	0.999 858	3.11	0.999 989 08	3.52	0.999 999 358	3.93	0.999 999 973
2.70	0.999 866	3.12	0.999 989 77	3.53	0.999 999 403	3.94	0.999 999 975
2.71	0.999 873	3.13	0.999 990 42	3.54	0.999 999 445	3.95	0.999 999 977
2.72	0.999 880	3.14	0.999 991 03	3.55	0.999 999 485	3.96	0.999 999 979
2.73	0.999 887	3.15	0.999 991 60	3.56	0.999 999 521	3.97	0.999 999 980
2.74	0.999 893	3.16	0.999 992 14	3.57	0.999 999 555	3.98	0.999 999 982
2.75	0.999 899	3.17	0.999 992 64	3.58	0.999 999 587	3.99	0.999 999 983
2.76	0.999 905	3.18	0.999 993 11	3.59	0.999 999 617		
2.77	0.999 910	3.19	0.999 993 56	3.60	0.999 999 644		

(from May & Sze, Fundamentals of Semiconductor Manufacturing, Wiley, 2004)

APPENDIX B

CUMULATIVE STANDARD NORMAL DISTRIBUTION

$$\Phi(z) = \int_{-\infty}^{z} \frac{1}{\sqrt{2\pi}}\, e^{-u^2/2}\, du$$

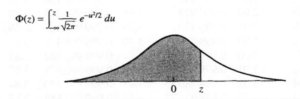

z	0.00	0.01	0.02	0.03	0.04	z
0.0	0.50000	0.50399	0.50798	0.51197	0.51595	0.0
0.1	0.53983	0.54379	0.54776	0.55172	0.55567	0.1
0.2	0.57926	0.58317	0.58706	0.59095	0.59483	0.2
0.3	0.61791	0.62172	0.62551	0.62930	0.63307	0.3
0.4	0.65542	0.65910	0.62276	0.66640	0.67003	0.4
0.5	0.69146	0.69497	0.69847	0.70194	0.70540	0.5
0.6	0.72575	0.72907	0.73237	0.73565	0.73891	0.6
0.7	0.75803	0.76115	0.76424	0.76730	0.77035	0.7
0.8	0.78814	0.79103	0.79389	0.79673	0.79954	0.8
0.9	0.81594	0.81859	0.82121	0.82381	0.82639	0.9
1.0	0.84134	0.84375	0.84613	0.84849	0.85083	1.0
1.1	0.86433	0.86650	0.86864	0.87076	0.87285	1.1
1.2	0.88493	0.88686	0.88877	0.89065	0.89251	1.2
1.3	0.90320	0.90490	0.90658	0.90824	0.90988	1.3
1.4	0.91924	0.92073	0.92219	0.92364	0.92506	1.4

Fundamentals of Semiconductor Manufacturing and Process Control,
By Gary S. May and Costas J. Spanos
Copyright © 2006 John Wiley & Sons, Inc.

z	0.00	0.01	0.02	0.03	0.04	z
1.5	0.93319	0.93448	0.93574	0.93699	0.93822	1.5
1.6	0.94520	0.94630	0.94738	0.94845	0.94950	1.6
1.7	0.95543	0.95637	0.95728	0.95818	0.95907	1.7
1.8	0.96407	0.96485	0.96562	0.96637	0.96711	1.8
1.9	0.97128	0.97193	0.97257	0.97320	0.97381	1.9
2.0	0.97725	0.97778	0.97831	0.97882	0.97932	2.0
2.1	0.98214	0.98257	0.98300	0.98341	0.98382	2.1
2.2	0.98610	0.98645	0.98679	0.98713	0.98745	2.2
2.3	0.98928	0.98956	0.98983	0.99010	0.99036	2.3
2.4	0.99180	0.99202	0.99224	0.99245	0.99266	2.4
2.5	0.99379	0.99396	0.99413	0.99430	0.99446	2.5
2.6	0.99534	0.99547	0.99560	0.99573	0.99585	2.6
2.7	0.99653	0.99664	0.99674	0.99683	0.99693	2.7
2.8	0.99744	0.99752	0.99760	0.99767	0.99774	2.8
2.9	0.99813	0.99819	0.99825	0.99831	0.99836	2.9
3.0	0.99865	0.99869	0.99874	0.99878	0.99882	3.0
3.1	0.99903	0.99906	0.99910	0.99913	0.99916	3.1
3.2	0.99931	0.99934	0.99936	0.99938	0.99940	3.2
3.3	0.99952	0.99953	0.99955	0.99957	0.99958	3.3
3.4	0.99966	0.99968	0.99969	0.99970	0.99971	3.4
3.5	0.99977	0.99978	0.99978	0.99979	0.99980	3.5
3.6	0.99984	0.99985	0.99985	0.99986	0.99986	3.6
3.7	0.99989	0.99990	0.99990	0.99990	0.99991	3.7
3.8	0.99993	0.99993	0.99993	0.99994	0.99994	3.8
3.9	0.99995	0.99995	0.99996	0.99996	0.99996	3.9

(from Montgomery, Intro to Statistical Quality Control, 3rd ed., Wiley, 1997)

$$\Phi(z) = \int_{-\infty}^{z} \frac{1}{\sqrt{2\pi}} e^{-u^2/2} \, du$$

z	0.05	0.06	0.07	0.08	0.09	z
0.0	0.51994	0.52392	0.52790	0.53188	0.53586	0.0
0.1	0.55962	0.56356	0.56749	0.57142	0.57534	0.1
0.2	0.59871	0.60257	0.60642	0.61026	0.61409	0.2
0.3	0.63683	0.64058	0.64431	0.64803	0.65173	0.3
0.4	0.67364	0.67724	0.68082	0.68438	0.68793	0.4
0.5	0.70884	0.71226	0.71566	0.71904	0.72240	0.5
0.6	0.74215	0.74537	0.74857	0.75175	0.75490	0.6
0.7	0.77337	0.77637	0.77935	0.78230	0.78523	0.7
0.8	0.80234	0.80510	0.80785	0.81057	0.81327	0.8
0.9	0.82894	0.83147	0.83397	0.83646	0.83891	0.9
1.0	0.85314	0.85543	0.85769	0.85993	0.86214	1.0
1.1	0.87493	0.87697	0.87900	0.88100	0.88297	1.1
1.2	0.89435	0.89616	0.89796	0.89973	0.90147	1.2
1.3	0.91149	0.91308	0.91465	0.91621	0.91773	1.3
1.4	0.92647	0.92785	0.92922	0.93056	0.93189	1.4
1.5	0.93943	0.94062	0.94179	0.94295	0.94408	1.5
1.6	0.95053	0.95154	0.95254	0.95352	0.95448	1.6
1.7	0.95994	0.96080	0.96164	0.96246	0.96327	1.7
1.8	0.96784	0.96856	0.96926	0.96995	0.97062	1.8
1.9	0.97441	0.97500	0.97558	0.97615	0.97670	1.9
2.0	0.97982	0.98030	0.98077	0.98124	0.98169	2.0
2.1	0.98422	0.98461	0.98500	0.98537	0.98574	2.1
2.2	0.98778	0.98809	0.98840	0.98870	0.98899	2.2
2.3	0.99061	0.99086	0.99111	0.99134	0.99158	2.3
2.4	0.99286	0.99305	0.99324	0.99343	0.99361	2.4
2.5	0.99461	0.99477	0.99492	0.99506	0.99520	2.5
2.6	0.99598	0.99609	0.99621	0.99632	0.99643	2.6
2.7	0.99702	0.99711	0.99720	0.99728	0.99736	2.7
2.8	0.99781	0.99788	0.99795	0.99801	0.99807	2.8
2.9	0.99841	0.99846	0.99851	0.99856	0.99861	2.9
3.0	0.99886	0.99889	0.99893	0.99897	0.99900	3.0
3.1	0.99918	0.99921	0.99924	0.99926	0.99929	3.1
3.2	0.99942	0.99944	0.99946	0.99948	0.99950	3.2
3.3	0.99960	0.99961	0.99962	0.99964	0.99965	3.3
3.4	0.99972	0.99973	0.99974	0.99975	0.99976	3.4
3.5	0.99981	0.99981	0.99982	0.99983	0.99983	3.5
3.6	0.99987	0.99987	0.99988	0.99988	0.99989	3.6
3.7	0.99991	0.99992	0.99992	0.99992	0.99992	3.7
3.8	0.99994	0.99994	0.99995	0.99995	0.99995	3.8
3.9	0.99996	0.99996	0.99996	0.99997	0.99997	3.9

APPENDIX C

PERCENTAGE POINTS OF THE χ^2 DISTRIBUTION[a]

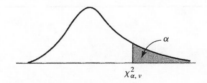

$\chi^2_{\alpha, v}$

					α				
v	0.995	0.990	0.975	0.950	0.500	0.050	0.025	0.010	0.005
1	0.00+	0.00+	0.00+	0.00+	0.45	3.84	5.02	6.63	7.88
2	0.01	0.02	0.05	0.10	1.39	5.99	7.38	9.21	10.60
3	0.07	0.11	0.22	0.35	2.37	7.81	9.35	11.34	12.84
4	0.21	0.30	0.48	0.71	3.36	9.49	11.14	13.28	14.86
5	0.41	0.55	0.83	1.15	4.35	11.07	12.38	15.09	16.75
6	0.68	0.87	1.24	1.64	5.35	12.59	14.45	16.81	18.55
7	0.99	1.24	1.69	2.17	6.35	14.07	16.01	18.48	20.28
8	1.34	1.65	2.18	2.73	7.34	15.51	17.53	20.09	21.96
9	1.73	2.09	2.70	3.33	8.34	16.92	19.02	21.67	23.59
10	2.16	2.56	3.25	3.94	9.34	18.31	20.48	23.21	25.19
11	2.60	3.05	3.82	4.57	10.34	19.68	21.92	24.72	26.76
12	3.07	3.57	4.40	5.23	11.34	21.03	23.34	26.22	28.30
13	3.57	4.11	5.01	5.89	12.34	22.36	24.74	27.69	29.82

[a]Adapted with permission from *Biometrika Tables for Statisticians*, Vol. 1, 3rd ed., by E. S. Pearson and H. O. Hartley, Cambridge University Press, Cambridge, 1966.

Fundamentals of Semiconductor Manufacturing and Process Control,
By Gary S. May and Costas J. Spanos
Copyright © 2006 John Wiley & Sons, Inc.

				α					
v	0.995	0.990	0.975	0.950	0.500	0.050	0.025	0.010	0.005
14	4.07	4.66	5.63	6.57	13.34	23.68	26.12	29.14	31.32
15	4.60	5.23	6.27	7.26	14.34	25.00	27.49	30.58	32.80
16	5.14	5.81	6.91	7.96	15.34	26.30	28.85	32.00	34.27
17	5.70	6.41	7.56	8.67	16.34	27.59	30.19	33.41	35.72
18	6.26	7.01	8.23	9.39	17.34	28.87	31.53	34.81	37.16
19	6.84	7.63	8.91	10.12	18.34	30.14	32.85	36.19	38.58
20	7.43	8.26	9.59	10.85	19.34	31.41	34.17	37.57	40.00
25	10.52	11.52	13.12	14.61	24.34	37.65	40.65	44.31	46.93
30	13.79	14.95	16.79	18.49	29.34	43.77	46.98	50.89	53.67
40	20.71	22.16	24.43	26.51	39.34	55.76	59.34	63.69	66.77
50	27.99	29.71	32.36	34.76	49.33	67.50	71.42	76.15	79.49
60	35.53	37.48	40.48	43.19	59.33	79.08	83.30	88.38	91.95
70	43.28	45.44	48.76	51.74	69.33	90.53	95.02	100.42	104.22
80	51.17	53.54	57.15	60.39	79.33	101.88	106.63	112.33	116.32
90	59.20	61.75	65.65	69.13	89.33	113.14	118.14	124.12	128.30
100	67.33	70.06	74.22	77.93	99.33	124.34	129.56	135.81	140.17

v = degrees of freedom.
(from Montgomery, 1997)

APPENDIX D

PERCENTAGE POINTS OF THE t DISTRIBUTION[a]

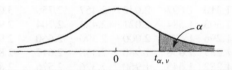

					α					
v	0.40	0.25	0.10	0.05	0.025	0.01	0.005	0.0025	0.001	0.0005
1	0.325	1.000	3.078	6.314	12.706	31.821	63.657	127.32	318.31	636.62
2	0.289	0.816	1.886	2.920	4.303	6.965	9.925	14.089	23.326	31.598
3	0.277	0.765	1.638	2.353	3.182	4.541	5.841	7.453	10.213	12.924
4	0.271	0.741	1.533	2.132	2.776	3.747	4.604	5.598	7.173	8.610
5	0.267	0.727	1.476	2.015	2.571	3.365	4.032	4.773	5.893	6.869
6	0.265	0.727	1.440	1.943	2.447	3.143	3.707	4.317	5.208	5.959
7	0.263	0.711	1.415	1.895	2.365	2.998	3.49	4.019	4.785	5.408
8	0.262	0.706	1.397	1.860	2.306	2.896	3.355	3.833	4.501	5.041
9	0.261	0.703	1.383	1.833	2.262	2.821	3.250	3.690	4.297	4.781
10	0.260	0.700	1.372	1.812	2.228	2.764	3.169	3.581	4.144	4.587
11	0.260	0.697	1.363	1.796	2.201	2.718	3.106	3.497	4.025	4.437
12	0.259	0.695	1.356	1.782	2.179	2.681	3.055	3.428	3.930	4.318
13	0.259	0.694	1.350	1.771	2.160	2.650	3.012	3.372	3.852	4.221

[a] Adapted with permission from *Biometrika Tables for Statisticians*, Vol. 1, 3rd ed., by E. S. Pearson and H. O. Hartley, Cambridge University Press, Cambridge, 1966.

Fundamentals of Semiconductor Manufacturing and Process Control,
By Gary S. May and Costas J. Spanos
Copyright © 2006 John Wiley & Sons, Inc.

v	0.40	0.25	0.10	0.05	0.025	α 0.01	0.005	0.0025	0.001	0.0005
14	0.258	0.692	1.345	1.761	2.145	2.624	2.977	3.326	3.787	4.140
15	0.258	0.691	1.341	1.753	2.131	2.602	2.947	3.286	3.733	4.073
16	0.258	0.690	1.337	1.746	2.120	2.583	2.921	3.252	3.686	4.015
17	0.257	0.689	1.333	1.740	2.110	2.567	2.898	3.222	3.646	3.965
18	0.257	0.688	1.330	1.734	2.101	2.552	2.878	3.197	3.610	3.992
19	0.257	0.688	1.328	1.729	2.093	2.539	2.861	3.174	3.579	3.883
20	0.257	0.687	1.325	1.725	2.086	2.528	2.845	3.153	3.552	3.850
21	0.257	0.686	1.323	1.721	2.080	2.518	2.831	3.135	3.527	3.819
22	0.256	0.686	1.321	1.717	2.074	2.508	2.819	3.119	3.505	3.792
23	0.256	0.685	1.319	1.714	2.069	2.500	2.807	3.104	3.485	3.767
24	0.256	0.685	1.318	1.711	2.064	2.492	2.797	3.091	3.467	3.745
25	0.256	0.684	1.316	1.708	2.060	2.485	2.787	3.078	3.450	3.725
26	0.256	0.684	1.315	1.706	2.056	2.479	2.779	3.067	3.435	3.707
27	0.256	0.684	1.314	1.703	2.052	2.473	2.771	3.057	3.421	3.690
28	0.256	0.683	1.313	1.701	2.048	2.467	2.763	3.047	3.408	3.674
29	0.256	0.683	1.311	1.699	2.045	2.462	2.756	3.038	3.396	3.659
30	0.256	0.683	1.310	1.697	2.042	2.457	2.750	3.030	3.385	3.646
40	0.255	0.681	1.303	1.684	2.021	2.423	2.704	2.971	3.307	3.551
60	0.254	0.679	1.296	1.671	2.000	2.390	2.660	2.915	3.232	3.460
120	0.254	0.677	1.289	1.658	1.980	2.358	2.617	2.860	3.160	3.373
∞	0.253	0.674	1.282	1.645	1.960	2.326	2.576	2.807	3.090	3.291

v = degrees of freedom.
(from Montgomery, 1997)

APPENDIX E

PERCENTAGE POINTS OF THE F DISTRIBUTION

Fundamentals of Semiconductor Manufacturing and Process Control,
By Gary S. May and Costas J. Spanos
Copyright © 2006 John Wiley & Sons, Inc.

$$F_{0.25, \nu_1, \nu_2}$$

$\nu_2 \backslash \nu_1$	Degrees of freedom for the numerator (ν_1)																		
	1	2	3	4	5	6	7	8	9	10	12	15	20	24	30	40	60	120	∞
1	5.83	7.50	8.20	8.58	8.82	8.98	9.10	9.19	9.26	9.32	9.41	9.49	9.58	9.63	9.67	9.71	9.76	9.80	9.85
2	2.57	3.00	3.15	3.23	3.28	3.31	3.34	3.35	3.37	3.38	3.39	3.41	3.43	3.43	3.44	3.45	3.46	3.47	3.48
3	2.02	2.28	2.36	2.39	2.41	2.42	2.43	2.44	2.44	2.44	2.45	2.46	2.46	2.46	2.47	2.47	2.47	2.47	2.47
4	1.81	2.00	2.05	2.06	2.07	2.08	2.08	2.08	2.08	2.08	2.08	2.08	2.08	2.08	2.08	2.08	2.08	2.08	2.08
5	1.69	1.85	1.88	1.89	1.89	1.89	1.89	1.89	1.89	1.89	1.89	1.89	1.88	1.88	1.88	1.88	1.87	1.87	1.87
6	1.62	1.76	1.78	1.79	1.79	1.78	1.78	1.78	1.77	1.77	1.77	1.76	1.76	1.75	1.75	1.75	1.74	1.74	1.74
7	1.57	1.70	1.72	1.72	1.71	1.71	1.70	1.70	1.70	1.69	1.68	1.68	1.67	1.67	1.66	1.66	1.65	1.65	1.65
8	1.54	1.66	1.67	1.66	1.66	1.65	1.64	1.64	1.63	1.63	1.62	1.62	1.61	1.60	1.60	1.59	1.59	1.58	1.58
9	1.51	1.62	1.63	1.63	1.62	1.61	1.60	1.60	1.59	1.59	1.58	1.57	1.56	1.56	1.55	1.54	1.54	1.53	1.53
10	1.49	1.60	1.60	1.59	1.59	1.58	1.57	1.56	1.56	1.55	1.54	1.53	1.52	1.52	1.51	1.51	1.50	1.49	1.48
11	1.47	1.58	1.58	1.57	1.56	1.55	1.54	1.53	1.53	1.52	1.51	1.50	1.49	1.49	1.48	1.47	1.47	1.46	1.45
12	1.46	1.56	1.56	1.55	1.54	1.53	1.52	1.51	1.51	1.50	1.49	1.48	1.47	1.46	1.45	1.45	1.44	1.43	1.42
13	1.45	1.55	1.55	1.53	1.52	1.51	1.50	1.49	1.49	1.48	1.47	1.46	1.45	1.44	1.43	1.42	1.42	1.41	1.40
14	1.44	1.53	1.53	1.52	1.51	1.50	1.49	1.48	1.47	1.46	1.45	1.44	1.43	1.42	1.41	1.41	1.40	1.39	1.38
15	1.43	1.52	1.52	1.51	1.49	1.48	1.47	1.46	1.46	1.45	1.44	1.43	1.41	1.41	1.40	1.39	1.38	1.37	1.36
16	1.42	1.51	1.51	1.50	1.48	1.47	1.46	1.45	1.44	1.44	1.43	1.41	1.40	1.39	1.38	1.37	1.36	1.35	1.34
17	1.42	1.51	1.50	1.49	1.47	1.46	1.45	1.44	1.43	1.43	1.41	1.40	1.39	1.38	1.37	1.36	1.35	1.34	1.33
18	1.41	1.50	1.49	1.48	1.46	1.45	1.44	1.43	1.42	1.42	1.40	1.39	1.38	1.37	1.36	1.35	1.34	1.33	1.32
19	1.41	1.49	1.49	1.47	1.46	1.44	1.43	1.42	1.41	1.41	1.40	1.38	1.37	1.36	1.35	1.34	1.33	1.32	1.30

Degrees of freedom for the denominator (ν_2)

v_2	$v_1=1$	2	3	4	5	6	7	8	9	10	12	15	20	24	30	40	60	120	∞
20	1.40	1.49	1.48	1.47	1.45	1.44	1.43	1.42	1.41	1.40	1.39	1.37	1.36	1.35	1.34	1.33	1.32	1.31	1.29
21	1.40	1.48	1.48	1.46	1.44	1.43	1.42	1.41	1.40	1.39	1.38	1.37	1.35	1.34	1.33	1.32	1.31	1.30	1.28
22	1.40	1.48	1.47	1.45	1.44	1.42	1.41	1.40	1.39	1.39	1.37	1.36	1.34	1.33	1.32	1.31	1.30	1.29	1.28
23	1.39	1.47	1.47	1.45	1.43	1.42	1.41	1.40	1.39	1.38	1.37	1.35	1.34	1.33	1.32	1.31	1.30	1.28	1.27
24	1.39	1.47	1.46	1.44	1.43	1.41	1.40	1.39	1.38	1.38	1.36	1.35	1.33	1.32	1.31	1.30	1.29	1.28	1.26
25	1.39	1.47	1.46	1.44	1.42	1.41	1.40	1.39	1.38	1.37	1.36	1.34	1.33	1.32	1.31	1.29	1.28	1.27	1.25
26	1.38	1.46	1.45	1.44	1.42	1.41	1.39	1.38	1.37	1.37	1.35	1.34	1.32	1.31	1.30	1.29	1.28	1.26	1.25
27	1.38	1.46	1.45	1.44	1.42	1.41	1.39	1.38	1.37	1.36	1.35	1.33	1.32	1.31	1.30	1.28	1.27	1.26	1.24
28	1.38	1.46	1.45	1.43	1.41	1.40	1.39	1.38	1.37	1.36	1.34	1.33	1.31	1.30	1.29	1.28	1.27	1.25	1.24
29	1.38	1.45	1.45	1.43	1.41	1.40	1.38	1.37	1.37	1.35	1.34	1.32	1.31	1.30	1.29	1.27	1.26	1.25	1.23
30	1.38	1.45	1.44	1.42	1.41	1.39	1.38	1.37	1.36	1.35	1.34	1.32	1.30	1.29	1.28	1.27	1.26	1.24	1.23
40	1.36	1.44	1.42	1.40	1.39	1.37	1.36	1.35	1.34	1.33	1.31	1.30	1.28	1.26	1.25	1.24	1.22	1.21	1.19
60	1.35	1.42	1.41	1.38	1.37	1.35	1.33	1.32	1.31	1.30	1.29	1.27	1.25	1.24	1.22	1.21	1.19	1.17	1.15
120	1.34	1.40	1.39	1.37	1.35	1.33	1.31	1.30	1.29	1.28	1.26	1.24	1.22	1.21	1.19	1.18	1.16	1.13	1.10
∞	1.32	1.39	1.37	1.35	1.33	1.31	1.29	1.28	1.27	1.25	1.24	1.22	1.19	1.18	1.16	1.14	1.12	1.08	1.00

Note: $F_{0.75,v_1,v_2} = 1/F_{0.25,v_2,v_1}$.

Source: Adapted with permission from *Biometrika Tables for Statisticians*, Vol. 1, 3rd ed., by E. S. Pearson and H. O. Hartley, Cambridge University Press, Cambridge, 1966.

(from Montgomery, 1997)

$$F_{0.10,v_1,v_2}$$

v_2 \ v_1	1	2	3	4	5	6	7	8	9	10	12	15	20	24	30	40	60	120	∞
1	39.86	49.50	53.59	55.83	57.24	58.20	58.91	59.44	59.86	60.19	60.71	61.22	61.74	62.00	62.26	62.53	62.79	63.06	63.33
2	8.53	9.00	9.16	9.24	9.29	9.33	9.35	9.37	9.38	9.39	9.41	9.42	9.44	9.45	9.46	9.47	9.47	9.48	9.49
3	5.54	5.46	5.39	5.34	5.31	5.28	5.27	5.25	5.24	5.23	5.22	5.20	5.18	5.18	5.17	5.16	5.15	5.14	5.13
4	4.54	4.32	4.19	4.11	4.05	4.01	3.98	3.95	3.94	3.92	3.90	3.87	3.84	3.83	3.82	3.80	3.79	3.78	3.76
5	4.06	3.78	3.62	3.52	3.45	3.40	3.37	3.34	3.32	3.30	3.27	3.24	3.21	3.19	3.17	3.16	3.14	3.12	3.10
6	3.78	3.46	3.29	3.18	3.11	3.05	3.01	2.98	2.96	2.94	2.90	2.87	2.84	2.82	2.80	2.78	2.76	2.74	2.72
7	3.59	3.26	3.07	2.96	2.88	2.83	2.78	2.75	2.72	2.70	2.67	2.63	2.59	2.58	2.56	2.54	2.51	2.49	2.47
8	3.46	3.11	2.92	2.81	2.73	2.67	2.62	2.59	2.56	2.54	2.50	2.46	2.42	2.40	2.38	2.36	2.34	2.32	2.29
9	3.36	3.01	2.81	2.69	2.61	2.55	2.51	2.47	2.44	2.42	2.38	2.34	2.30	2.28	2.25	2.23	2.21	2.18	2.16
10	3.29	2.92	2.73	2.61	2.52	2.46	2.41	2.38	2.35	2.32	2.28	2.24	2.20	2.18	2.16	2.13	2.11	2.08	2.06
11	3.23	2.86	2.66	2.54	2.45	2.39	2.34	2.30	2.27	2.25	2.21	2.17	2.12	2.10	2.08	2.05	2.03	2.00	1.97
12	3.18	2.81	2.61	2.48	2.39	2.33	2.28	2.24	2.21	2.19	2.15	2.10	2.06	2.04	2.01	1.99	1.96	1.93	1.90
13	3.14	2.76	2.56	2.43	2.35	2.28	2.23	2.20	2.16	2.14	2.10	2.05	2.01	1.98	1.96	1.93	1.90	1.88	1.85
14	3.10	2.73	2.52	2.39	2.31	2.24	2.19	2.15	2.12	2.10	2.05	2.01	1.96	1.94	1.91	1.89	1.86	1.83	1.80
15	3.07	2.70	2.49	2.36	2.27	2.21	2.16	2.12	2.09	2.06	2.02	1.97	1.92	1.90	1.87	1.85	1.82	1.79	1.76
16	3.05	2.67	2.46	2.33	2.24	2.18	2.13	2.09	2.06	2.03	1.99	1.94	1.89	1.86	1.84	1.81	1.78	1.75	1.72
17	3.03	2.64	2.44	2.31	2.22	2.15	2.10	2.06	2.03	2.00	1.96	1.91	1.86	1.84	1.81	1.78	1.75	1.72	1.69
18	3.01	2.62	2.42	2.29	2.20	2.13	2.08	2.04	2.00	1.98	1.93	1.89	1.84	1.81	1.78	1.75	1.72	1.69	1.66
19	2.99	2.61	2.40	2.27	2.18	2.11	2.06	2.02	1.98	1.96	1.91	1.86	1.81	1.79	1.76	1.73	1.70	1.67	1.63

Degrees of freedom for the numerator (v_1)

Degrees of freedom for the denominator (v_2)

430

20	2.97	2.59	2.38	2.25	2.16	2.09	2.04	2.00	1.96	1.94	1.89	1.84	1.79	1.77	1.74	1.71	1.68	1.64	1.61
21	2.96	2.57	2.36	2.23	2.14	2.08	2.02	1.98	1.95	1.92	1.87	1.83	1.78	1.75	1.72	1.69	1.66	1.62	1.59
22	2.95	2.56	2.35	2.22	2.13	2.06	2.01	1.97	1.93	1.90	1.86	1.81	1.76	1.73	1.70	1.67	1.64	1.60	1.57
23	2.94	2.55	2.34	2.21	2.11	2.05	1.99	1.95	1.92	1.89	1.84	1.80	1.74	1.72	1.69	1.66	1.62	1.59	1.55
24	2.93	2.54	2.33	2.19	2.10	2.04	1.98	1.94	1.91	1.88	1.83	1.78	1.73	1.70	1.67	1.64	1.61	1.57	1.53
25	2.92	2.53	2.32	2.18	2.09	2.02	1.97	1.93	1.89	1.87	1.82	1.77	1.72	1.69	1.66	1.63	1.59	1.56	1.52
26	2.91	2.52	2.31	2.17	2.08	2.01	1.96	1.92	1.88	1.86	1.81	1.76	1.71	1.68	1.65	1.61	1.58	1.54	1.50
27	2.90	2.51	2.30	2.17	2.07	2.00	1.95	1.91	1.87	1.85	1.80	1.75	1.70	1.67	1.64	1.60	1.57	1.53	1.49
28	2.89	2.50	2.29	2.16	2.06	2.00	1.94	1.90	1.87	1.84	1.79	1.74	1.69	1.66	1.63	1.59	1.56	1.52	1.48
29	2.89	2.50	2.28	2.15	2.06	1.99	1.93	1.89	1.86	1.83	1.78	1.73	1.68	1.65	1.62	1.58	1.55	1.51	1.47
30	2.88	2.49	2.28	2.14	2.03	1.98	1.93	1.88	1.85	1.82	1.77	1.72	1.67	1.64	1.61	1.57	1.54	1.50	1.46
40	2.84	2.44	2.23	2.09	2.00	1.93	1.87	1.83	1.79	1.76	1.71	1.66	1.61	1.57	1.54	1.51	1.47	1.42	1.38
60	2.79	2.39	2.18	2.04	1.95	1.87	1.82	1.77	1.74	1.71	1.66	1.60	1.54	1.51	1.48	1.44	1.40	1.35	1.29
120	2.75	2.35	2.13	1.99	1.90	1.82	1.77	1.72	1.68	1.65	1.60	1.55	1.48	1.45	1.41	1.37	1.32	1.26	1.19
∞	2.71	2.30	2.08	1.94	1.85	1.77	1.72	1.67	1.63	1.60	1.55	1.49	1.42	1.38	1.34	1.30	1.24	1.17	1.00

Note: $F_{0.90,v_1,v_2} = 1/F_{0.10,v_2,v_1}$.

$$F_{0.05, \nu_1, \nu_2}$$

ν_2 \ ν_1	1	2	3	4	5	6	7	8	9	10	12	15	20	24	30	40	60	120	∞
1	161.4	199.5	215.7	224.6	230.2	234.0	236.8	238.9	240.5	241.9	243.9	245.9	248.0	249.1	250.1	251.1	252.2	253.3	254.3
2	18.51	19.00	19.16	19.25	19.30	19.33	19.35	19.37	19.38	19.40	19.41	19.43	19.45	19.45	19.46	19.47	19.48	19.49	19.50
3	10.13	9.55	9.28	9.12	9.01	8.94	8.89	8.85	8.81	8.79	8.74	8.70	8.66	8.64	8.62	8.59	8.57	8.55	8.53
4	7.71	6.94	6.59	6.39	6.26	6.16	6.09	6.04	6.00	5.96	5.91	5.86	5.80	5.77	5.75	5.72	5.69	5.66	5.63
5	6.61	5.79	5.41	5.19	5.05	4.95	4.88	4.82	4.77	4.74	4.68	4.62	4.56	4.53	4.50	4.46	4.43	4.40	4.36
6	5.99	5.14	4.76	4.53	4.39	4.28	4.21	4.15	4.10	4.06	4.00	3.94	3.87	3.84	3.81	3.77	3.74	3.70	3.67
7	5.59	4.74	4.35	4.12	3.97	3.87	3.79	3.73	3.68	3.64	3.57	3.51	3.44	3.41	3.38	3.34	3.30	3.27	3.23
8	5.32	4.46	4.07	3.84	3.69	3.58	3.50	3.44	3.39	3.35	3.28	3.22	3.15	3.12	3.08	3.04	3.01	2.97	2.93
9	5.12	4.26	3.86	3.63	3.48	3.37	3.29	3.23	3.18	3.14	3.07	3.01	2.94	2.90	2.86	2.83	2.79	2.75	2.71
10	4.96	4.10	3.71	3.48	3.33	3.22	3.14	3.07	3.02	2.98	2.91	2.85	2.77	2.74	2.70	2.66	2.62	2.58	2.54
11	4.84	3.98	3.59	3.36	3.20	3.09	3.01	2.95	2.90	2.85	2.79	2.72	2.65	2.61	2.57	2.53	2.49	2.45	2.40
12	4.75	3.89	3.49	3.26	3.11	3.00	2.91	2.85	2.80	2.75	2.69	2.62	2.54	2.51	2.47	2.43	2.38	2.34	2.30
13	4.67	3.81	3.41	3.18	3.03	2.92	2.83	2.77	2.71	2.67	2.60	2.53	2.46	2.42	2.38	2.34	2.30	2.25	2.21
14	4.60	3.74	3.34	3.11	2.96	2.85	2.76	2.70	2.65	2.60	2.53	2.46	2.39	2.35	2.31	2.27	2.22	2.18	2.13
15	4.54	3.68	3.29	3.06	2.90	2.79	2.71	2.64	2.59	2.54	2.48	2.40	2.33	2.29	2.25	2.20	2.16	2.11	2.07
16	4.49	3.63	3.24	3.01	2.85	2.74	2.66	2.59	2.54	2.49	2.42	2.35	2.28	2.24	2.19	2.15	2.11	2.06	2.01
17	4.45	3.59	3.20	2.96	2.81	2.70	2.61	2.55	2.49	2.45	2.38	2.31	2.23	2.19	2.15	2.10	2.06	2.01	1.96
18	4.41	3.55	3.16	2.93	2.77	2.66	2.58	2.51	2.46	2.41	2.34	2.27	2.19	2.15	2.11	2.06	2.02	1.97	1.92
19	4.38	3.52	3.13	2.90	2.74	2.63	2.54	2.48	2.42	2.38	2.31	2.23	2.16	2.11	2.07	2.03	1.98	1.93	1.88

Degrees of freedom for the numerator (ν_1)

Degrees of freedom for the denominator (ν_2)

20	4.35	3.49	3.10	2.87	2.71	2.60	2.51	2.45	2.39	2.35	2.28	2.20	2.12	2.08	2.04	1.99	1.95	1.90	1.84
21	4.32	3.47	3.07	2.84	2.68	2.57	2.49	2.42	2.37	2.32	2.25	2.18	2.10	2.05	2.01	1.96	1.92	1.87	1.81
22	4.30	3.44	3.05	2.82	2.66	2.55	2.46	2.40	2.34	2.30	2.23	2.15	2.07	2.03	1.98	1.94	1.89	1.84	1.78
23	4.28	3.42	3.03	2.80	2.64	2.53	2.44	2.37	2.32	2.27	2.20	2.13	2.05	2.01	1.96	1.91	1.86	1.81	1.76
24	4.26	3.40	3.01	2.78	2.62	2.51	2.42	2.36	2.30	2.25	2.18	2.11	2.03	1.98	1.94	1.89	1.84	1.79	1.73
25	4.24	3.39	2.99	2.76	2.60	2.49	2.40	2.34	2.28	2.24	2.16	2.09	2.01	1.96	1.92	1.87	1.82	1.77	1.71
26	4.23	3.37	2.98	2.74	2.59	2.47	2.39	2.32	2.27	2.22	2.15	2.07	1.99	1.95	1.90	1.85	1.80	1.75	1.69
27	4.21	3.35	2.96	2.73	2.57	2.46	2.37	2.31	2.25	2.20	2.13	2.06	1.97	1.93	1.88	1.84	1.79	1.73	1.67
28	4.20	3.34	2.95	2.71	2.56	2.45	2.36	2.29	2.24	2.19	2.12	2.04	1.96	1.91	1.87	1.82	1.77	1.71	1.65
29	4.18	3.33	2.93	2.70	2.55	2.43	2.35	2.28	2.22	2.18	2.10	2.03	1.94	1.90	1.85	1.81	1.75	1.70	1.64
30	4.17	3.32	2.92	2.69	2.53	2.42	2.33	2.27	2.21	2.16	2.09	2.01	1.93	1.89	1.84	1.79	1.74	1.68	1.62
40	4.08	3.23	2.84	2.61	2.45	2.34	2.25	2.18	2.12	2.08	2.00	1.92	1.84	1.79	1.74	1.69	1.64	1.58	1.51
60	4.00	3.15	2.76	2.53	2.37	2.25	2.17	2.10	2.04	1.99	1.92	1.84	1.75	1.70	1.65	1.59	1.53	1.47	1.39
120	3.92	3.07	2.68	2.45	2.29	2.17	2.09	2.02	1.96	1.91	1.83	1.75	1.66	1.61	1.55	1.55	1.43	1.35	1.25
∞	3.84	3.00	2.60	2.37	2.21	2.10	2.01	1.94	1.88	1.83	1.75	1.67	1.57	1.52	1.46	1.39	1.32	1.22	1.00

Note: $F_{0.95,v_1,v_2} = 1/F_{0.05,v_2,v_1}$.

$$F_{0.25, v_1, v_2}$$

Degrees of freedom for the numerator (v_1)

v_2	1	2	3	4	5	6	7	8	9	10	12	15	20	24	30	40	60	120	∞
1	647.8	799.5	864.2	899.6	921.8	937.1	948.2	956.7	963.3	968.6	976.7	984.9	993.1	997.2	1001.0	1006.0	1010.0	1014.0	1018.0
2	38.51	39.00	39.17	39.25	39.30	39.33	39.36	39.37	39.39	39.40	39.41	39.43	39.45	39.46	39.46	39.47	39.48	39.49	39.50
3	17.44	16.04	15.44	15.10	14.88	14.73	14.62	14.54	14.47	14.42	14.34	14.25	14.17	14.12	14.08	14.04	13.99	13.95	13.90
4	12.22	10.65	9.98	9.60	9.36	9.20	9.07	8.98	8.90	8.84	8.75	8.66	8.56	8.51	8.46	8.41	8.36	8.31	8.26
5	10.01	8.43	7.76	7.39	7.15	6.98	6.85	6.76	6.68	6.62	6.52	6.43	6.33	6.28	6.23	6.18	6.12	6.07	6.02
6	8.81	7.26	6.60	6.23	5.99	5.82	5.70	5.60	5.52	5.46	5.37	5.27	5.17	5.12	5.07	5.01	4.96	4.90	4.85
7	8.07	6.54	5.89	5.52	5.29	5.12	4.99	4.90	4.82	4.76	4.67	4.57	4.47	4.42	4.36	4.31	4.25	4.20	4.14
8	7.57	6.06	5.42	5.05	4.82	4.65	4.53	4.43	4.36	4.30	4.20	4.10	4.00	3.95	3.89	3.84	3.78	3.73	3.67
9	7.21	5.71	5.08	4.72	4.48	4.32	4.20	4.10	4.03	3.96	3.87	3.77	3.67	3.61	3.56	3.51	3.45	3.39	3.33
10	6.94	5.46	4.83	4.47	4.24	4.07	3.95	3.85	3.78	3.72	3.62	3.52	3.42	3.37	3.31	3.26	3.20	3.14	3.08
11	6.72	5.26	4.63	4.28	4.04	3.88	3.76	3.66	3.59	3.53	3.43	3.33	3.23	3.17	3.12	3.06	3.00	2.94	2.88
12	6.55	5.10	4.47	4.12	3.89	3.73	3.61	3.51	3.44	3.37	3.28	3.18	3.07	3.02	2.96	2.91	2.85	2.79	2.72
13	6.41	4.97	4.35	4.00	3.77	3.60	3.48	3.39	3.31	3.25	3.15	3.05	2.95	2.89	2.84	2.78	2.72	2.66	2.60
14	6.30	4.86	4.24	3.89	3.66	3.50	3.38	3.29	3.21	3.15	3.05	2.95	2.84	2.79	2.73	2.67	2.61	2.55	2.49
15	6.20	4.77	4.15	3.80	3.58	3.41	3.29	3.20	3.12	3.06	2.96	2.86	2.76	2.70	2.64	2.59	2.52	2.46	2.40
16	6.12	4.69	4.08	3.73	3.50	3.34	3.22	3.12	3.05	2.99	2.89	2.79	2.68	2.63	2.57	2.51	2.45	2.38	2.32
17	6.04	4.62	4.01	3.66	3.44	3.28	3.16	3.06	2.98	2.92	2.82	2.72	2.62	2.56	2.50	2.44	2.38	2.32	2.25
18	5.98	4.56	3.95	3.61	3.38	3.22	3.10	3.01	2.93	2.87	2.77	2.67	2.56	2.50	2.44	2.38	2.32	2.26	2.19
19	5.92	4.51	3.90	3.56	3.33	3.17	3.05	2.96	2.88	2.82	2.72	2.62	2.51	2.45	2.39	2.33	2.27	2.20	2.13

Degrees of freedom for the denominator (v_2)

20	5.87	4.46	3.86	3.51	3.29	3.13	3.01	2.91	2.84	2.77	2.68	2.57	2.46	2.41	2.35	2.29	2.22	2.16	2.09
21	5.83	4.42	3.82	3.48	3.25	3.09	2.97	2.87	2.80	2.73	2.64	2.53	2.42	2.37	2.31	2.25	2.18	2.11	2.04
22	5.79	4.38	3.78	3.44	3.22	3.05	2.93	2.84	2.76	2.70	2.60	2.50	2.39	2.33	2.27	2.21	2.14	2.08	2.00
23	5.75	4.35	3.75	3.41	3.18	3.02	2.90	2.81	2.73	2.67	2.57	2.47	2.36	2.30	2.24	2.18	2.11	2.04	1.97
24	5.72	4.32	3.72	3.38	3.15	2.99	2.87	2.78	2.70	2.64	2.54	2.44	2.33	2.27	2.21	2.15	2.08	2.01	1.94
25	5.69	4.29	3.69	3.35	3.13	2.97	2.85	2.75	2.68	2.61	2.51	2.41	2.30	2.24	2.18	2.12	2.05	1.98	1.91
26	5.66	4.27	3.67	3.33	3.10	2.94	2.82	2.73	2.65	2.59	2.49	2.39	2.28	2.22	2.16	2.09	2.03	1.95	1.88
27	5.63	4.24	3.65	3.31	3.08	2.92	2.80	2.71	2.63	2.57	2.47	2.36	2.25	2.19	2.13	2.07	2.00	1.93	1.85
28	5.61	4.22	3.63	3.29	3.06	2.90	2.78	2.69	2.61	2.55	2.45	2.34	2.23	2.17	2.11	2.05	1.98	1.91	1.83
29	5.59	4.20	3.61	3.27	3.04	2.88	2.76	2.67	2.59	2.53	2.43	2.32	2.21	2.15	2.09	2.03	1.96	1.89	1.81
30	5.57	4.18	3.59	3.25	3.03	2.87	2.75	2.65	2.57	2.51	2.41	2.31	2.20	2.14	2.07	2.01	1.94	1.87	1.79
40	5.42	4.05	3.46	3.13	2.90	2.74	2.62	2.53	2.45	2.39	2.29	2.18	2.07	2.01	1.94	1.88	1.80	1.72	1.64
60	5.29	3.93	3.34	3.01	2.79	2.63	2.51	2.41	2.33	2.27	2.17	2.06	1.94	1.88	1.82	1.74	1.67	1.58	1.48
120	5.15	3.80	3.23	2.89	2.67	2.52	2.39	2.30	2.22	2.16	2.05	1.94	1.82	1.76	1.69	1.61	1.53	1.43	1.31
∞	5.02	3.69	3.12	2.79	2.57	2.41	2.29	2.19	2.11	2.05	1.94	1.83	1.71	1.64	1.57	1.48	1.39	1.27	1.00

Note: $F_{0.95, v_1, v_2} = 1/F_{0.05, v_2, v_1}$.

$F_{0.01,\nu_1,\nu_2}$

$\nu_2 \backslash \nu_1$	1	2	3	4	5	6	7	8	9	10	12	15	20	24	30	40	60	120	∞
1	4052.0	4999.5	5403.0	5625.0	5764.0	5859.0	5928.0	5982.0	6022.0	6056.0	6106.0	6157.0	6209.0	6235.0	6261.0	6287.0	6313.0	6339.0	6366.0
2	98.50	99.00	99.17	99.25	99.30	99.33	99.36	99.37	99.39	99.40	99.42	99.43	99.45	99.46	99.47	99.47	99.48	99.49	99.50
3	34.12	30.82	29.46	28.71	28.24	27.91	27.67	27.49	27.35	27.23	27.05	26.87	26.69	26.00	26.50	26.41	26.32	26.22	26.13
4	21.20	18.00	16.69	15.98	15.52	15.21	14.98	14.80	14.66	14.55	14.37	14.20	14.02	13.93	13.84	13.75	13.65	13.56	13.46
5	16.26	13.27	12.06	11.39	10.97	10.67	10.46	10.29	10.16	10.05	9.89	9.72	9.55	9.47	9.38	9.29	9.20	9.11	9.02
6	13.75	10.92	9.78	9.15	8.75	8.47	8.26	8.10	7.98	7.87	7.72	7.56	7.40	7.31	7.23	7.14	7.06	6.97	6.88
7	12.25	9.55	8.45	7.85	7.46	7.19	6.99	6.84	6.72	6.62	6.47	6.31	6.16	6.07	5.99	5.91	5.82	5.74	5.65
8	11.26	8.65	7.59	7.01	6.63	6.37	6.18	6.03	5.91	5.81	5.67	5.52	5.36	5.28	5.20	5.12	5.03	4.95	4.86
9	10.56	8.02	6.99	6.42	6.06	5.80	5.61	5.47	5.35	5.26	5.11	4.96	4.81	4.73	4.65	4.57	4.48	4.40	4.31
10	10.04	7.56	6.55	5.99	5.64	5.39	5.20	5.06	4.94	4.85	4.71	4.56	4.41	4.33	4.25	4.17	4.08	4.00	3.91
11	9.65	7.21	6.22	5.67	5.32	5.07	4.89	4.74	4.63	4.54	4.40	4.25	4.10	4.02	3.94	3.86	3.78	3.69	3.60
12	9.33	6.93	5.95	5.41	5.06	4.82	4.64	4.50	4.39	4.30	4.16	4.01	3.86	3.78	3.70	3.62	3.54	3.45	3.36
13	9.07	6.70	5.74	5.21	4.86	4.62	4.44	4.30	4.19	4.10	3.96	3.82	3.66	3.59	3.51	3.43	3.34	3.25	3.17
14	8.86	6.51	5.56	5.04	4.69	4.46	4.28	4.14	4.03	3.94	3.80	3.66	3.51	3.43	3.35	3.27	3.18	3.09	3.00
15	8.68	6.36	5.42	4.89	4.56	4.32	4.14	4.00	3.89	3.80	3.67	3.52	3.37	3.29	3.21	3.13	3.05	2.96	2.87
16	8.53	6.23	5.29	4.77	4.44	4.20	4.03	3.89	3.78	3.69	3.55	3.41	3.26	3.18	3.10	3.02	2.93	2.84	2.75
17	8.40	6.11	5.18	4.67	4.34	4.10	3.93	3.79	3.68	3.59	3.46	3.31	3.16	3.08	3.00	2.92	2.83	2.75	2.65
18	8.29	6.01	5.09	4.58	4.25	4.01	3.84	3.71	3.60	3.51	3.37	3.23	3.08	3.00	2.92	2.84	2.75	2.66	2.57
19	8.18	5.93	5.01	4.50	4.17	3.94	3.77	3.63	3.52	3.43	3.30	3.15	3.00	2.92	2.84	2.76	2.67	2.58	2.49

Degrees of freedom for the numerator (ν_1)

Degrees of freedom for the denominator (ν_2)

20	8.10	5.85	4.94	4.43	4.10	3.87	3.70	3.56	3.46	3.37	3.23	3.09	2.94	2.86	2.78	2.69	2.61	2.52	2.42
21	8.02	5.78	4.87	4.37	4.04	3.81	3.64	3.51	3.40	3.31	3.17	3.03	2.88	2.80	2.72	2.64	2.55	2.46	2.36
22	7.95	5.72	4.82	4.31	3.99	3.76	3.59	3.45	3.35	3.26	3.12	2.98	2.83	2.75	2.67	2.58	2.50	2.40	2.31
23	7.88	5.66	4.76	4.26	3.94	3.71	3.54	3.41	3.30	3.21	3.07	2.93	2.78	2.70	2.62	2.54	2.45	2.35	2.26
24	7.82	5.61	4.72	4.22	3.90	3.67	3.50	3.36	3.26	3.17	3.03	2.89	2.74	2.66	2.58	2.49	2.40	2.31	2.21
25	7.77	5.57	4.68	4.18	3.85	3.63	3.46	3.32	3.22	3.13	2.99	2.85	2.70	2.62	2.54	2.45	2.36	2.27	2.17
26	7.72	5.53	4.64	4.14	3.82	3.59	3.42	3.29	3.18	3.09	2.96	2.81	2.66	2.58	2.50	2.42	2.33	2.23	2.13
27	7.68	5.49	4.60	4.11	3.78	3.56	3.39	3.26	3.15	3.06	2.93	2.78	2.63	2.55	2.47	2.38	2.29	2.20	2.10
28	7.64	5.45	4.57	4.07	3.75	3.53	3.36	3.23	3.12	3.03	2.90	2.75	2.60	2.52	2.44	2.35	2.26	2.17	2.06
29	7.60	5.42	4.54	4.04	3.73	3.50	3.33	3.20	3.09	3.00	2.87	2.73	2.57	2.49	2.41	2.33	2.23	2.14	2.03
30	7.56	5.39	4.51	4.02	3.70	3.47	3.30	3.17	3.07	2.98	2.84	2.70	2.55	2.47	2.39	2.30	2.21	2.11	2.01
40	7.31	5.18	4.31	3.83	3.51	3.29	3.12	2.99	2.89	2.80	2.66	2.52	2.37	2.29	2.20	2.11	2.02	1.92	1.80
60	7.08	4.98	4.13	3.65	3.34	3.12	2.95	2.82	2.72	2.63	2.50	2.35	2.20	2.12	2.03	1.94	1.84	1.73	1.60
120	6.85	4.79	3.95	3.48	3.17	2.96	2.79	2.66	2.56	2.47	2.34	2.19	2.03	1.95	1.86	1.76	1.66	1.53	1.38
∞	6.63	4.61	3.78	3.32	3.02	2.80	2.64	2.51	2.41	2.32	2.18	2.04	1.88	1.79	1.70	1.59	1.47	1.32	1.00

Note: $F_{0.99, v_1, v_2} = 1/F_{0.01, v_2, v_1}$.

APPENDIX F

FACTORS FOR CONSTRUCTING VARIABLES CONTROL CHARTS

| | Chart for Averages | | | Chart for Standard Deviations | | | | | | | Chart for Ranges | | | | | | |
| | Factors for Control Limits | | | Factors for Center Line | | Factors for Control Limits | | | | Factors for Center Line | | | Factors for Control Limits | | | |
Observations in Sample, n	A	A_2	A_3	c_4	$1/c_4$	B_3	B_4	B_5	B_6	d_2	$1/d_2$	d_3	D_1	D_2	D_3	D_4
2	2.121	1.880	2.659	0.7979	1.2533	0	3.267	0	2.606	1.128	0.8865	0.853	0	3.686	0	3.267
3	1.732	1.023	1.954	0.8862	1.1284	0	2.568	0	2.276	1.693	0.5907	0.888	0	4.358	0	2.575
4	1.500	0.729	1.628	0.9213	1.0854	0	2.266	0	2.088	2.059	0.4857	0.880	0	4.698	0	2.282
5	1.342	0.577	1.427	0.9400	1.0638	0	2.089	0	1.964	2.326	0.4299	0.864	0	4.918	0	2.115
6	1.225	0.483	1.287	0.9515	1.0510	0.030	1.970	0.029	1.874	2.534	0.3946	0.848	0	5.078	0	2.004
7	1.134	0.419	1.182	0.9594	1.04230	0.118	1.882	0.113	1.806	2.704	0.3698	0.833	0.204	5.204	0.076	1.924
8	1.061	0.373	1.099	0.9650	1.0363	0.185	1.815	0.179	1.751	2.847	0.3512	0.820	0.388	5.306	0.136	1.864
9	1.000	0.337	1.032	0.9693	1.0317	0.239	1.761	0.232	1.707	2.970	0.3367	0.808	0.547	5.393	0.184	1.816
10	0.949	0.308	0.975	0.9727	1.0281	0.284	1.716	0.276	1.669	3.078	0.3249	0.797	0.687	5.469	0.223	1.777
11	0.905	0.285	0.927	0.9754	1.0252	0.321	1.679	0.313	1.637	3.173	0.3152	0.787	0.811	5.535	0.256	1.744
12	0.866	0.266	0.886	0.9776	1.0229	0.354	1.646	0.346	1.610	3.258	0.3069	0.778	0.922	5.594	0.283	1.717
13	0.832	0.249	0.850	0.9794	1.0210	0.382	1.618	0.374	1.585	3.336	0.2998	0.770	1.025	5.647	0.307	1.693
14	0.802	0.235	0.817	0.9810	1.0194	0.406	1.594	0.399	1.563	3.407	0.2935	0.763	1.118	5.696	0.328	1.672
15	0.775	0.223	0.789	0.9823	1.0180	0.428	1.572	0.421	1.544	3.472	0.2880	0.756	1.203	5.741	0.347	1.653
16	0.750	0.212	0.763	0.9835	1.0168	0.448	1.552	0.440	1.526	3.532	0.2831	0.750	1.282	5.782	0.363	1.637
17	0.728	0.203	0.739	0.9845	1.0157	0.466	1.534	0.458	1.511	3.588	0.2787	0.744	1.356	5.820	0.378	1.622
18	0.707	0.194	0.718	0.9854	1.0148	0.482	1.518	0.475	1.496	3.640	0.2747	0.739	1.424	5.856	0.391	1.608

(continued overleaf)

Observations in Sample, n	Chart for Averages			Chart for Standard Deviations						Chart for Ranges						
	Factors for Control Limits			Factors for Center Line		Factors for Control Limits				Factors for Center Line			Factors for Control Limits			
	A	A_2	A_3	c_4	$1/c_4$	B_3	B_4	B_5	B_6	d_2	$1/d_2$	d_3	D_1	D_2	D_3	D_4
19	0.688	0.187	0.698	0.9862	1.0140	0.497	1.503	0.490	1.483	3.689	0.2711	0.734	1.487	5.891	0.403	1.597
20	0.671	0.180	0.680	0.9869	1.0133	0.510	1.490	0.504	1.470	3.735	0.2677	0.729	1.549	5.921	0.415	1.585
21	0.655	0.173	0.663	0.9876	1.0126	0.523	1.477	0.516	1.459	3.778	0.2647	0.724	1.605	5.951	0.425	1.575
22	0.640	0.167	0.647	0.9882	1.0119	0.534	1.466	0.528	1.448	3.819	0.2618	0.720	1.659	5.979	0.434	1.566
23	0.626	0.162	0.633	0.9887	1.0114	0.545	1.455	0.539	1.438	3.858	0.2592	0.716	1.710	6.006	0.443	1.557
24	0.612	0.157	0.619	0.9892	1.0109	0.555	1.445	0.549	1.429	3.895	0.2567	0.712	1.759	6.031	0.451	1.548
25	0.600	0.153	0.606	0.9896	1.0105	0.565	1.435	0.559	1.420	3.931	0.2544	0.708	1.806	6.056	0.459	1.541

For $n > 25$

$$A = \frac{3}{\sqrt{n}}, \quad A_3 = \frac{3}{c_4\sqrt{n}}, \quad c_4 \simeq \frac{4(n-1)}{4n-3},$$

$$B_3 = 1 - \frac{3}{c_4\sqrt{2(n-1)}}, \quad B_4 = 1 + \frac{3}{c_4\sqrt{2(n-1)}},$$

$$B_5 = c_4 - \frac{3}{\sqrt{2(n-1)}}, \quad B_6 = c_4 + \frac{3}{\sqrt{2(n-1)}}.$$

(from Montgomery)

440

INDEX

Abstraction, of modern semiconductor
    manufacturing processes, 8
Acceptability regions, in design centering,
    172–173
Acceptance charts, 206, 207–208
Additive model, 242–243
Adhesion promoter, 41
Advanced process control, 333–373
    multivariate control with complete model
        adaptation, 351–359
    run-by-run control, 333–335, 335–351
    statistical process control and, 333
    supervisory control, 359–373
Air filters, 99
Air monitoring, in cleanrooms, 99–100
Alarms
    false and missed, 141, 182
    feedforward control and, 358
    in multivariate control, 352–354
Algorithmic methods, for diagnostic checking,
    380, 381–391
Alternative hypothesis, 140
    one-sided, 141, 142
Aluminum, in NMOS fabrication, 70
Aluminum interconnect, 6
Amplifier applications, for BiCMOS
    technology, 74–75
Analysis of variance (ANOVA), 232–249, 301.
    See also ANOVA entries
    described, 232
    with exponential models, 286

randomized block experiments with,
    240–244
in single-parameter models, 276–277
sums of squares in, 232–234
in Taguchi method, 266–268
in two-parameter regression models, 279
in two-way designs, 245–249
Angle-resolved scatterometers, 96–97
ANOVA diagnostics, 237–240. See also
    Analysis of variance (ANOVA)
ANOVA table, 234–240, 282, 283
    formats of, 234, 235
    full, 235
    geometric interpretation of, 235–237, 238
    hypothesis testing and, 234–235
    randomized block experiments and,
        241–242
    in single-parameter models, 276–277
Applied Materials 7800 barrel reactor, 338,
    343, 344
Approximations, of probability distributions,
    132–133
Architecture, PEDX, 396–397
ARIMA models. See Integrated autoregressive
    moving-average (ARIMA) models
Array diagnostic monitor (ADM), 105
Arrays, orthogonal, 264–265. See also Matrices
Arsenic, as dopant, 52
Artificial intelligence (AI), 273. See also
    Computational intelligence; Intelligent
    entries
Assembly line, 4

Fundamentals of Semiconductor Manufacturing and Process Control,
By Gary S. May and Costas J. Spanos
Copyright © 2006 John Wiley & Sons, Inc.

Assignable causes
  diagnosing, 380
  of yield loss, 149
Associational links, with PIES, 393–394
Associational strengths, with PIES, 393–394
Atmospheric pressure chemical vapor
  deposition (APCVD), 60
Atomic force microscopy (AFM), of patterned
  thin films, 93–95
Attachment methods, for integrated circuits,
  79–80
Attributes, 186
  control charts for, 186–195
Augmented response surface model, 293–294,
  295
Autocorrelated measurements, 221–222
Autocorrelation function, 221
Automated recipe generation, 356–357
Automated test equipment (ATE), 106
Automatic internal analysis, with MERLIN,
  388
Autoregressive models, first- and second-order,
  222
Autoregressive moving-average (ARMA)
  models, 222–223, 223–224
Average
  correction factor for, 235
  grand, 197, 199, 232, 236, 237, 238
  sample, 123, 183
  weighted, 213–214
Average run length (ARL), 184
  control alarms and, 354
  cusum charts and, 210–211
  for fraction nonconforming chart, 191–192
  for $\bar{x}$ chart, 200, 201
Average sample size, for fraction
  nonconforming chart, 189–191
Axial points, 294
  in central composite design, 260, 261

Backpropagation (BP) algorithm, 310–314
Backscattering, in e-beam lithography, 45
Baking, in Taguchi method, 263, 264
Ball-grid arrays (BGAs), 77–78
Barrel susceptors, 60
Base–base shorts, monitoring, 102–103
Base implantation, 65, 66
Base region, forming, 65, 66
Basic probability mass distribution (BPMD), in
  Dempster–Shafer theory, 406–408
Batch operations/processes
  in semiconductor manufacturing, 19, 20
  single-variable control methods for,
    335–343, 344

Batch uniformity
  gradual controller operation mode for,
    338–339
  rapid controller operation mode for, 343,
    344
Beachy, M., 402
Beam blanking plates, in e-beam lithography,
  43–44
Bell System Technical Journal, Taguchi method
  in, 263
Benzocyclobutene (BCB), 366
Bernoulli trials, 125, 128, 132–133
Beryllium, in X-ray lithography, 46
Between-lot statistics, 199–200
Between-sample variability, 199
Between-treatment sum of squares, 233, 234
Bhatikar, S., 400, 401
Bias, in implantation monitoring, 117
Bias terms, with run-by-run control, 350–351
BiCMOS (bipolar CMOS) technology, 62,
  74–75
Binary strings, in genetic algorithms, 323
Binomial distribution, 125–127. *See also*
    Negative binomial yield model
  negative, 128
  normal approximation to, 132–133
  Poisson approximation to, 132
  Poisson distribution as limiting form of, 128
  Poisson yield model and, 152
Bipolar chips, shmoo plots for, 107
Bipolar device structures, with BiCMOS
  technology, 74–75
Bipolar junction transistors, fabrication of, 62
Bipolar technology, 62, 63–66
  PIES and, 395
Bipolar transistors, fabrication of, 11–12
Bivariate process control, 215–217
Blanket etches, 47
Blanket thin films, metrology of, 85–92
Block effect (β_i), 242
Block experiments, randomized, 240–244
Blocking, 240–242
  fractional factorial experimental designs
    and, 256
  of two-level factorial design, 254–255
Boat, 20
Boron
  as dopant, 52
  in gate engineering, 73–74
Boron ions, implanting, 65
Bose–Einstein statistics, 153–154
Bottom numbers
  in gradual controller operation mode, 339
  in Yates algorithm, 258

BP neural networks, 310–314
Bridgman technique, 5t, 6

Capacitance manometers, 111
Catastrophic yield, 148
Causal chains, in PIES knowledge base, 392–393, 394
Caused-by links, with PIES, 393–394
Causes links, with PIES, 393–394
c chart, 186. *See also* Defect chart (c chart)
Cell maps, 107
Centerline
    for control charts, 182, 183, 184
    for defect chart, 193
    for defect density chart, 193–195
    for exponentially weighted moving-average charts, 214
    for fraction nonconforming chart, 187, 188
    for moving-average charts, 212
    in multivariate process variability, 220
    for s and $\bar{x}$ charts, 203–204
    for variable control chart, 197–199
Centerpoints, in central composite design, 260, 261
Central composite circumscribed (CCC) design, 294, 295, 297
Central composite design (CCD), 260–261
Ceramic DIP (CerDIP), 77
Chain scission, 45
"Change in mean" (CIM) effect, evolutionary operation and, 303
Chanstop, 64
Charged coupled devices (CCDs), in interferometry, 88
Charges, at silicon–silicon dioxide interface, 33
Chemical etching, 50
Chemical industry, designed experiments in, 5
Chemical–mechanical polishing (CMP), 5t, 6, 61, 62
    planarization monitoring in, 118
    shallow trench and, 73
Chemical vapor deposition (CVD), 20, 31, 58, 60–61. *See also* CVD entries
    fuzzy logic in modeling, 317–318
    physical vapor deposition versus, 61
    in two-level factorial design example, 251–255
    in two-way experimental design example, 245–249
Chillers, 111
Chips
    FABRICS software for, 167–171
    VLASIC software for, 162–167
    on wafers, 62, 63

Chip-scale packages (CSPs), 78–79
Chi-square (χ^2) distribution, 134,138
    F distribution and, 135–136
    in multivariate process control, 217, 218
    t distribution and, 134–135
Chromosomes, in genetic algorithms, 323, 324, 326
Circuits, computer-integrated manufacturing of, 7–8
Cleanrooms
    air monitoring in, 99–100
    classes of, 35
    as examples of linear model with nonzero intercept, 280–283
    in photolithography, 34–35
Closed-loop recirculation system, 111
CMOS (complementary MOSFET) technology, 5t, 6, 51, 62, 66–74
CMOS fabrication sequence, 70–74
CMOS inverter, 70–71
CMOS process flow, 83
CMOS structures, with BiCMOS technology, 74–75
Collector contact chain, 103, 104
Collimator, sputtering through, 59, 60
Comb–meander–comb structure, 102
Competitive Semiconductor Manufacturing program, 175
Complementary error function (erfc), 54
Complete model adaptation
    multivariate process control with, 351–359
    supervisory control using, 359–364, 365, 366, 367, 368, 369
Complex index of refraction (N), 90
Compromise recipes, with model update algorithm, 354
Computational intelligence, 310. *See also* Artificial intelligence (AI); Intelligent entries
Computer-aided design (CAD), 7
    in photolithographic masking, 38
Computer-integrated manufacturing of circuits (IC-CIM), 7–8, 12–14
Computer numeric control (CNC), 4
Computers
    in manufacturing, 4
    in semiconductor manufacturing, 12–14
Concatenated multiparameter fixed-point coding, in genetic algorithms, 323–324
Concentration gradient, in diffusion, 53
Condenser lenses, in e-beam lithography, 43
Confidence intervals, 137–140

Confidence intervals, (*continued*)
  defined, 137
  for difference between two means, known
    variance, 138
  for difference between two means, unknown
    variances, 138–139
  for mean with known variance, 137
  for mean with unknown variance, 137
  for ratio of two variances, 139–140
  for variance, 137–138
Confidence level (C), in maintenance diagnosis,
  409
Confirmation step, with PIES, 394
Conformance quality, of semiconductor
  manufacturing, 17
Confounding patterns
  with block effect, 254, 255
  in fractional factorial experimental designs,
    256–257
Conjugate direction, in Powell's algorithm, 319
Constants, with MERLIN, 385, 387
Constant-surface-concentration diffusion, 54
Constant-total-dopant diffusion, 54
Constraints, in simplex method, 322, 323
Contact printing, 35, 36
Containment, in fuzzy logic, 315
Contamination
  defined, 156
  in IC fabrication, 98–102
Continuity equation, one-dimensional, 53
Continuous-flow manufacturing, of integrated
  circuits, 18–19, 19–21
Continuous probability distributions, 124,
  128–132
Contour plots, 295, 300
Contraction, in simplex method, 320, 321
Contrast, 252
  in fractional factorial experimental designs,
    256–257
  geometric representation of, 253
Control alarms, 352–354
Control charts, 5, 181–182
  for attributes, 186–195
  basics of, 182–184
  cusum, 208–212
  for defect density, 186, 193–195
  for defects, 186, 193
  exponentially weighted moving-average,
    213–215
  for fraction nonconforming, 186, 187–192
  moving-average, 212–215
  patterns in, 184–186
  rational subgroups and, 199–200

  for single-variable control methods,
    336–337
  standardized, 191
  for variables, 195–215
Control ellipse, in multivariate process control,
  217–220
Control factors, in Taguchi method, 263,
  266
Control region, in multivariate process control,
  216, 218
Convergence, in simplex method, 322
Copper interconnect, 5t, 6–7
Correction factor for the average, 235
Correlation measure of verification (ρ), with
  MERLIN, 389
Cost, of semiconductor manufacturing, 15–16
Cost function, in fault diagnosis, 383, 384
Covariance
  with autocorrelation function, 221
  in multivariate process control, 217–220
Covariance matrix
  in multivariate process control, 219–220
  in principal-component analysis, 309
Crisp sets, 314–315
Critical area (A_c), 150, 157–158
Critical area integral, 158
Critical dimension (CD), 95, 96, 97, 98
  lithography operation monitoring and, 117
  for photolithographic machines, 361–362,
    364
  in shadow printing, 36
Crosier's multivariate cusum scheme, control
  alarms in, 353–354
Crossover, in genetic algorithms, 323,
  324–325, 326
Crystal, in quartz crystal monitor, 91
Crystal growth techniques, 5–6
Cumulative distribution function
  in maintenance diagnosis, 409
  standard normal, 133
Cumulative normal distribution, 130
Cusum (cumulative sum) charts, 208–212,
  336–337
  average run length and, 210–211
  Shewhart control charts and, 208, 209
  tabular, 210
  for variance, 211–212
Cusum scheme, control alarms in, 353–354
CVD barrel reactor, 400–401
CVD diagnosis, pattern recognition in,
  400–402. *See also* Chemical vapor
  deposition (CVD)
Cyclic behavior, in control charts, 184, 185
Czochralski technique, 5t, 6

Data analysis, in Taguchi method, 266–268
Data transformations, 244, 246–249
　variance-stabilizing, 248
Decision interval (H), 210
　in online diagnosis, 412
Deep junctions, diffusion and, 51
Defect chart (c chart), 186, 193
Defect density, 156–157
　in masks, 39
Defect density chart (u chart), 186, 193–195
Defect diameter, 156, 157
Defect inspection, 98–102
Defects
　defined, 156
　Murphy's yield integral and, 152–154
　negative binomial yield model and,
　　154–155, 157
　Poisson yield model and, 151–152
　size distribution of, 156, 157
　software for estimating, 162–167
　yield and, 150
Defining relation, of fractional factorial
　　experimental designs, 257
Degrees of freedom, 134, 135, 235
　in single-parameter models, 276
　with sums of squares, 233
　in two-parameter regression models, 279
Delta function (δ), 152, 153, 155
Demagnification ratio (M), in photolithography,
　36
Dempster–Shafer theory, 403, 406–408, 410
Dependence links, with MERLIN, 387
Dependent variables, 273–274
Deposition. See also Chemical vapor deposition
　　(CVD)
　gradual controller operation mode for,
　　338–339
　in integrated circuit fabrication, 58–61
　rapid controller operation mode for, 343,
　　344
Depth of focus (DoF), in photolithography, 38
Design centering, yield and, 171–174. See also
　　Design for manufacturability
Designed experiments, 5, 228. See also
　　Statistical experimental design
Design for manufacturability, 149, 161, 172.
　　See also Design centering
Design matrix, 251, 256
Design quality, of semiconductor
　　manufacturing, 17
Design yield, 148
Developer, lithography operation monitoring
　　and, 117
Development, in Taguchi method, 263, 264

Deviation graph, with MERLIN, 387, 388,
　390–391
Device–component objects, with MERLIN,
　385
Device diagnosis, with MERLIN, 384
Diagnosis. See also Diagnostic checking;
　　Diagnostic systems
　inline, 413–414
　maintenance, 408–409
　online, 409–413
Diagnostic checking, 379–414. See also
　　Diagnostic systems
　algorithmic methods for, 380, 381–391
　ANOVA, 237–240
　expert systems for, 380, 391–398
　hybrid methods of, 402–414
　need for, 379–380
　neural network approaches to, 381,
　　398–402
　of randomized block experiment models,
　　243–244
　in response surface methodology, 292–293
Diagnostic reasoner, with PIES, 393–394,
　394–395
Diagnostic systems, 380–381
　Hippocrates, 381–384
　MERLIN, 384–391
Diamond saw, in die separation, 76
Diamond scribe, in die separation, 76
Die separation, in IC packaging, 76
Die yield, 148,ielectric function, 89–90, 91
Differential pressure mass flowmeter, 110
Differential term, 215
Diffraction
　in e-beam lithography, 45
　photoresists and, 41
　in proximity printing, 35
Diffusion
　defined, 53
　impurity doping via, 51–52, 52–56
　in p–n junction fabrication, 10, 11
Diffusion coefficient (D), 28–29, 53, 54
Diffusion length, 56
Diffusion techniques, 5t, 6
Diffusivity, 53
Dimensionality reduction, via
　　principal-component analysis, 306–309
Direct costs, in semiconductor manufacturing,
　16
Discrete-parts manufacturing, of integrated
　　circuits, 19, 21
Discrete probability distributions, 124–128
Discrete time index (t), 335
Dispersion model, 91

Distributions
  comparing, 229–231
  external reference, 229–231
Disturbances, 381, 383–384
  detecting, 352–354
Divisors, in Yates algorithm, 258
Dodge, Harold, 5
Doping, 51–58
Dot diagrams, 238–240, 244
  with normal probability plots, 259
Double-comb structure, 102
Double-exponential forecasting filter (DEFF),
    with predictor–corrector controller, 345
Drain-induced barrier lowering, in gate
    engineering, 73
DRAM (dynamic random access memory), 5t, 6
  production of, 7
  simulating with VLASIC software, 163–167
Drift
  in cusum charts, 209
  gradual controller operation mode and,
    337–339
  in process control, 351
Drive-in diffusion, 56
Dry etching, 5t, 6, 48–51
  wet chemical etching versus, 48–49
Dry oxidation, 9–10
  growth kinetics of, 27–31
  thin oxide growth during, 31, 32
Drytek quad reactor, 402
Dual-inline package (DIP), 77
Dummy wafers, 83
Dust particles
  mask damage from, 35
  in photolithography, 34–35

Economics, of semiconductor manufacturing,
    15–16
Economy of control actions, in process control,
    351
Effective dielectric function (ε), 90, 91
Effects, 245. See also Interaction effects; Main
    effects
  with normal probability plots, 259–260
  in Yates algorithm, 258
Eigenvalues, in principal-component analysis,
    306–307, 309
Eigenvectors, in principal-component analysis,
    306–307, 309
Electrical linewidth measurement, of patterned
    thin films, 98
Electrical testing, 102–107
  of integrated circuits, 75
Electrochemical anodization, 26

Electrodes
  in parallel-plate diode system, 51
  in sputtering, 59
Electron-beam evaporation, 59
Electron-beam (e-beam) lithographic systems,
    38, 43–45. See also Lithography
  advantages and disadvantages of, 44
Electron beams, in scanning electron
    microscopy, 95–96
Electron cyclotron resonance (ECR) systems,
    59
Electron gun, in e-beam lithography, 43
Electronic packaging hierarchy, 75–76
Electronics industry
  computer-integrated manufacturing in, 7–8
  semiconductor devices in, 2–3
Electron resist, 38
Electron scattering, in e-beam lithography, 45
Ellipse. See Control ellipse
Ellipsometry
  measuring film thickness via, 88–91
  with predictor–corrector controller,
    343–345
Emissivity, extrinsic and intrinsic, 109
Emitter region, forming, 65, 66
Endpoint detector, 50
Endpoint trace, 114
Enhancement-mode n-channel device, 69
Epitaxial growth, 5t, 6
  with chemical vapor deposition, 60–61
  fuzzy logic in modeling, 317–318
  PIES diagnosis of, 395
Equality, in fuzzy logic, 315
Equation objects, with MERLIN, 385, 387
Equipment costs, in semiconductor
    manufacturing, 16
Equipment diagnosis. See Diagnostic checking
Equipment state measurements, in process
    monitoring, 83, 107–118
Error function, complementary, 54
Errors. See also Experimental error entries;
    Standard error (SE)
  in evolutionary operation, 302–303, 304
  forecast, 214–215
  in neural networks, 312–313, 314
  types I and II, 140, 182, 184
Error variance, response surface methodology
    and, 293–294
Estimates, pooled, 142
Estimation, 136–140
  of global yield loss, 159
  for malfunction alarms, 353
  of parametric yield, 160–161
  shift, 340–342

in two-parameter regression models, 279–280
Estimators, 136–137
Etching, 396, 397
  dry, 5t, 6, 48–51
  as linear regression example, 274–275, 276–277
  in photolithography, 34, 47–51
  plasma, 49–50, 86, 88
  in $p-n$ junction fabrication, 9, 11
  predictor–corrector controller for, 345–346, 347
  wet chemical, 47–48, 48–49
Etch rates, 48
  with dry etching, 48–49, 50
Etch tools, with dry etching, 50
Evans, Oliver, 4
Evaporation, 59
Evidential reasoning, 403, 406–408
Evolutionary operation (EVOP), 301–306
Expansion, in simplex method, 320, 321
Experiential knowledge, with MERLIN, 386
Experimental error (ε_{ti}), 242
Experimental error variance, response surface methodology and, 293–294
Experiments. See also Statistical experimental design
  designed, 5, 228
  randomized block, 240–244
Expert systems
  for diagnostic checking, 380, 391–398
  hybrid, 403–414
Exponential distribution, 131–132
Exponentially weighted moving-average (EWMA) control charts, 213–215, 337–338
Exponentially weighted moving-average gradual model, for run-by-run control, 343
Exponential models, 285–286
Exposure, in Taguchi method, 263, 264
Exposure response curves, for photoresists, 40–41
Exposure tools
  lithography operation monitoring and, 117
  in photolithography, 35–38
Expressions, with MERLIN, 385
External reference distribution, 229–231
Extinction coefficient, 90
Extrema counting, 87
Extrinsic emissivity, 109

Fabrication facilities ("fabs"), in semiconductor manufacturing, 14

FABRICS (FABRication of Integrated Circuits Simulator), 161–162, 167–171, 383
Factorial experimental designs, 245, 246, 249–261
  advanced, 260–261
  analyzing factorials in, 257–260, 261
  fractional, 256–257
  response surface methodology and, 290, 293, 297
  in Taguchi method, 264–265
  two-level, 250–255
Factors, 245
  in factorial experimental designs, 249–240
Fail points, in design centering, 172
Failure(s)
  interval between, 132
  mean time to, 131
  PIES diagnosis of, 395
Failure cases, with PIES, 393
Failure rate, 131
  in maintenance diagnosis, 408, 409
False alarm, 141, 182
Faraday cup, 117
Fault analysis, 163
Fault combinations, 151t
Fault correlation matrix (FCM), 382
Fault diagnosis, 383
Fault patterns, in online diagnosis, 410
Fault ranking, in maintenance diagnosis, 409
Faults. See also Process faults
  defined, 156
  parametric yield and, 159–161
  in semiconductor manufacturing, 147
  software for estimating, 162–167
  in wafers, 148
F distribution, 135–136
Feedback control
  alarms for, 353–354
  for automated recipe generation, 356–357
  full model adaptation of, 354–356
  in process control, 351
  in run-by-run control, 333–334
Feedforward control, 358–359, 364, 365, 366, 367, 368, 369
  in process control, 351
  in supervisory control, 334
Feedforward neural networks, 310
Fick's diffusion equation (law), 53–54
Field oxides, 26, 68
Film dielectric function, 91
Film formation
  in integrated circuit fabrication, 11–12, 62–63

Films, nonuniformity of, 245–246. *See also*
    Thin films
Film thickness
    measuring via ellipsometry, 88–91
    measuring via interferometry, 85–88
Filtering operations, 163
Filters
    air, 99
    forecast, 345
Final testing, 106–107
Final testing yield, 148
First-order autoregressive model, 222
First-order autoregressive moving-average
    (ARMA) models, 222–223
First-order integrated autoregressive moving-
    average (ARIMA) model, 222–223
Fitness, in genetic algorithms, 323, 324
Fixed-oxide charge, 33
Fixed-point coding, in genetic algorithms,
    323–324
Flipchip bonding, of integrated circuits, 79, 80
Flipchip-mounted packages, 78
Flowmeters, 110
Fluorine, in gate engineering, 73–74
Flux analysis, in plasma operation monitoring,
    112
Ford, Henry, 4
Forecast filter, with predictor–corrector
    controller, 345
Forecasts, of process mean, 214
Foreign particles, defects due to, 150. *See also*
    Particles
Forgetting factor (w_{kk}), with model update
    algorithm, 355–356
Fourier transform infrared (FTIR) spectroscopy,
    in plasma operation monitoring, 115
Four-point probe, measuring resistivity via, 92
Fractional factorial experimental designs,
    256–257
    constructing, 256–257
    resolution in, 257
Fraction nonconforming chart (p chart), 186,
    187–192
    average run length for, 191–192
    design of, 188–189
    operating characteristic curve for, 191–192
    sample size for, 188–189
    variable sample size for, 189–191
Frame of discernment, in Dempster–Shafer
    theory, 406, 407
F ratio, 246
Freedom, degrees of, 134, 135, 235, 276, 279
Fresnel equations, 89–90
Fresnel reflection coefficients, 90–91

Full ANOVA table, 235
Full-wafer interferometry, 88
Functional test structures, 105
Functional yield, 148
    components of, 156–159
    models for, 149–155
    parametric yield versus, 160
    software for simulating, 162–167
Function minimization/maximization, 318–320
Furnaces
    oxidation, 26, 27
    vertical oxidation, 32
Fused-silica plate, as standard mask substrate,
    38–39
Fuzzy cell, 316
Fuzzy logic, 273, 310, 314–318
Fuzzy rules, 316–317
Fuzzy sets, 314–315

Gain factor, in process control, 351
Gamma function (Γ), 134, 154–155
Gamma probability density function, 154–155
Gaseous dopant sources, 52, 54
Gas flow, monitoring, 110–111, 112
Gate engineering, in CMOS circuits, 73–74
Gate formation, in NMOS fabrication, 69
Gate oxides, 26
    in NMOS fabrication, 68–69
Gaussian distribution, 153
    in diffusion, 56
    in ion implantation, 57–58
Gauss–Seidel algorithm, for automated recipe
    generation, 356–357
GEM standard, 13–14
Generalized sample variance, 220–221
Generalized SPC approach, 335, 336–337. *See
    also* Statistical process control (SPC)
Generation, of blocking, 254, 255
Generic equipment model (GEM), 13–14
Generic particle counts, 100–101
Generic rules, with PEDX, 397
Genetic "code," in genetic algorithms, 324
Genetic optimization algorithms (GAs), 318,
    323–325, 326, 327
    in intelligent supervisory control, 369–373
Geometric distribution, 128
Geometric interpretation of ANOVA table,
    235–237, 238
Geometric moving-average (GMA) control
    charts, 213. *See also* Exponentially
    weighted moving-average (EWMA)
    control charts
Geometric representation of contrasts, 253
Global controllers, supervisory control and,
    360–361, 364, 365, 366, 367, 368, 369

Global defects, 158
Global optimization, 325
Global yield loss, 158–159
  negative binomial yield model and, 159
Gold, in IC attachment, 79
Grade of membership, 315
Gradient descent approach, in neural networks, 311
Gradual controller operation mode, 335, 336, 337–339
Grand average, 197, 199, 232
  ANOVA table and, 236, 237, 238
Grossberg layer, with process control neural network, 398–399
Gross world product (GWP), electronics industry and, 2–3
Growth kinetics, of silicon oxidation, 27–31
Gyvez, J. Piñeda de, 151

Hard faults, 381
Hard yield, 148
Hazard rate, in maintenance diagnosis, 408
Helium, in X-ray lithography, 46
Hessian matrix, 318–319
Hexamethylene–disiloxane (HMDS), 41
High-atomic-number materials, in X-ray lithography masks, 46
High-density bipolar transisto chain, 102, 103
High-density plasma (HDP) etching, 49
High-efficiency particulate air (HEPA) filters, 99
"Hill climbing" approach, to process optimization, 318
Himmel, C., 314
Hippocrates diagnostic system, 381–384
Holland, John, 323
Hopfield networks, 310
Horizontal susceptors, 60
Hotelling's T^2 statistic, 220
  in inline diagnosis, 413
Hybrid diagnostic checking methods, 402–414
Hybrid expert systems, 403–414
Hybrid optimization methods, 325, 326, 327
Hydrofluoric acid (HF), in $p-n$ junction fabrication, 11
Hypergeometric distribution, 124–125
Hypersphere, in parametric yield optimization, 173
Hypothesis testing, 140–144
  ANOVA table and, 234–235
  control charts and, 182, 183
  defined, 140
  with MERLIN, 388–389
  in online diagnosis, 410

statistical, 231
statistical experimental design for, 229–231
Hypothesization step, with PIES, 394

IC-CIM systems, 12–14
Identification (ID), in Yates algorithm, 258
Image cross sections, for photoresists, 40, 41
Immersion etching, 47
Implantation, monitoring, 117. See also Ion implantation
Implication step, with PIES, 394
Implicit optimization functions, 319
Impurities, in $p-n$ junction fabrication, 10, 11
Impurity doping, 51–58
  in IC fabrication, 63
In-control processes, 184
Independent distributions, 133
Independent variables, 273–274
Index of refraction (n), 85–86
  complex, 90
Indicator variable, 280
Indirect costs, in semiconductor manufacturing, 16
Inert gases, thin oxide growth in, 31. See also Helium
Inference mechanism, with MERLIN, 387–390
Information flow, in semiconductor manufacturing, 12–14
Infrared (IR) spectroscopy, in plasma operation monitoring, 115
Inline diagnosis, 413–414
Inline monitoring techniques, 100
Input bounds algorithm, with run-by-run control, 346–348
Input–output pairs, in neural networks, 312–313, 313–314
Input ranges, for photolithographic machines, 361–363
Input resolution, with run-by-run control, 348, 349
Inputs, to manufacturing, 1–2
Input setting matrix (X), with model update algorithm, 354–355
Input weights, with run-by-run control, 348–350
In situ particle monitoring (ISPM), 100
Inspections, 84
Insulator image, 41–43
Integral term, 215
Integrated autoregressive moving-average (ARIMA) models, 222–223
Integrated circuits (ICs), 5t, 6
  attachment methods for, 79–80
  computer-integrated manufacturing of, 7–8

Integrated circuits (ICs), (*continued*)
   described, 61–62
   fabrication of, 11–12
   faults in, 147
   manufacture of, 1–2, 7–8
   modeling yield of, 147, 148–149, 149–176
   monitoring fabrication of, 82–118
   packaging of, 75–80
   pattern transfer in fabricating, 41–43
   photolithography in fabricating, 34–46
   planar fabrication technology in
      manufacturing, 25
   planarization of, 61, 62
   plasma reactor technology and, 50
   process integration in fabricating, 61–80
   thin films in fabricating, 26
   unit processes in fabricating, 25–61
Intelligent modeling techniques, 310–318
   fuzzy logic, 310, 314–318
   neural networks, 310–314
Intelligent supervisory control, 364–373
Interaction effects, 245
   in two-level factorial design, 251–252
Interactions
   in fractional factorial experimental designs,
      256–257
   in two-level factorial experimental design,
      250, 251–252
Interchangeable parts, 3–4
Interconnect test structures, 101–102
Interface-trapped charge, 33
Interference, optical, 85–86
Interferograms, 86
Interferometry
   full-wafer, 88
   measuring film thickness via, 85–88
Interlevel causes, with PIES, 393–394
Internal analysis, with MERLIN, 388
Internal functions, with fuzzy logic, 316–317
Interval between failures, 132
Interval estimator, 136–137
Intralevel causalities, with PIES, 395
Intralevel causes, with PIES, 393–394
Intrinsic emissivity, 109
Ion distribution, in ion implantation, 57–58
Ion implantation, 5t, 6, 42–43, 63–64, 65. *See*
     also Implantation
   in gate engineering, 73–74
   impurity doping via, 51, 52, 56–58
   inspecting, 84
   in $p–n$ junction fabrication, 10, 11
Ion implantor, 56–57
Ionization gauge, 110

Isolation, in CMOS circuits, 72–73
ISO-Z bipolar process, 395

Japan, electronics industry in, 7, 8
Junction depth, 92
Junctions, in thermocouple, 109

Key measurement points, in process
     monitoring, 83
Kim, B., 403
Kim, T., 364, 366
Kinetics, of silicon oxidation, 27–31
Knowledge base, with PIES, 391–392,
   393–394
Knowledge editor, with PIES, 393–394
Knowledge representation, with MERLIN, 384,
   385–387
Kohonen self-organizing network, 399, 400

Labor, in manufacturing, 3–4
Labor costs, in semiconductor manufacturing,
   16
Lagrange multipliers, in principal-component
   analysis, 307
Lam Rainbow reactive-ion etching system,
   223–224
Lanthanum hexaboride (LaB_6), 43
Latchup, in CMOS circuits, 72
Lateral oxide isolation region, forming, 64
Lateral straggle ($\sigma_\perp$), of an ion, 57
Layers, in neural networks, 310–312, 313,
   314
Leakage current, in gate engineering, 73
Learning, by neural networks, 310–314
Learning curve, yield, 174–176
Least-squares approach, to thin-film metrology,
   87–88
Least squares method, 272, 273, 275
   with exponential models, 285–286
   with multivariate models, 283
   response methodology and, 340–342
   response surface methodology and,
     291–292
   with run-by-run control, 348
Level 1 packaging, of integrated circuits, 79
Levels. *See also* Two-level factorial design
   in factorial experimental designs, 249
   of packaging, 75–76, 79
   in PIES knowledge base, 392–393, 394
   in Taguchi method, 265
Likely state, with PIES, 393–394
Linear models, 283
   with nonzero intercept, 280–283
Linear rate constant (B/A), 30–31

Linear regression, 274–275
  analysis, 272–273
Linewidth, 95
  lithography operation monitoring and, 117
Liquid dopant sources, 52–53
Lithographic exposure tools, 35–38
Lithography, 5t, 6. *See also* Electron-beam
    (e-beam) lithographic systems;
    Photolithography; X-ray lithography
    (XRL)
  in IC fabrication, 63
Lithography operations, monitoring, 116–117.
    See also Photolithography
Local controllers, supervisory control and,
    359–361, 364, 365, 366, 367, 368, 369
Local defects, 158, 162
Local optimization, 325
Local oxidation of silicon (LOCOS), 64, 68
Logic devices, functional testing of, 106
Long-throw sputtering, 59–60
Loss function, in Taguchi method, 262
Low-atomic number materials, in X-ray
    lithography masks, 46
Lower control limit (LCL), 182, 183, 184
  for acceptance charts, 207–208
  for defect chart, 193
  for defect density chart, 193–195
  for exponentially weighted moving-average
    charts, 214
  in feedforward control, 358, 359
  for fraction nonconforming chart, 187, 188,
    189
  for moving-average charts, 212
  in multivariate process variability, 220–221
  for s and $\bar{x}$ charts, 203–204
  for variable control chart, 197–199, 200
Lower cusum (C^-), 210
Lower natural tolerance limit (LNTL), 206
Lower specification limit (LSL), 205, 206
  for acceptance charts, 207–208
  in feedforward control, 358, 359
  for modified charts, 207
  in variable control charts, 196
Low pressure chemical vapor deposition
    (LPCVD), 60, 61, 297

Magnetically enhanced RIE (MERIE), 49
Magnetron sputtering, 59, 60
Mahajan, R., 400, 401
Main effects, 245, 253
  defined, 251
  in Taguchi method, 265
  in two-level factorial design, 251
Maintenance diagnosis, 408–409

Malfunction alarms, 352–353
Malfunctions, diagnosing, 379–381
Management, scientific, 4
Manometers, capacitance, 111
Manufacturing. *See also* Semiconductor
    manufacturing
  assembly line in, 4
  computers in, 4
  defined, 1, 2
  of integrated circuits, 7–8
  interchangeable parts in, 3–4
  major historical milestones in, 3t
  quality control in, 3–5
  standardization in, 4
  statistical process control in, 5
Manufacturing line monitors, 82–83
Manufacturing science, 7–8
Manufacturing yield, 148
Mask damage
  from dust particles, 35
  in shadow printing, 36
Masks, 38–39
  defect density in, 39
  for dry etching, 48–49
  in e-beam lithography, 43–44
  exposure tools and, 35, 36, 37
  in X-ray lithography, 46
Mask substrate, standard, 38–39
Mass flow controller, 110, 111
Mass flowmeters, 110
Mass production, 4
Mass spectroscopy. *See* Residual gas analysis
    (RGA)
Matching algorithm, with PEDX, 397
Matrices. *See also* Hessian matrix; Orthogonal
    arrays; Weight matrices
  for diagnosis, 382
  with model update algorithm, 354–356
  in principal-component analysis, 308–309
Matrix algebra, for multivariate models,
    283–284
Maximum variance, in principal-component
    analysis, 306–307
Maxwell–Boltzmann distribution, Poisson yield
    model and, 152
May, Gary S., 366
Maybe state, with PIES, 393
MC1530 amplifier, with FABRICS software,
    169–171
Mean-centered matrix, in principal-component
    analysis, 308–309
Meander structure, 102
Mean etch rate (MER), predicting, 345
Mean square, 233, 234

Mean time to failure, 131
Mean value (μ), 123, 125, 127, 129, 130, 242.
    See also Sample mean
    control charts for, 195–199
    estimating, 136
    forecasting, 214
    multivariate control of, 217–220
    tests on, with known variance, 141–142
    tests on, with unknown variance, 142–143
Measurement plan, in diagnosis, 382
Measurements, with MERLIN, 386–387
Measure of association, with MERLIN, 388
Membership functions
    with fuzzy logic, 317
    in set theory, 315–316
Memory products, functional testing of, 106
MERLIN (MEasurement ReLational
        INterpreter) diagnostic system, 384–391
Metal films, in $p–n$ junction fabrication, 10, 11
Metallization
    with chemical vapor deposition, 61
    in NMOS fabrication, 70
    in $p–n$ junction fabrication, 10, 11
Metallorganic chemical vapor deposition
        (MOCVD), 5t, 6
Metal–oxide–semiconductors. *See* MOS entries
Metals, in X-ray lithography masks, 46
Metrology
    of blanket thin films, 85–92
    of patterned thin films, 93–98
Metrology equipment, 82–83
Microelectronic industry, reactive-ion etching
        in, 51
Microstrips, parametric yield of, 160–161
Minimization problem, in fault diagnosis, 383
Missed alarm, 141, 182
Mixtures, in control charts, 184
Mobile ionic charge, 33–34
Model-based SPC, 223–224. *See also*
        Statistical process control (SPC)
    process disturbance detection via, 352–354
Model term matrix (T), with model update
        algorithm, 354–355
Model update algorithm, for feedback control,
        354–356
Modified charts, 206–207
Molecular-beam epitaxy (MBE), 5t, 6
Monitoring. *See* Air monitoring; Evolutionary
        operation (EVOP); Process monitoring;
        Product monitoring; Work-in-progress
        (WIP) monitoring
Monitor wafers, 83, 92
Monolithic microprocessors, 5t, 6
Monotonicity, in fuzzy logic, 315

Monte Carlo simulation. *See also* VLASIC
        (VLSI LAyout Simulator for Integrated
        Circuits)
    with FABRICS software, 169–170
    parametric yield estimation via, 160–161
    software for, 162
    supervisory control using, 360–361
MOS devices, oxide quality in, 33
MOSFETs (metal–oxide–semiconductor
        field-effect transistors), fabrication of,
        11–12, 26, 31, 51, 56, 62, 63; BiCMOS
        (bipolar CMOS) technology; CMOS
        (complementary MOSFET) technology;
        NMOS (n-channel MOSFET)
        technology; PMOS (p-channel
        MOSFET) technology
MOSFET technology, 6
MOS process, in CMOS fabrication, 70
Moving-average charts, 212–215
    basic, 212–213
    exponentially weighted, 213–215
Moving-average models, autoregressive,
        222–223
Multiparameter fixed-point coding, in genetic
        algorithms, 323–324
Multivariate process control, 215–221
    with complete model adaptation, 351–359
    described, 215–217
    of means, 217–220
    run-by-run, 343–346, 347
Multivariate process variability, 220–221
Multivariate regression models, 283–285
Multiway principal-component analysis
        (MPCA), 308–309
Murphy, B. T., 152–153
Murphy's yield integral, 152–154
Must state, with PIES, 393
Mutation, in genetic algorithms, 323, 324–325,
        326

Natural tolerance limits (NTLs), 205, 206
Negative binomial distribution, 128
Negative binomial yield model, 154–155, 157
    global yield loss and, 159
Negative electron resists, 45
Negative resists, 40, 41. *See also* Photoresist
Network-based time-series (NTS) models, 409,
        410–411
Neural networks, 273, 310–314
    for diagnostic checking, 381, 398–402
    in intelligent supervisory control, 369–373
    in PECVP optimization, 326
Neurons, in neural networks, 310, 311, 312,
        313

Nitride layer, 64
NMOS fabrication sequence, 67–70
NMOS (n-channel MOSFET) technology, 6, 67
    in CMOS fabrication, 70, 71
Noise
    gradual controller operation mode and, 337
    malfunction alarms and, 353
    in Taguchi method, 262, 263, 264
Noise factors, in Taguchi method, 263
Nominal vales, in Taguchi method, 262
Nonadditivity, 244
    transformable, 247
Nonlinear regression models, 285–287
Nonuniformity of films, 245–246
Nonzero intercept, linear model with, 280–283
Normal approximation to binomial distribution,
    132–133
Normal distribution, 129–130, 259
    cumulative, 130
    sampling from, 133–136
    standard, 130
    t distribution and, 135
Normal equations, in two-parameter regression
    models, 277–279
Normality, in fuzzy logic, 315
Normal probability paper, 259
Normal probability plots, 258–260, 261
Norm body, in parametric yield optimization,
    173
n^+-buried layer, for BiCMOS technology, 75
n^+-collector region, forming, 65, 66
n^+-emitter region, forming, 65, 66
n^+-polysilicon gates, 73
$n–p–n$ bipolar transistors, in integrated
    circuits, 63–66, 67
npn transistors, 72
NPSOL nonlinear optimization software
    package, 345–346
n tub, 71
n-type dopants, 52
n-type epitaxial layer, forming, 64
Null hypothesis, 140
    ANOVA table and, 234–235
    randomized block experiments and,
        241–242
    sums of squares and, 234
Numerical aperture (NA), in photolithography,
    37–38

Observability, in diagnosis, 382, 384
ODOS (one-dimensional orthogonal search)
    acceptability region approximation, 172,
    173
Okabe, T., 154

One-dimensional continuity equation, 53
One-sided alternative hypothesis, 141, 142
One-sided cusum, 211
Online diagnosis, 409–413
On-target ARL, 354
Operating characteristic (OC) curve
    for defect density chart, 194, 195
    for fraction nonconforming chart, 191–192
    for R chart, 202
    for $\bar{x}$ chart, 200–202
Operating costs, in semiconductor
    manufacturing, 16
Operational amplifiers (op-amps), diagnosing
    faults in, 398–399
Optical emission spectroscopy (OES)
    PEDX and, 396, 398
    in plasma operation monitoring, 114–115
Optical interference, 85–86
Optical lithography. *See* Photolithography
Optical metrology, 85
Optical particle counters, in cleanroom air
    monitoring, 99–100
Optical path length, 85
Optimization
    of plasma-enhanced chemical vapor
        deposition, 326–327
    Powell's algorithm for, 318–320, 325, 326,
        327
    with process models, 318–327
Optimization functions, implicit, 319
Orthogonal arrays, in Taguchi method, 264–265
Oscillator applications, for BiCMOS
    technology, 74–75
Oscillators, ring, 104, 105
Out-of-control processes, 184
    rules defining, 185–186
Output discrepancy matrix (Δz), with model
    update algorithm, 355–356
Outputs, from manufacturing, 1–2
Output weights, with run-by-run control,
    350–351
Oxidation
    dry, 9–10, 31, 32
    in IC fabrication, 26–34
    wet, 9–10
Oxidation furnaces, 26, 27
    vertical, 32
Oxide isolation, 64, 66
Oxide layers, in NMOS fabrication, 68–70
Oxide masking, 5t, 6
Oxide quality
    in thermal oxidation, 33–34
    in wet oxidation, 33
Oxide-trapped charge, 33

Packaging
  of integrated circuits, 75–80
  levels of, 75–76
  types of, 77–79
Pancake susceptors, 60
Paperless yield models, 176
Parabolic rate constant (B), 30, 32
Parallel-plate diode system, 51
Parameter design, in Taguchi method, 262
Parametric yield, 148, 149, 159–161
  global yield loss and, 159
  optimizing, 173–174
  software for simulating, 167–171
Pareto distribution, 399
Parsons, John, 4
Partial-pressure analysis, in plasma operation
  monitoring, 112
Particle counts, 100–101
  as example of linear model with nonzero
    intercept, 280–283
  as example of nonlinear regression model,
    286–287
Particle/defect inspection, 98–102
Particles, defects due to, 150
Pascal distribution, 128
Pass points, in design centering, 172
Path of steepest ascent, 293
Patterned masks, in $p-n$ junction fabrication,
  9, 10, 11
Patterned thin films, metrology of, 93–98
Pattern recognition, in CVD diagnosis,
  400–402
Patterns, in control charts, 184–186
Pattern transfer, in integrated circuit fabrication,
  41–43
p chart, 186. *See also* Fraction nonconforming
  chart (p chart)
PEDX (plasma etch diagnosis expert) system,
  395–398
Performance bins, 149
P-glass, 67, 69, 70
Phase, in evolutionary operation, 302
Phase changes, 85
Phase shifts, 89
Phosphorus, as dopant, 52, 53
Phosphorus-doped silicon dioxide (P-glass), 67,
  69, 70
Photoactive compound (PAC) concentration, for
  photolithographic machines, 361–363
Photolithographic machines, acceptable input
  ranges of, 361–363
Photolithography. *See also* Lithography;
  Lithography operations
  e-beam lithography versus, 43

in integrated circuit fabrication, 11–12,
  34–46
multivariate control of, 351–352, 354–356
in $p-n$ junction fabrication, 9, 10–11
supervisory control of, 361–364, 365, 366,
  367, 368, 369
X-ray lithography versus, 45–46
Photomask patterns, dust particles and, 34–35
Photoresist, 5t, 6, 9, 10–11, 34, 39–41, 64, 65
  electron resists versus, 45
  lithography operation monitoring and, 117
  in NMOS fabrication, 68
  in pattern transfer process, 41–43
  in Taguchi method, 263, 264
Physical etching, 50
Physical vapor deposition (PVD), 58, 59–60
  chemical vapor deposition versus, 61
PIES (parametric interpretation expert system),
  391–395
Piñeda de Gyvez, J., 151
Pin-grid array (PGA), 77, 78
π (parallel) polarization component, 89
Planar fabrication technology, 25
Planarization
  in integrated circuit fabrication, 61, 62
  monitoring, 118
Planar models, response surface methodology
  and, 290–292, 292–293
Plasma-assisted etching, 49
Plasma-enhanced chemical vapor deposition
  (PECVD), 26
  optimization of, 326–327
Plasma etching, 49–50. *See also* PEDX
  (plasma etch diagnosis expert) system
  monitoring, 86, 88, 111–116
  neural network modeling and, 314
  as response surface methodology example,
    294–301
  in Taguchi method, 263, 264
Plasma etch monitoring, 86
Plasma operations, monitoring, 111–116
Plasma reactor technology, 50
Plasmas, 49
Plasma spray deposition, 59
Plasma Therm 700 series RIE, 403
Plastic DIP, 77
Plausibility, in Dempster–Shafer theory, 406,
  408
Player piano, 4
Plots, for displaying test results, 107. *See also*
  Control charts; Cusum (cumulative sum)
  charts; Dot diagrams; Normal
  probability plots; Regression charts;
  Shewhart control charts

Plummer, J., 398, 399–400
PMOS (*p*-channel MOSFET) technology, 6, 67
    in CMOS fabrication, 70, 71
p–n junctions, fabrication of, 9–11
pnp transistors, 72
Point-based acceptability region approximation,
    172, 173
Point estimator, 136
Poisson approximation to binomial distribution,
    132
Poisson distribution, 127–128
    with defect chart, 193
    with defect density chart, 194
    exponential distribution and, 132
Poisson yield model, 151–152
    probability density function for, 153
Polarization, in ellipsometry, 88–91
Poly(butene-1 sulfone) (PBS), as electron resist,
    45
Polyglycidylmethacrylate–coethylacrylate
    (COP), as electron resist, 45
Polyhedra
    in parametric yield optimization, 173–174
    in simplicial acceptability region
        approximation, 172, 173
Polymers
    electron resists as, 45
    in negative resists, 40
Poly(methyl methacrylate) (PMMA), as
    electron resist, 45
Polynomial models, 273, 274
    response surface methodology and, 290,
        293–294, 295
Polynomial regression models, 283
Polysilicon, etching of, 297–301, 396, 397
Polysilicon bridging, 102–103
Polysilicon deposition rate, regression charts
    for, 287–289
Polysilicon gate process, 5t, 6
Polysilicon gates, predictor–corrector controller
    for etching, 345–346, 347
Pooled ANOVA, in Taguchi method, 266–268
Pooled estimates, 142
Positive electron resists, 45
Positive resists, 39, 40. *See also* Photoresist
"Postbaking," of wafers, 41
Powell's algorithm, 318–320, 325, 326, 327
p[+]-channel stop, 64
p[+]-polysilicon gates, 73
p[+] regions, in BiCMOS technology, 75
p polarization, 89, 97
Precision of estimates, in two-parameter
    regression models, 279–280
Predeposition diffused layer, 56

Predictability, with MERLIN, 388
Prediction equation, 335
Predictor–corrector controller (PCC), for
    run-by-run control, 343–346, 347
Pressure, monitoring, 109–110, 112
Pressure by temperature by flowrate interaction,
    252
Pressure by temperature interaction, 252
Principal-component analysis (PCA), 273,
    306–309
    multiway, 308–309
Principal components, 273
Printed circuit boards (PCBs), 19
    attaching ICs to, 79
    IC packaging and, 77
Probability. *See also* Normal probability plots
    in multivariate process control, 215–217
    of shifts, 342
Probability density function (pdf), 129, 153,
    161
    in FABRICS simulations, 168
    gamma, 154–155
Probability distributions, 123–133
    continuous, 124, 128–132
    defined, 123–124
    discrete, 124–128
    useful approximations of, 132–133
Probably state, with PIES, 393–394
Probe testing yield, 148
Process capability, 204–206
Process capability ratio (PCR), 205, 206
Process control, multivariate, 215–221,
        351–359. *See also* Advanced process
        control; Statistical process control
        (SPC)
Process control monitors (PCMs), 101–102
Process control neural network (PCNN),
    398–400
Process diagnosis. *See also* Diagnostic checking
    defined, 381
    with MERLIN, 384
Process disturbances, 381, 383–384
    detecting, 352–354
Processes. *See also* Batch operations/processes;
        Process monitoring; Semiconductor
        processing
    abstraction of, 8
    in-control and out-of-control, 184, 185–186
    in manufacturing, 1–2
    optimizing, 318–327
    unit, 9–11, 25–61
Process faults, 381
Process flow, 83
Process integration, in IC fabrication, 61–80

Process mean, 204
  forecasting, 214
Process models, 22, 272–327
  applications of, 272–273
  evolutionary operation monitoring of,
    301–306
  fuzzy logic and, 316–318
  intelligent, 310–318
  in optimizing processes, 318–327
  principal-component analysis with,
    306–309
  response surface methods for, 289–301
  via regression modeling, 273–289
Process monitoring, 82–118
  described, 82–83
  equipment state measurements in, 83,
    107–118
  key measurement points in, 83
  wafer state measurements in, 82–83,
    84–107
Process optimization
  genetic algorithms for, 318, 323–325, 326,
    327
  hybrid methods for, 325, 326, 327
  simplex method for, 318, 320–323, 325,
    326, 327
Process organization, in semiconductor
  manufacturing, 14–15
Process sequences, in semiconductor
  manufacturing, 11–12
Process-specific rules, with PEDX, 397
Process variables, in Taguchi method, 262,
  263–264
Product flow, in semiconductor manufacturing,
  14–15
Product monitoring, for particle defects,
  100–102
Product variability, 122–123
Profilometers, 93
Profilometry, of patterned thin films, 93
Projected range (R_p), of an ion, 57, 58
Projected straggle (σ_p), of an ion, 57
Projection printing, 35, 36–38
Propagation algorithm, with MERLIN, 388
Proportional–integral–differential (PID)
  approach, 214–215
Proportional term, 214
Proximity effect, in e-beam lithography, 45
Proximity printing, 35, 36
p tub, 71, 72
p-type dopants, 52
p-type silicon substrate, for BiCMOS
  technology, 75
Pure chemical etching, 50

Pyrometers, 109
Pyrometry, 109

QMS chamber. *See* Quadrupole mass
  spectrometer (QMS)
Quad flatpack (QFP), 77, 78
Quadratic forms, minimizing, 319–320
Quadratic loss function, in Taguchi method, 262
Quad reactor, 402
Quadrupole mass spectrometer (QMS), in
  plasma operation monitoring, 112–113
Qualifiers, with MERLIN, 390
Quality. *See also* Oxide quality
  defined, 122
  of semiconductor manufacturing, 15, 17
  statistics and, 122–123
Quality characteristics, 123
Quality control, in manufacturing, 3–5
Quality improvement, 123
  in semiconductor manufacturing, 17
Quality of conformance, 123
Quartz crystal monitor, measuring deposition
  rate via, 91–92

Radiofrequency (RF) heating, 59
Radiofrequency monitoring, 116
Randomization, standard error and, 252
Randomized block experiments, 240–244
  described, 240–242
  diagnostic checking of, 243–244
  mathematical model in, 242–243
Random number generators (RNGs), with
  FABRICS software, 168, 169–170
Random sampling, 133, 134, 135, 136
Random variables, 124, 128, 132, 133,
  134–135, 135–136
  confidence intervals and, 137–140
  in FABRICS simulations, 167–171
Range (R), 197
  control charts for, 195–199, 199–200,
    202
  of an ion, 57, 58
Range matrix (D), with model update
  algorithm, 355–356
Rapid controller operation mode, 335, 336,
  337, 339–343, 344
Raster scan systems
  for e-beam lithography, 44–45
  for X-ray lithography, 46
Rational subgroups, control charts and,
  199–200
Raw materials, in manufacturing, 1–2
Raw sample matrix, in principal-component
  analysis, 308–309

R chart, 195–199, 199–200
  operating characteristic curve for, 202
Reactive-ion-beam etching, 49
Reactive-ion etching (RIE), 49, 50, 51,
    223–224
Real-time control, in semiconductor
    manufacturing, 14
Reasoning, evidential, 406–408
Recipes
  automated generation of, 356–357
  intelligent generation of, 369–373
  with model update algorithm, 354
  optimization, 318, 325, 326, 327
  in run-by-run control, 333–334, 335
Redistribution diffusion, 56
Reduction projection lithography, 37
Redundancy technique, in testing memory
    products, 106
Reference junction, in thermocouple, 109
Reference value(s) (*K*)
  cusum charts and, 210–211
  with MERLIN, 387
  in online diagnosis, 411–412
Reflection, in simplex method, 320, 321
Reflectometry, 85
Region objects, with PEDX, 396–397
Regions, with PEDX, 396
Registration, of exposure tools, 35
Regression, with model update algorithm, 354
Regression analysis, linear, 272–273
Regression charts, 287–289
Regression equation, 274
Regression modeling
  multivariate, 283–285
  nonlinear, 285–287
  process models via, 273–289
  regression charts, 287–289
  single-parameter, 274–277
  two-parameter, 277–283
Regular simplex, 320
Reitman, E., 402, 403
Relative range, 196
Relevant reference set, 230
Reliability, in semiconductor manufacturing,
    15, 18
Reproduction, in genetic algorithms, 323, 324
Residual gas analysis (RGA), in plasma
    operation monitoring, 112–113
Residuals
  ANOVA diagnostics and, 238–240
  in randomized block experiments, 242, 243,
    244
  in single-parameter models, 275–276
  in single-variable control methods, 336–337

in two-parameter regression models,
    277–279
  in two-way designs, 247
Resistance, sheet, 98
Resistance heating, 59
Resistance thermometer devices (RTDs),
    111–112
Resistance thermometers, 110–111
Resistivity, four-point probe measurement of,
    92
Resists. *See also* Photoresist
  electron-beam, 43, 44, 45, 46
  X-ray, 46
Resolution
  of exposure tools, 35, 37–38
  in fractional factorial experimental designs,
    257, 296
Response
  evolutionary operation and, 303
  in factorial experimental designs, 249
  by rapid controller operation mode,
    339–343, 344
Response surface, 272, 273
Response surface methodology (RSM),
    289–301
  with predictor–corrector controller,
    345–346, 347
Retrograde well, in CMOS circuits, 72
Ring oscillator, 104, 105
Romig, Harry, 5
Root-cause diagnosis, with MERLIN, 384
Root mean square (RMS) error, in neural
    networks, 314
Rotatability, in central composite design,
    260–261
R/S Discover software package, 298
RS/Explore software package, 286–287
Rule-based reasoning, with PEDX, 395,
    396–397
Rules, for control charts, 185–186, 208
Run-by-run basis, in controlling semiconductor
    manufacturing, 14
Run-by-run (RbR) control, 333–335
  with constant term adaptation, 335–351
  practical considerations in, 346–351
Runs
  in control charts, 184
  regression charts and, 287–289

Sachs run-by-run control architecture, 335, 336,
    338, 343
Sample autocorrelation function, 221
Sample average, 123, 183. *See also* Sample
    mean

Sample fraction nonconforming, 126–127, 187
Sample mean, 133, 135, 136. *See also* Mean value (μ); Sample average
  confidence intervals and, 137–140
  control charts for, 195–199, 202–204
Samples, 133. *See also* Sampling entries
Sample size
  for fraction nonconforming chart, 188–189
  malfunction alarms and, 353
  $\bar{x}$ chart and, 200–201
Sample standard deviation (*s*), 123
  control charts for, 202–204
Sample variance (s^2), 123, 133, 135, 136. *See also* Variance (σ^2)
  confidence intervals and, 137–140
  generalized, 220–221
Sampling
  in multivariate process control, 217–220
  from normal distribution, 133–136
Sampling distribution, 134
Scanning electron microscopy (SEM), 297
  of patterned thin films, 95–96
Scatterometers, 96–97
Scatterometry, of patterned thin films, 96–97
Scatterplots, 299, 362, 363
Scatter signatures, 97
s chart, 195–199, 202–204
Schemata, in genetic algorithms, 325
Scientific management, 4
Screening, 295, 296
Scribing, 76
Seasonal ARIMA (SARIMA) model, 223
Second-order autoregressive model, 222
SECS-I protocol, 13
SECS-II messages, 13–14
Seebeck effect, 109
Seeds, R. B., 153–154
Seeds exponential yield model, 153–154, 155
Self-aligned polysilicon gate process, 5t, 6
SEMATECH consortium, 8
Semiconductor devices, in electronics industry, 2–3
Semiconductor equipment communications standard (SECS) protocol, 13–14
Semiconductor manufacturing, 1–22
  advanced process control for, 334–335
  computers in, 4
  described, 1–2
  goals of, 15–18
  history of, 2–8
  information flow in, 12–14
  integrated circuits, 7–8
  modeling yield of, 147–176
  modern, 8–15

process and equipment diagnosis in, 379–414
process organization in, 14–15
process sequences in, 11–12
quality control in, 3–5
statistical process control in, 5
systems for, 18–21
Semiconductor processing, 5–7
  major historical milestones in, 5t
  wet chemical etching in, 47–48
Semiconductors, impurity doping of, 51–58
Sensing junction, in thermocouple, 109
Series-type chains, 103
Sets, in Dempster–Shafer theory, 406–408
Set theory, fuzzy logic and, 314–315
Shadow printing, in photolithography, 35, 36
Shallow junctions, ion implantation and, 51
Shallow trench, in CMOS circuits, 73
Sheet resistance, measuring, 98
Shewhart, Walter, 5, 182
Shewhart control charts, 182, 208, 209, 223, 336–337. *See also* Control charts
  in statistical process control, 221
Shifts
  in control charts, 184
  in cusum charts, 209, 210
  diagnosing, 379–380, 384
  rapid controller operation mode and, 339–343, 344
Shmoo plot, 107
σ(perpendicular) polarization component, 89
Signal factors, in Taguchi method, 263, 266
Signal-to-noise ratio (*SN*), in Taguchi method, 264, 266
Signal-to-symbol transformation, with PEDX, 395, 396, 397
Signatures, in scatterometry, 97
Significance, statistical, 229, 231
Silicon, thermal oxidation of, 27–34
Silicon carbide, in X-ray lithography masks, 46
Silicon deposition
  gradual controller operation mode for, 338–339
  rapid controller operation mode for, 343, 344
Silicon dioxide (SiO_2)
  high-quality, 9–11
  thermal oxidation of silicon to, 27–34
Silicon nitride deposition, in NMOS fabrication, 68
Silicon–silicon dioxide interface
  during oxide growth, 27–31
  oxide quality at, 33

Silicon wafers, 9–11. *See also* Wafers
   pattern transfer to, 41–43
Simplex, regular, 320
Simplex method, 318, 320–323, 325, 326, 327
Simplicial acceptability region approximation,
   172, 173
Simulation tools, in semiconductor
   manufacturing, 15
Single-parameter regression models, 274–277
Single-variable methods, for run-by-run control,
   335–343, 344
Single-wafer operations, in semiconductor
   manufacturing, 19, 20–21
Single-workpiece operations, in semiconductor
   manufacturing, 19, 20–21
Singular value decomposition, in principal-
   component analysis, 309
Sinusoidal regression models, 283
"Soft-baking," 41
Soft faults, 159, 381
Software
   for multivariate regression models, 284
   nonlinear optimization, 345–346
   for nonlinear regression models, 286–287
   yield simulation, 162–171
Soft yield, 148
Solvents, for photoresists, 39, 40
Source and drain formation, in NMOS
   fabrication, 69–70
Spanos, Costas J., 383
SPC automation, 176. *See also* Statistical
   process control (SPC)
Specification limits (SLs), 204, 205, 206
   in variable control charts, 196
Spectroscopic ellipsometry (SE), 88
Spectroscopic measurements, 87
SPICE (simulation program with integrated
   circuit emphasis), 169, 399–400
s polarization, 89, 97
Spray etching, 47–48
Spread, 123
Sputter etching, 50
Sputtering, 59
   through collimator, 59, 60
Standard cubic centimeter per minute (sccm),
   110
Standard deviation (Φ), 129, 200, 203–204,
   252. *See also* Standard error (SE)
   control charts for, 195–199
   evolutionary operation and, 304
   in multivariate process control, 217–220
   sample, 123
Standard error (SE). *See also* Standard
   deviation (σ)

in feedforward control, 358–359
   response surface methodology and,
      292–293, 293–294
   in single-parameter models, 276
   in two-level factorial design, 252–254
Standardization, in manufacturing, 4
Standardized control chart, 191
Standardized matrix (Y), with model update
   algorithm, 355–356
Standard normal cumulative distribution
   function (Φ), 133
Standard normal distribution, 130
Standard order, in analyzing factorials, 258
Standard sputtering, 59
Stapper, C. H., 154
Star points, 294
Static random access memory (SRAM), with
   BiCMOS technology, 74
Statistical Control Quality Handbook (Western
   Electric), control chart rules in,
      185–186
Statistical experimental design, 22, 228–268
   analysis of variance in, 232–249
   applications of, 228–229
   comparing distributions in, 229–231
   factorial designs, 245, 246, 249–261
   statistical process control and, 228–229
   Taguchi method of, 262–268
Statistical hypotheses, 140
   testing, 140–144
Statistical hypothesis test, 231
Statistical process control (SPC), 5, 22, 122,
   181–224. *See also* Generalized SPC
   approach; Model-based SPC; SPC
   automation
   advanced process control and, 333
   control charts for, 181–182, 182–184,
      184–186, 186–195, 195–215
   with correlated process data, 221–224
   defined, 181–182
   experimental design and, 228–229
   model-based, 223–224
   multivariate, 215–221
Statistical significance, 229, 231
Statistical tests, in feedforward control,
   358–359
Statistics
   between-lot and within-lot, 199–200
   fundamentals of, 122–144
Steepest ascent path, 293
Stefan–Boltzmann relationship, 109
Step-and-repeat projection lithography,
   36, 37
Step-and-scan projection lithography, 37

Steppers
feedforward control for, 358, 367
input settings for, 362
Stepwise regression, with model update
algorithm, 354, 355
Straggle, of an ion, 57
Stratification, in control charts, 184
Strengths, with PIES, 393–394
String manipulation, in genetic algorithms,
323–324
Structural failure, PIES diagnosis of, 395
Subgroups, rational, 199–200
Subject values, with MERLIN, 387
Submicrometer resist geometries, 44
Sums of squares
in analysis of variance, 232–234, 235,
236
with fuzzy logic, 317
for nonlinear regression models, 286
in single-parameter models, 276
total, 235
in two-parameter regression models, 279
Supervisory control, 359–373
intelligent, 364–373
run-by-run control as, 334
in semiconductor manufacturing, 14
using complete model adaptation, 359–364,
365, 366, 367, 368, 369
Support
in Dempster–Shafer theory, 406
in online diagnosis, 411, 413
Surface-mount packages, 77
Surface profilometers, 93
Surfscans, 100, 101
Susceptors, 109
with chemical vapor deposition, 60–61
Symmetry, in fuzzy logic, 315
Symptom correlation matrix (SCM), 382
Symptoms, 381
Symptom vector, 382
System design, in Taguchi method, 262

Tabular cusum chart, 210
Taguchi, Genichi, 262
Taguchi method, 262–268
data analysis in, 266–268
orthogonal arrays in, 264–265
process variables in, 262, 263–264
signal-to-noise ratio in, 264, 266
Tape-automated bonding (TAB), of integrated
circuits, 79
Taylor, Frederick, 4
Taylor series, 318–319

t distribution, 134–135
Technology
in manufacturing, 3–5, 25–80
semiconductor processing, 5–7
Temperature. *See also* Thermal entries
diffusion and, 53–54
monitoring, 109, 111–112
in thermal oxidation of silicon, 30–31, 32
Temperature sensors, 110–111
Test structures
electrical, 102–105
functional, 105
for wafer defects, 101–102
Testing
final, 106–107
hypothesis, 140–144, 182, 183, 229–231,
234–235, 388–389, 410
of integrated circuits, 75
Thermal conductivity gauges, 110
Thermal mass flowmeter, 110
Thermal operations, monitoring, 109–111. *See
also* Temperature entries
Thermal oxidation, 26
oxide quality in, 33–34
of silicon, 27–31
thin oxide growth in, 31–32
Thermocompression bonding, 79
Thermocouples, 109
Thermoelectric effect, 109
Thermopile, 109
Thin-film deposition, inspecting, 84
Thin films. *See also* Blanket thin films; Film
entries; Patterned thin films
in IC fabrication, 26, 58–61
Thin oxide growth, in thermal oxidation of
silicon, 31–32
3-sigma (3σ) control charts, 183
3-sigma (3σ) control limits, 185–186, 193, 194
Three-transistor DRAM cell, simulating with
VLASIC software, 163–167
Thresholding, with PIES, 394
Threshold voltage rolloff, in gate engineering,
73
Throughput
of e-beam systems, 44
exposure tools and, 35
in semiconductor manufacturing, 15
Time-series diagnosis, 402–403, 404, 405
Time-series modeling, in statistical process
control, 221–223
Time-to-yield metric, 174–176
Tolerance design, in Taguchi method, 262

Tool attachments, 4
Tool signatures, 402–403
Tool variables, in equipment state
   measurements, 108–109
Top numbers
   in gradual controller operation mode, 339
   in Yates algorithm, 258
Toshiba study (1986), 7
Total dopant diffusion, 55
Total reflection coefficients, 89
Total sum of squares, 235
Transceiver applications, for BiCMOS
   technology, 74–75
Transformable nonadditivity, 247
Transformation of data, 244, 246–249
Transformation of response variable, 244
Transistor chains, 102–105
Transistors, fabrication of, 11–12
Traps, at silicon–silicon dioxide interface, 33
Treatment effect (τ_i), 242
Trench isolation technology, 5t, 6
Trends
   in control charts, 184, 185
   in cusum charts, 209
   regression charts and, 287–289
Trial control limits, for fraction nonconforming
   chart, 188
Triangular Murphy yield model, 153
Tungsten, in IC fabrication, 61
Turret lathe, 4
Twin tub, 71
Two-level factorial design, 250–255
   blocking of, 254–255
   interaction effects in, 251–252
   main effects in, 251
   standard error in, 252–254
Two-level IC-CIM architecture, 13–14
Two-parameter regression models, 277–283
   with nonzero intercept, 280–283
Two-sided cusum, 211, 212
2-sigma (2σ) control limits, 186
Two-way ANOVA, 245
Two-way designs, analysis of variance in,
   245–249
Type I error, 140, 182, 184
Type II error, 140, 182

u chart, 186, 193–195. *See also* Defect density
   chart (u chart)
Ultra-high-efficiency particulate air (ULPA)
   filters, 99
Ultrasonic bonding, 79
Ultraviolet (UV) radiation, in p–n junction
   fabrication, 10–11

Unacceptability regions, in design centering,
   172
Uniform density function, 152–153
Unique fault combinations, 151t
United States, electronics industry in, 7, 8
Unit processes
   in IC fabrication, 25–61
   in semiconductor manufacturing, 9–11
Unreliable performance, diagnosing, 379–380
Upper control limit (UCL), 182, 183, 184
   for acceptance charts, 207–208
   for defect chart, 193
   for defect density chart, 193–195
   for exponentially weighted moving-average
     charts, 214
   in feedforward control, 358, 359
   for fraction nonconforming chart, 187, 188,
     189
   malfunction alarms and, 353
   for moving-average charts, 212
   in multivariate process variability, 220
   for s and $\bar{x}$ charts, 203–204
   for variable control chart, 197–199, 200
Upper cusum (C^+), 210
Upper natural tolerance limit (UNTL), 206
Upper specification limit (USL), 205, 206
   for acceptance charts, 207–208
   in feedforward control, 358, 359
   for modified charts, 207
   in variable control charts, 196

Validity, standard error and, 252
Value measure, with MERLIN, 389
Van der Pauw structure, 98
van der Waals electrostatic force, 94
Vapor-phase epitaxy (VPE), 60
Variability. *See also* Variation
   between-sample and within-sample, 199
   defined, 123
   multivariate control of, 220–221
   product, 122–123
   of semiconductor manufacturing, 17
Variable objects, with MERLIN, 385, 386
Variables. *See also* Principal components
   for automated recipe generation, 356–357
   control charts for, 195–215
   dependent, 273–274
   independent, 273–274
   indicator, 280
   with MERLIN, 385, 387
   screening of, 295, 296
   transformation of, 244
Variable sample size, for fraction
   nonconforming chart, 189–191

Variance (σ^2), 123, 125, 127, 129, 130. *See also*
    Sample variance (s^2)s
  confidence intervals and, 137–140
  cusum charts for, 211–212
  estimating, 136
  evolutionary operation and, 304
  mean tests with known, 141–142
  mean tests with unknown, 142–143
  in online diagnosis, 410
  in principal-component analysis, 306–307,
    307–309
  tests on, 143–144
Variance–covariance matrix
  malfunction alarms and, 353
  for multivariate models, 284
Variance-stabilizing data transformations, 248
Variation, diagnosing, 379–380. *See also*
    Variability
Vector algebra
  for multivariate models, 283–284
  for principal-component analysis, 306, 309
Vector geometry
  ANOVA table and, 236–237, 238
  for randomized block ANOVA models,
    242–243
Vector of correction term coefficients ($\Delta\gamma$),
    with model update algorithm, 356
Vectors
  for automated recipe generation, 356–357
  in CVD diagnosis, 401
Vector scan systems
  for e-beam lithography, 44–45
  for X-ray lithography, 46
Vernier caliper, 4
Vertical oxidation furnaces, 32
Very likely state, with PIES, 393–394
Via chain structure, 103–105
Via diameter, 183
Virtual manufacturing environment, in
    semiconductor manufacturing, 15
Viscosity, in Taguchi method, 263, 264
VLASIC (VLSI LAyout Simulator for
    Integrated Circuits), 162–167

Wafers. *See also* Silicon wafers
  chemical vapor deposition onto, 60–61
  in cleanroom air monitoring, 99
  defect density on, 156–157
  die separation on, 76
  dry etching of, 48–51
  electrical testing of, 102–107
  equipment state measurements and,
    107–118
  exposure tools and, 35, 36, 37

faults in, 148
global yield loss on, 159
in IC fabrication, 61–63
implantation monitoring for, 117
impurity diffusion into, 52–56
in integrated circuit fabrication, 11–12
intelligent manufacture of, 369–373
lithography operation monitoring and,
    116–117
monitor, 83, 92
monitoring for particle defects, 100–102
in NMOS fabrication, 67–70
with $n-p-n$ bipolar transistors, 63–66
pattern transfer to, 41–43
physical vapor deposition onto, 59–60
planarization of, 61, 62
processing cost of, 15–16
scribing of, 76
silicon deposition on, 338–339, 343, 344
software for estimating defects on, 162–167
wet chemical etching of, 47–48
yield of, 148
Wafer state characterization, 84–85
Wafer state measurements, in process
    monitoring, 82–83, 84–107
Wafer yield losses, 148
  assignable causes of, 149
Waveforms, 86
  in RF monitoring, 116
Wavelength (λ), 85, 87, 90–91
  electron, 95
  in photolithography, 37–38
Weibull distribution, in maintenance diagnosis,
    408–409
Weighted average, 213–214
Weighting schemes, with run-by-run control,
    348–351
Weight matrices
  with fuzzy logic, 316–317
  in neural networks, 310–311, 312, 313
  with run-by-run control, 349–350
Weights, for automated recipe generation, 357
Well formation, in CMOS circuits, 72
Western Electric rules, for control charts,
    185–186, 208
Wet chemical etching, 47–48
  dry etching versus, 48–49
Wet oxidation, 9–10
  growth kinetics of, 27–31
  oxide quality during, 33
Whitney, Eli, 3
Windowing technique, for global yield loss
    estimation, 159
Window size, with model update algorithm, 355

Wire bonding, of integrated circuits, 79
Within-lot statistics, 199–200
Within-sample variability, 199
Within-treatment sum of squares, 233, 234
Witness plates, 99
Workcells, 369–373
    in semiconductor manufacturing, 14–15
Work-in-progress (WIP) monitoring, 13

$\bar{x}$ charts, 195, 196–199, 199–200
    average run length for, 200, 201
    in bivariate process control, 216
    operating characteristic curve for, 200–202
X-ray lithography (XRL), 45–46. *See also*
        Lithography

Y_0 factor, 159
Yates algorithm, for analyzing factorials, 258
Yield, 122
    analysis of variance and, 232, 234–235,
        235–236
    assessing in IC fabrication, 102
    critical area and, 150
    defects and, 150
    defined, 147, 148–149
    design centering and, 171–174
    evolutionary operation and, 301, 303
    functional, 148, 149–155, 156–159, 160,
        162–167
    mask, 39

modeling, 22, 147–176
    with multivariate regression models,
        284–285
    Murphy's integral of, 152–154
    negative binomial model of, 154–155
    parametric, 148, 149, 159–161, 167–171,
        173–174
    Poisson model of, 151–152, 153
    randomized block experiments and,
        241–242
    as response surface methodology example,
        289–294, 295
    Seeds exponential model of, 153–154, 155
    of semiconductor manufacturing, 15, 16,
        17–18, 22
    simulating, 160–161, 161–171
    in statistical experimental design, 229–231
    time-to-yield and, 174–176
    types of, 148
Yield groups, 176
Yield improvement, in semiconductor
        manufacturing, 17–18
Yield learning, in semiconductor
        manufacturing, 17–18
Yield learning coefficient, 175, 176
Yield learning curve, 174–176
Yield loss, 148
    assignable causes of, 149

Zadeh, Lotfi, 314, 315

Printed and bound by CPI Group (UK) Ltd, Croydon, CR0 4YY

16/04/2025

14658596-0005